AF307732

Klaus Hahn
Sibylle Fischer
Isky Gordon

Atlas of Bone Scintigraphy in the Developing Paediatric Skeleton

The Normal Skeleton, Variants and Pitfalls

In Collaboration with
J. Guillet, A. Piepsz, I. Roca and M. Wioland

Under the Auspices of the Paediatric Task Group
of the European Association of Nuclear Medicine

Foreword by David L. Gilday

With 283 Figures in 888 Separate Illustrations

Springer-Verlag
Berlin Heidelberg New York
London Paris Tokyo
Hong Kong Barcelona
Budapest

Prof. Dr. med. KLAUS HAHN
SIBYLLE FISCHER

Klinikum der Johannes Gutenberg-Universität
Klinik und Poliklinik für Nuklearmedizin
Postfach 39 60, D-55101 Mainz, Germany

ISKY GORDON, M.D.

The Hospitals for Sick Children
Department of Radiology
Great Ormond Street
London WC1N 3JH, United Kingdom

ISBN-13:978-3-642-84947-3   e-ISBN-13:978-3-642-84945-9
DOI: 10.1007/978-3-642-84945-9

Library of Congress Cataloging-in-Publication Data. Hahn, K. (Klaus) Altlas of bone scintigraphy in
the developing paediatric skeleton: the normal skeleton, variants, and pitfalls/Klaus Hahn, Sibylle
Fischer, Isky Gordon: in collaboration with J. Guillet... [et al.]. p.   cm.   "Under the auspices of the
Paediatric Task Group of the European Association of Nuclear Medicine."
ISBN-13:978-3-642-84947-3   1. Bones – Radionuclide imaging – Atlases. 2. Bone
diseases in children – Radionuclide imaging – Atlases. 3. Human skeleton – Abnormalities – Diagnosis
– Atlases. I. Fischer, Sibylle, 1961 –. II. Gordon, Isky. III. European Association of Nuclear Medicine. Paediatric Task Group. IV. Title. [DNLM: 1. Bone and Bones – radionuclide imaging – atlases.
2. Bone Diseases – radionuclide imaging – atlases. 3. Bone Diseases – in infancy & childhood – atlases.
4. Bone Diseases – in adolescence – atlases. WE 17 H148a 1993] RJ482.B65H34 1993
618.92'7107575 – dc20   DNLM/DLC for Library of Congress   93-5678 CIP

Production editor: Meike Seeker, Heidelberg
Cover design: Erich Kirchner, Heidelberg
Reproduction of the figures: Gustav Dreher GmbH, Stuttgart
Typesetting: Data conversion by Fotosatz-Service Köhler OHG, Würzburg-Heidingsfeld

21/3130-5 4 3 2 1 0 – Printed on acid-free paper

Foreword

Since the introduction of technetium − 99 m polyphosphate in 1972 by Dr. M. Subramanian, bone scintigraphy has become an integral part of the evaluation of paediatric musculoskeletal disorders. Using the current high-resolution gamma cameras and technetium −99 m MDP or DPD, the quality of images that we interpret is very high. From the very earliest days, there has been confusion over normal bone physiology as depicted by the bone scintigram in paediatric patients. This has resulted in a number of difficulties in detecting abnormalities, especially near the physes (growth zones). Primary examples of abnormalities that might be confused with normal activity are osteomyelitis, bucket handle fractures of the long bones, and neuroblastoma and leukaemic metastases.

Early in the course of interpreting bone scintigrams in children, we realized that there was a significant difference in the appearance of the physes in the first years of life compared to that in the more mature child. The growth zone is globular at birth and becomes discoid later in childhood. The process is completed in about 2 years. The maturation of the physes varies considerably, especially between males and females. As the growth zone fuses, there is a blurring of the normal linear pattern until eventually the adult appearance emerges. The time at which this occurs varies according to the child's degree of physical activity and its state of health and nutrition. As you will see in this atlas, there is an obvious general pattern to the closure of the growth zones. It is extremely important to understand this information so that one can determine whether there is premature closure secondary to focal disease or therapy. It is extremely important too to be aware of the presence of normal apophyses, sychondroses and sutures.

An important observation is that a child who ceases to use a limb for even as little as 24 h, particularly one who has not reached skeletal maturity, will suffer a 50% reduction of bone activity in that limb. This can be due simply to pain, even if it is not related to a musculoskeletal problem. In such circumstances, it is necessary to question the child or family carefully to determine whether or not there is a reason for the reduction in the use of that limb.

The purpose of this atlas is to familiarize diagnosticians with the normal appearance of the skeleton in children of ages from the newborn to the young adult. A good understanding of this progression of skeletal development is extremely important, since one may otherwise misinterpret normal structures as lesions and, vice versa, miss abnormalities by thinking they are normal. As the majority of paediatric bone scintigrams are interpreted by nonpaediatric nuclear physicians, the availability of this reference atlas should improve the care of children.

The Hospital for Sick Children,
Toronto/Canada

DAVID L. GILDAY

Contents

Introduction

Introduction

The paediatric skeleton is in the process of maturing from birth to adult-hood. This maturation involves the growth of every bone, whereby this growth is principally located at the epiphyseal plates. Since the ^{99m}Tc bone scanning agents are absorbed onto the hydroxy apatite, the bone scan reflects active bone turnover. In the developing skeleton, the appearances of the rapidly growing areas, are therefore, different at different stages of maturation.

The members of the Paediatric Task Group of the European Association of Nuclear Medicine (EANM) are all regularly consulted for their opinion on bone scans obtained in their respective countries. It became obvious that only physicians who were experienced in paediatric nuclear medicine could recognise immediately the appearances of "normal" bone scan. The crucial importance of maintaining high-quality bone images was another feature which emerged as being important. This was the stimulus that led the Paediatric Task Group to produce an atlas of bone scintigraphy in the developing paediatric skeleton as one of its collaborative projects.

To produce an atlas of the scintigraphic features of the usual developing skeleton, over 1700 bone scans considered normal were collected from the following Peadiatric Nuclear Medicine Centres:

- Hospital for Sick Children, Department of Radiology, London, Great Britain (Dr. Isky Gordon).
- Service de Biophysique, Centre Hospitalier d'Agen, Agen, France (Dr. Jacques Guillet).
- Klinik und Poliklinik für Nuklearmedizin der Universität Mainz, Mainz, Germany (Professor Klaus Hahn).
- Hospital St. Pierre, Akademisch Ziekenhuis, Department of Radiology, Brussels, Belgium (Professor Amy Piepsz).
- Hospital General Vall d'Hebron, Department of Nuclear Medicine, Barcelona, Spain (Dr. Isabel Roca).
- Assistance Publique, Hospiteaux de Paris, Saint-Antoine, Service de Medicine Nucleaire, Paris, France (Dr. Michel Wioland).

The authors selected representative images from the pool of images collected by S. Fischer in the Department of Nuclear Medicine in Mainz, Germany.

Patient Inclusion Criteria

No "normal" children underwent bone scans, so the scans included here were rather from children who had undergone bone scintigraphy for clinical reasons but in whom it was considered highly probable that the skeleton was normal. The Paeadiatric Task Group of the EANM defined the following groups of children for inclusion:

1. Focal symptoms thought to be due to bone pathology in a healthy child, if the child did not have any skeletal pathology on follow-up, and if the cause of the symptoms was found to lie outside the selekton. This group underwent bone scans mainly for suspected infection or trauma.

2. Generalized symptoms: The bone scintigraphy was carried out if diagnosis was not clear. Only those children were included, who had undergone bone scans within 72 h after the onset of symptoms. All these children were ambulatory and not confined to bed prior to the bone scan. There was no alteration of skeletal metabolism or radionuclide uptake. Clinical follow-up in all these children ensured that no children were included who were later found to have any malignancy or bone pathology.

3. Known malignancy: Children with certain solid tumours, such as rhabdomyosarcoma, teratoma, Wilms' and yolk sac tumours were included only if the bone scan had been carried out prior to the administration of any chemo-therapeutic agents. These patients were ambulatory prior to and at the time of bone scintigraphy.

4. Localized benign bone tumours: Only those parts of the skeleton not affected by the benign bone tumours were included.

5. Suspected fracture: If whole body images were obtained in those children with suspected single acute fracture, the images from the non-affected areas were included.

6. Children with localized pain, but without certain evidence of bone disease: most of these children were suffering from backache.

7. Children with ichthyosis.

In age, the "normal children" ranged from neonates to 22-years-olds.

Materials and Methods

Technique. The radiopharmaceuticals ^{99m}Tc-methylene diphosphonate (MDP; from various companies) or ^{99m}Tc-2,3-dicarboxypropane-1, 1-diphosphonic acid (DPD; Behring). The dose used followed the recommendations of the Paediatric Task Group of the EANM (Table 1).

Blood Pool Images. Static gamma camera images were completed within 5 min of the injection. The minimum acquired count rate was 100 000 counts. The maximum time per image was 3 min. Using whole body scans, imaging was started directly after the injection of the radioisotope using a scan speed of 30 cm/min.

Bone Scan Images. The images were acquired no sooner than 3h post injection, with a minimum count rate of 50 000 counts for hands and feet; 100 000 counts for the knee joint; and 200 000 counts for the skull. From 250 000 to 500 000 counts were acquired for images of the pelvis and the remaining skeleton. Static images were acquired with an Elscint, Picker or Siemens gamma camera, with the child lying directly on top of the head of the gamma camera (Figs. 1, 2).

Whole body scans were acquired with a speed of 8 cm/min in children up to the age of 8 years; 10 cm/min in children ranging from 8 to 12 years; and a scan speed of 12 cm/min in children ranging from 12 to 16 years;

Introduction

Table 1

Paediatric Task Group

Fraction of Adult Administered Activity

3 kg = 0.1	22 kg = 0.50	42 kg = 0.78
4 kg = 0.14	24 kg = 0.53	44 kg = 0.80
6 kg = 0.19	26 kg = 0.56	46 kg = 0.82
8 kg = 0.23	28 kg = 0.58	48 kg = 0.85
10 kg = 0.27	30 kg = 0.62	50 kg = 0.88
12 kg = 0.32	32 kg = 0.65	52-54 kg = 0.90
14 kg = 0.36	34 kg = 0.68	56-58 kg = 0.92
16 kg = 0.40	36 kg = 0.71	60-62 kg = 0.96
18 kg = 0.44	38 kg = 0.73	64-66 kg = 0.98
20 kg = 0.46	40 kg = 0.76	68 kg = 0.99

Recommended Adult and Minimum Amounts in MBq

Radiopharmaceutical	Adult	Minimum
Tc99m DTPA (Kidney)	200	20
Tc99m DMSA	100	15
Tc99m MAG3	70	15
Tc99m Pertechnetate (Cystography)	20	20
Tc99m MDP	500	40
Tc99m COLLOID (Liver/Spleen)	80	15
Tc99m COLLOID (Marrow)	300	20
Tc99m SPLEEN (Denatured R B C)	40	20
Tc99m R B C (Blood Pool)	800	80
Tc99m ALBUMIN (Cardiac)	800	80
Tc99m Pertechnetate (First Pass)	500	80
Tc99m MAA / Microspheres	80	10
Tc99m Pertechnetate (Ectopic Gastric)	150	20
Tc99m Colloid (Gastric Reflux)	40	10
Tc99m IDA (Biliary)	150	20
Tc99m Pertechnetate (Thyroid)	80	10
Tc99m HMPAO (Brain)	740	100
Tc99m HMPAO (W B C)	500	40
I-123 HIPPURAN	75	10
I-123 (Thyroid)	20	3
I-123 Amphetamine (Brain)	185	18
I-123 mIBG	200	35
I-131 mIBG	80	35
GALLIUM 67	80	10

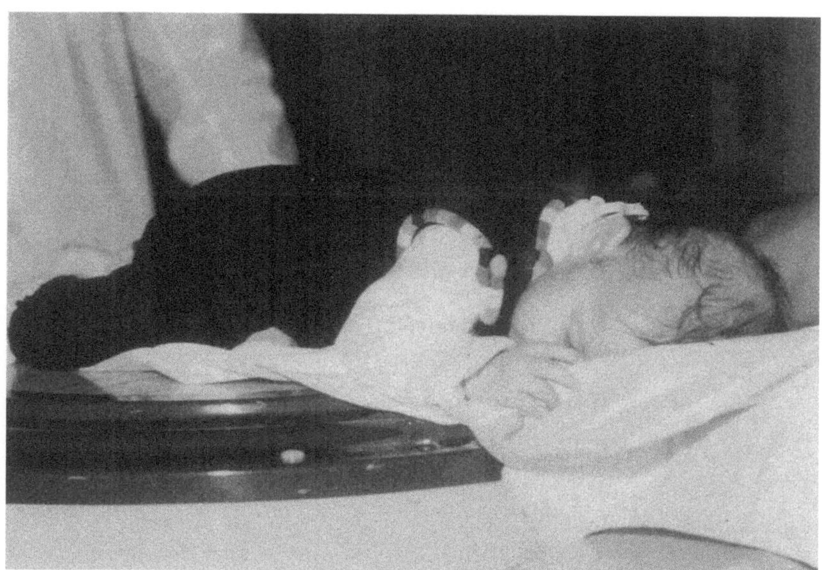

Fig.

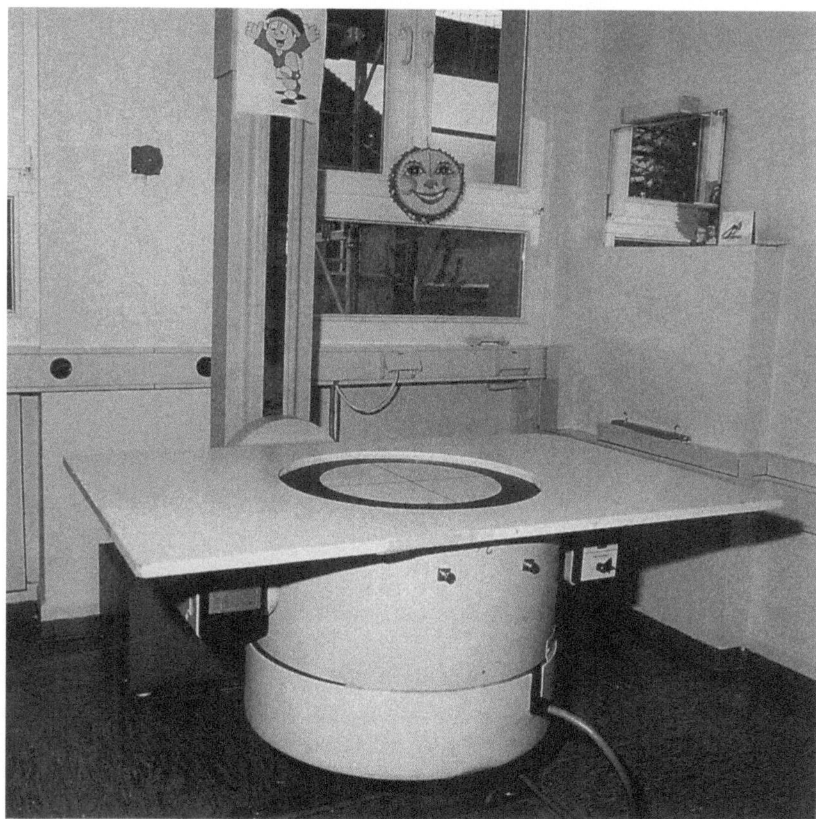

Fig.

beyond this age, the scan speed was 15 cm/min. Whole body scans were required on an Elscint or Siemens system.

Sedation. There was no routine sedation. If required, mild sedation with Dominal (prothipendyl) or Dormicum (midazolam) was used.

Immobilisation. Various techniques were used for immobilisation, as shown in Figs. 3 and 4.

Fig. 3

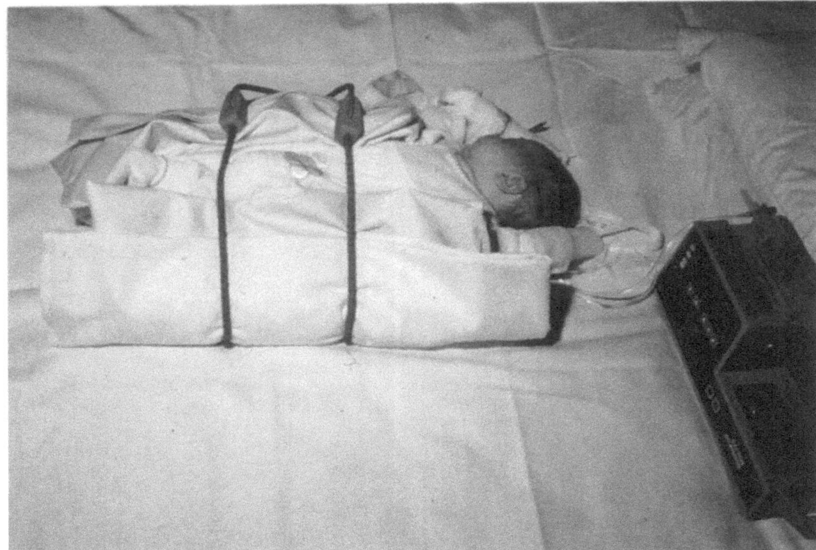

Fig. 4

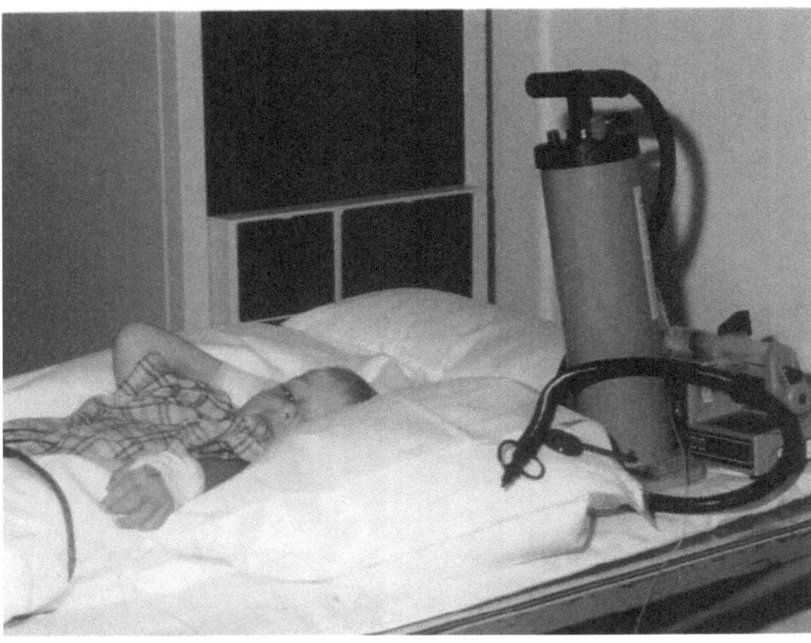

Hard-copy Recording. The primary images were recorded on X-ray film or photographic paper. All selected images were prepared in the same filmlaboratory in the Klinik für Nuklearmedizin, University Hospital, Mainz, Germany (head of department, Mrs. A. Keuchel).

Chapters 1–18 show blood pool images and bone scans of all parts of the skeleton in different age groups. In the hips and knees, different pathologies occur at different ages, and the scintigraphic features are clinically important. For this reason, Chaps. 19 and 20 are concerned with these areas, whereby most of the images shown here can also be seen in the Chaps. 1–18 under the appropriate age grouping.

1: Age 0–6 Months

Fig. 1. Posterior view of skull, upper limbs and thorax

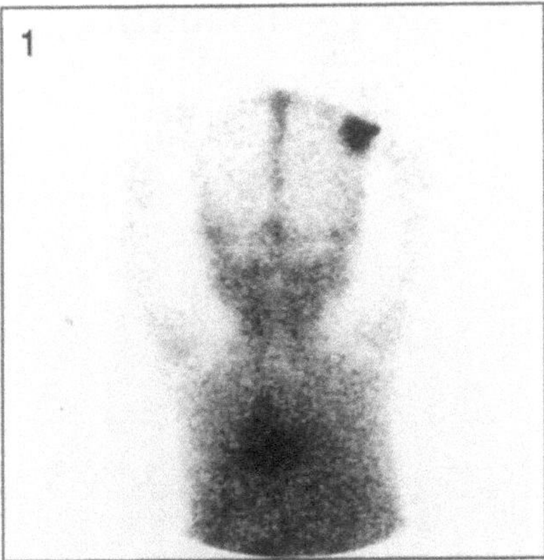

Fig. 2. Posterior view of spine, pelvis and lower limbs

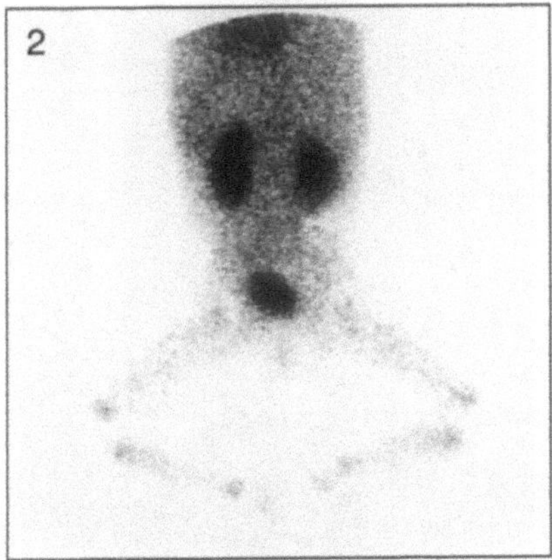

Technical Comments
– Note the sagittal suture in the skull in Fig. 1
– Note extravasation of isotope at the site of injection in the scalp in Fig. 1

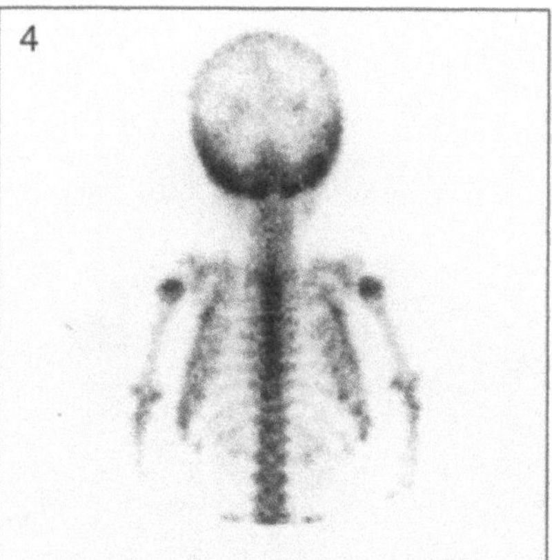

Fig. 4. Posterior view of skull, thorax and upper limbs

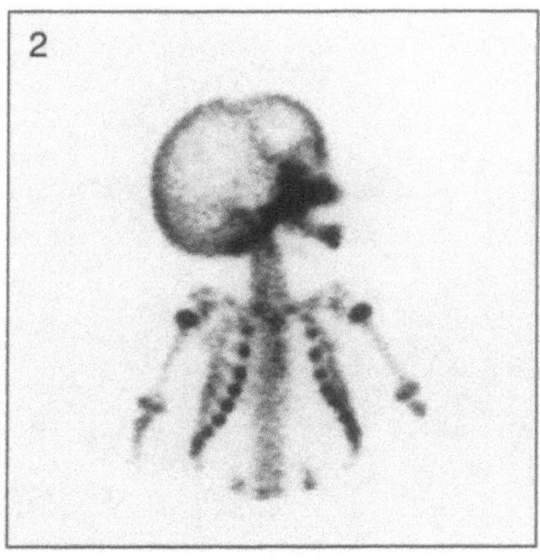

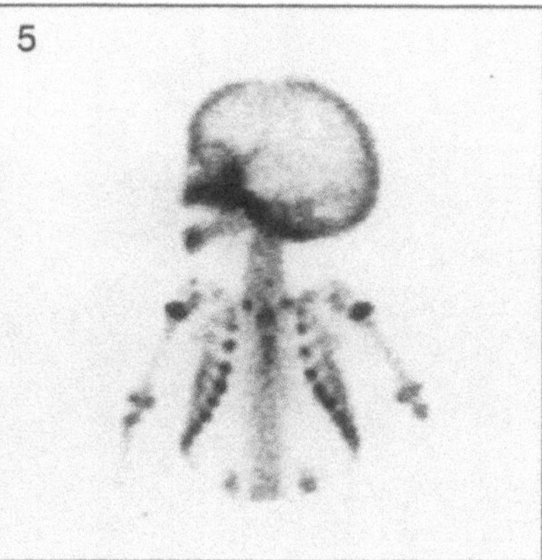

Fig. 2. Right lateral view of skull and anterior view of thorax and upper limbs

Fig. 5. Left lateral view of skull and anterior view of thorax and upper limbs

Technical Comments
- The coronal suture is seen in Figs. 2 and 5, note the difference in uptake between the left and right sides
- The transverse sinus is clearly seen in Fig. 4
- Note the increased activity at the growing ends of the bone, i.e. costo-chondral junctions and the epiphyseal plates of the upper humeri
- The indistinct lateral aspects of the ribs in Fig. 4 are due to "shine through" from the costo-chondral junctions

► **Potential Pitfalls**
- Decreased activity is noted in the region of the anterior fontanelle in Figs. 2 and 5. This should not be mistaken for a depressed fracture of the skull
- Posterior thorax shows increased activity laterally in Fig. 4 which should not be mistaken for multiple rib fractures. This increased activity is not in line with the posterior ribs and is due to the increased activity from the costo-chondral junctions

Fig. 1. Anterior view of skull and thorax

Fig. 4. Posterior view of skull, thorax and upper limbs

Fig. 2. Right lateral view of skull, right upper limb and anterior view of thorax

Fig. 5. Left lateral view of skull, left upper limb and anterior view of thorax

Fig. 3. Anterior view of thorax, pelvis and upper limbs

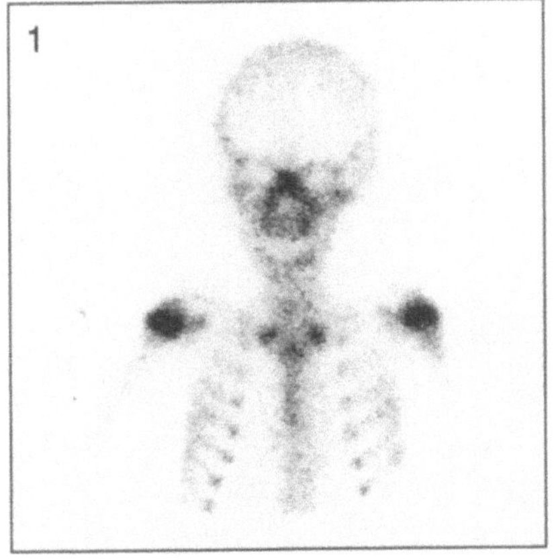
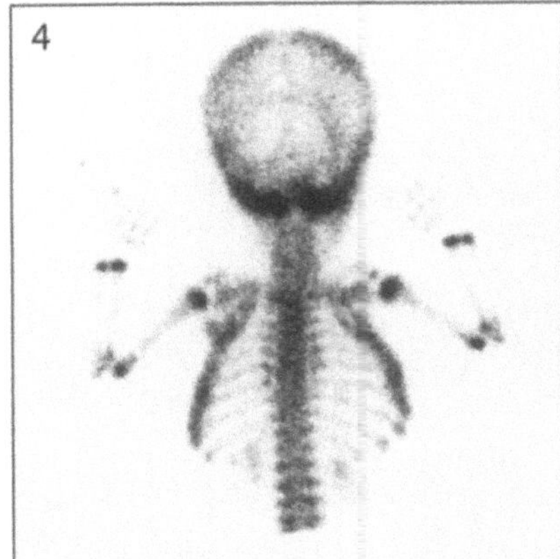
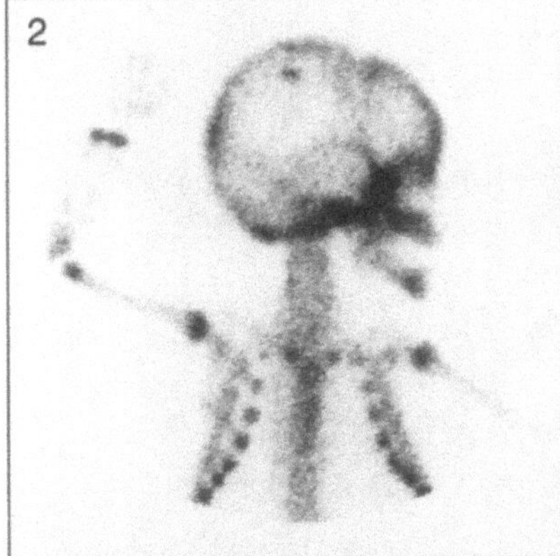
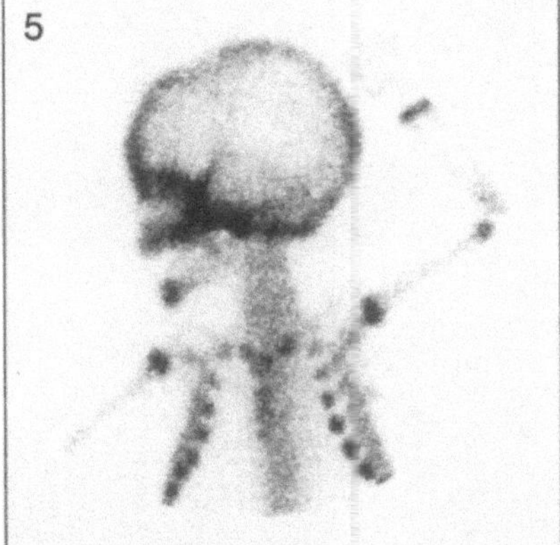
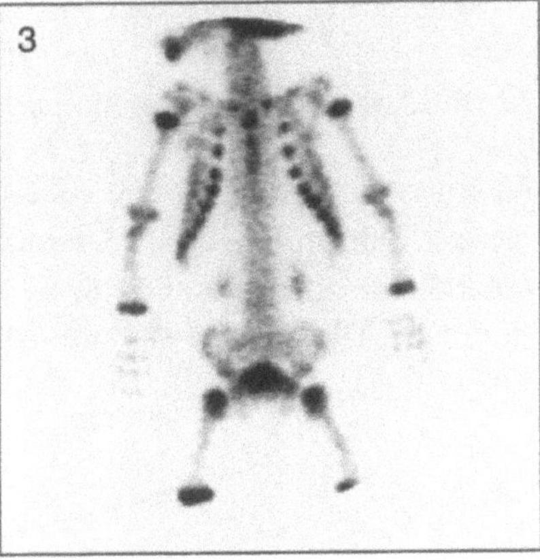

Technical Comments

- At this age the upper limbs may be imaged either with the skull (Figs. 2 and 5) or thorax (Fig. 3). Good positioning allows clear separation of the distal radius from the ulna
- Note extravasation of isotope at the site of injection in the scalp in Fig. 2

► Potential Pitfalls

- Decreased activity in the region of the anterior fontanelle in Figs. 2 and 5 should not be mistaken for a depressed fracture of the skull
- Increased activity seen laterally in Fig. 4 should not be mistaken for multiple rib fractures. This increased activity is not in line with the posterior ribs and is due to the increased activity in the region of the costo-chondral junctions. Also see p. 9

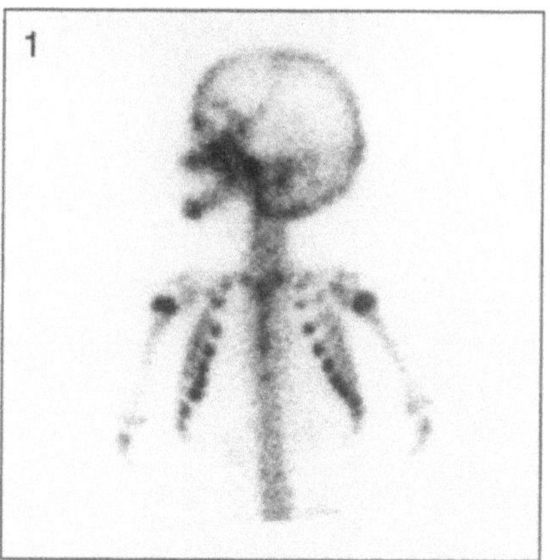

Fig. 1. Left lateral view of skull and anterior view of thorax

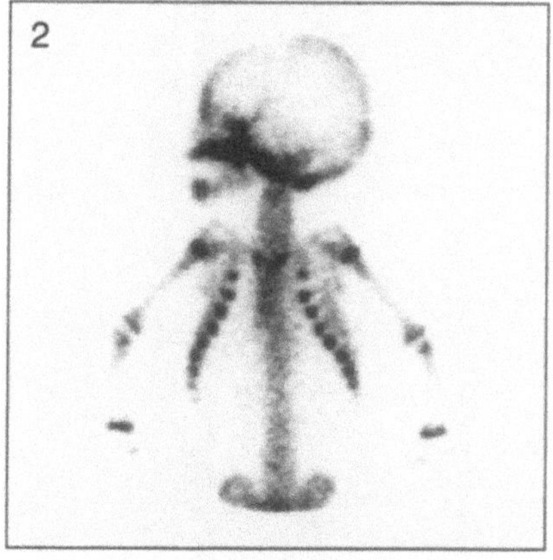

Fig. 2. Left lateral view of skull, anterior view of thorax and upper limbs

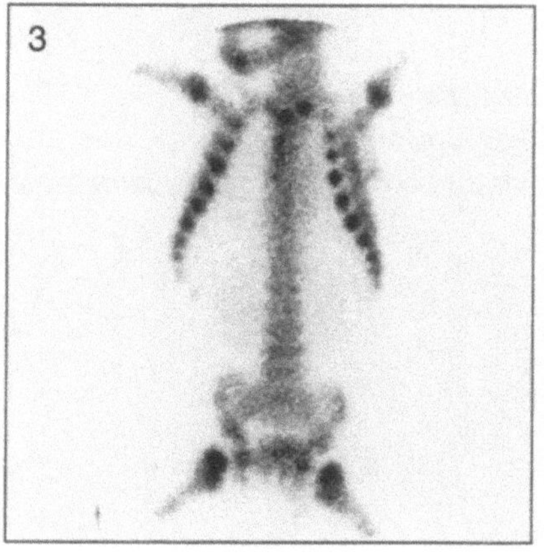

Fig. 3. Anterior view of thorax, spine and pelvis

Technical Comment
- Note the clarity of the lower lumbar spine on the anterior view in Fig. 3

▶ **Potential Pitfall**
- Note the difference in activity in the coronal sutures in Figs. 1 and 2. This variation makes the diagnosis of suture fusion difficult

Fig. 1. Right anterior oblique
view of thorax and pelvis

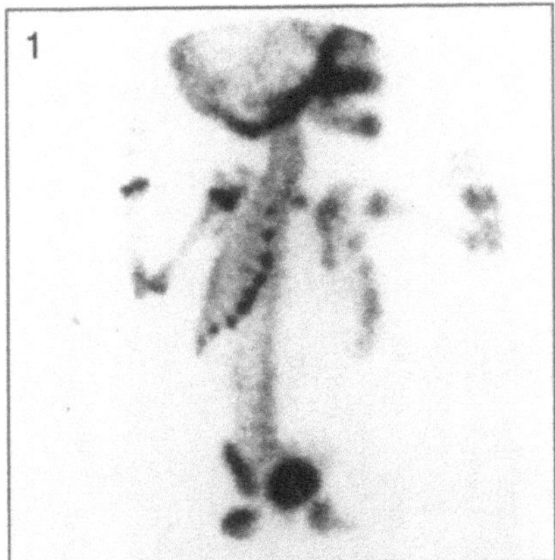

Fig. 2. Left anterior oblique view
of thorax and pelvis

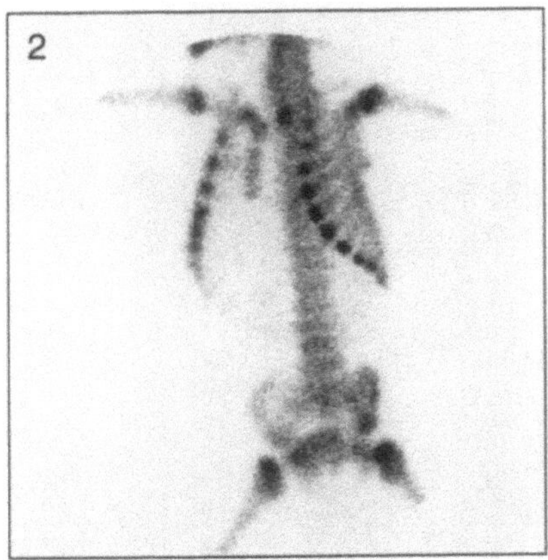

Technical Comments
- The sternum is well seen in both images
- Note the high activity in the full bladder in Fig. 1
- The anterior view of pelvis should not be obtained at the same time as
  the oblique views of the thorax

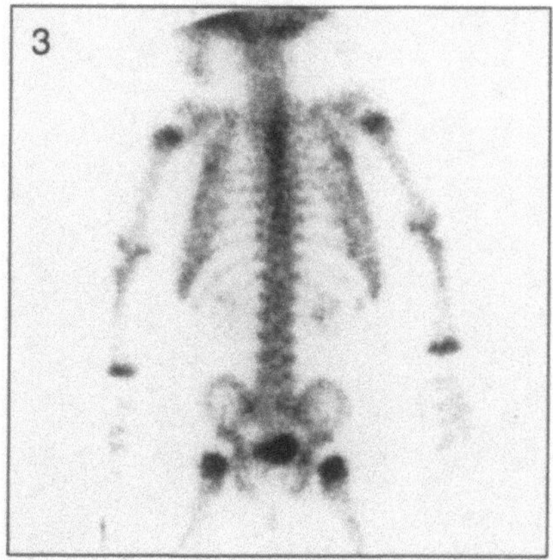

Fig. 1. Posterior view of thorax, spine, pelvis and upper limbs

Fig. 2. Posterior view of thorax, spine, pelvis and upper limbs

Fig. 3. Posterior view of thorax, spine, pelvis and upper limbs

Technical Comments
- The epiphysis of the scapula is seen in Fig. 2
- Note the difference in renal excretion, in Fig. 2 both kidneys are clearly seen, this is within normal limits
- Note lack of differentiation between distal radius and ulna due to poor positioning of the hands in all images

▶ **Potential Pitfall**
- Note the loss of clarity of the ribs in the axilla on all three images. This is due to the increased activity at the costo-chondral junctions and is not pathological

Fig. 1. Anterior view of spine and pelvis

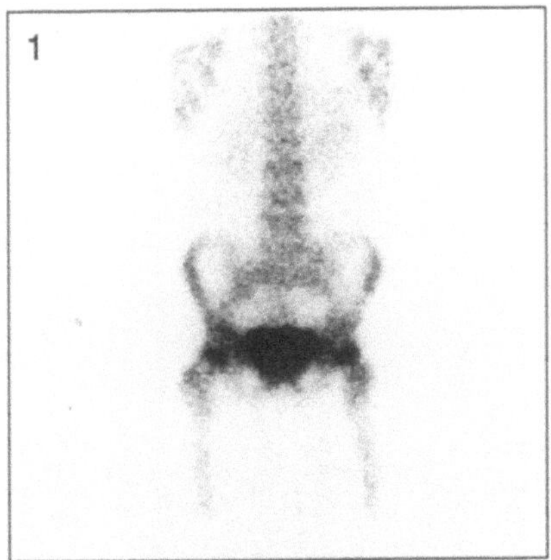

Fig. 2. Anterior view of pelvis and lower limbs

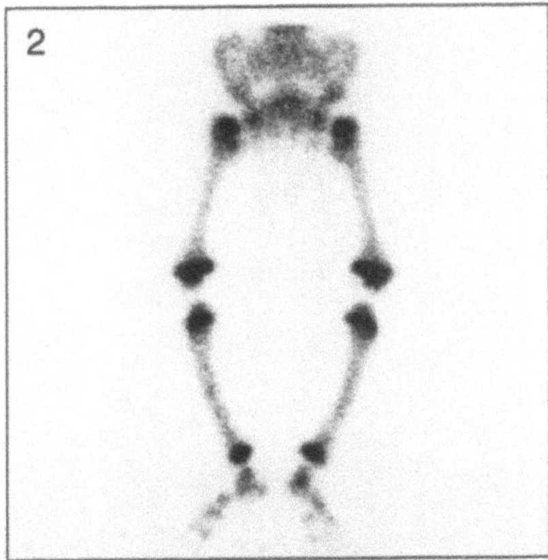

Fig. 3. Anterior view of pelvis and lower limbs

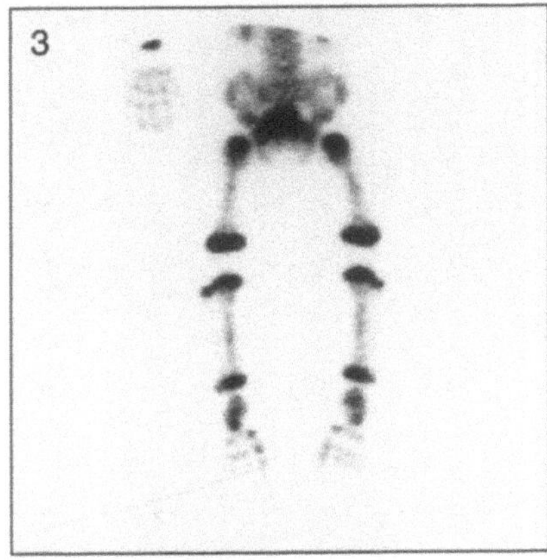

Technical Comments

- The full bladder in Figs. 1 and 3 precludes adequate visualization of the hips
- Note in Fig. 2 that the feet are facing laterally whilst in Fig. 3 the toes are facing medially, "the radiographic neutral position". This is the reason that the fibulae are clearly seen in Fig. 3 but not in Fig. 2

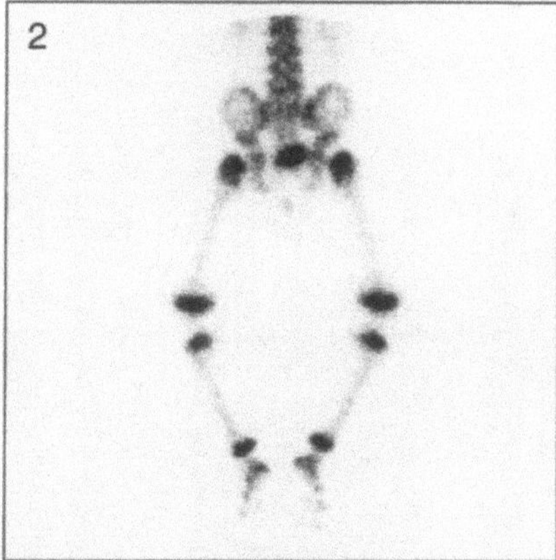

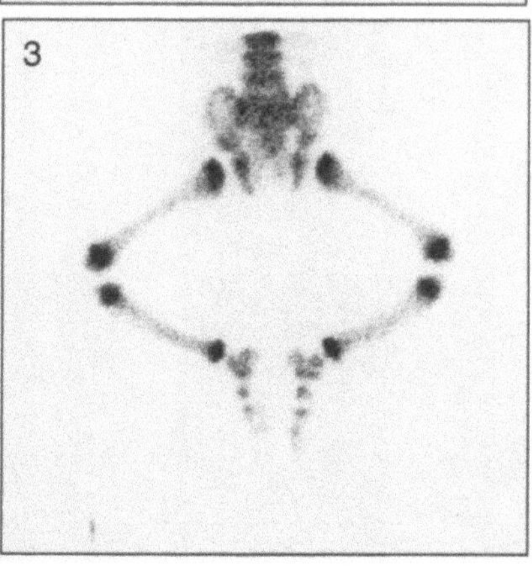

Fig. 1. Posterior view of thorax, spine and pelvis

Fig. 2. Posterior view of pelvis and lower limbs

Fig. 3. Posterior view of pelvis and lateral view of lower limbs

Technical Comments

– Urine contamination of the nappy below the pelvis is seen in Fig. 2
– Fig. 3 shows the projection which is required for images of the feet, this position is not ideal for imaging the knees (see Chap. 19: Knees)

Fig. 1. Posterior view of thorax, spine, pelvis and lower limbs

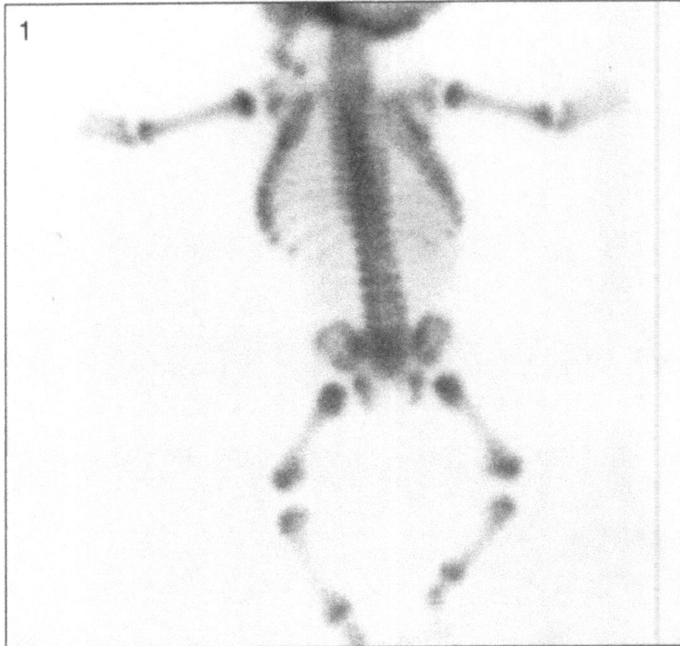

Fig. 2. Left lateral view of skull and posterior view of upper limbs, thorax, spine and pelvis

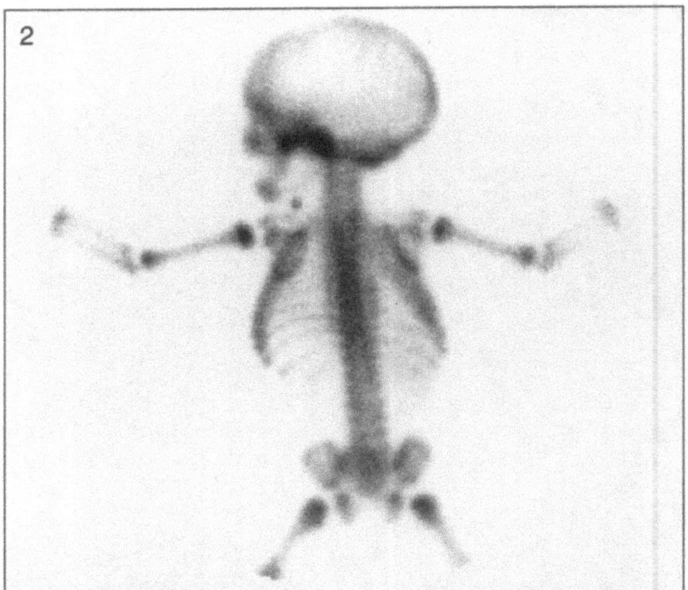

Technical Comments

- It is tempting to obtain whole body imaging on a large field of view gamma camera at this age. The quality of the images is unacceptable
- Spot images, with appropriate patient positioning and use of magnification, is preferable

2: Age 6–12 Months

Fig. 1. Posterior view of skull and thorax

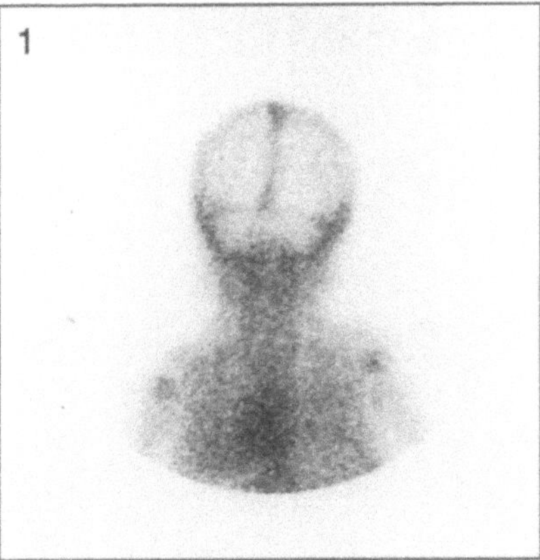

Fig. 2. Left lateral view of skull and posterior view of thorax

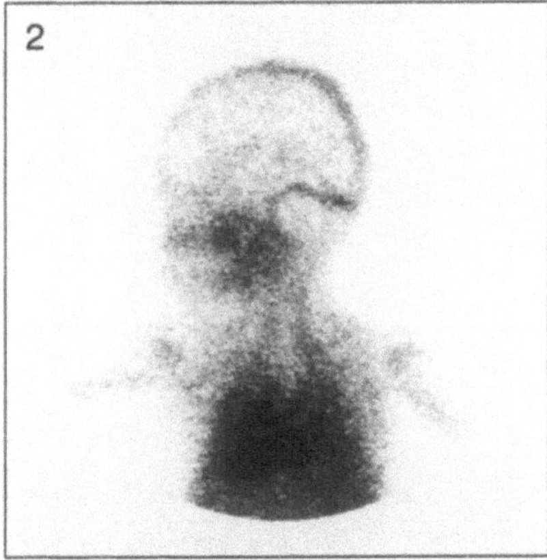

Fig. 3. Right lateral view of skull and posterior view of thorax

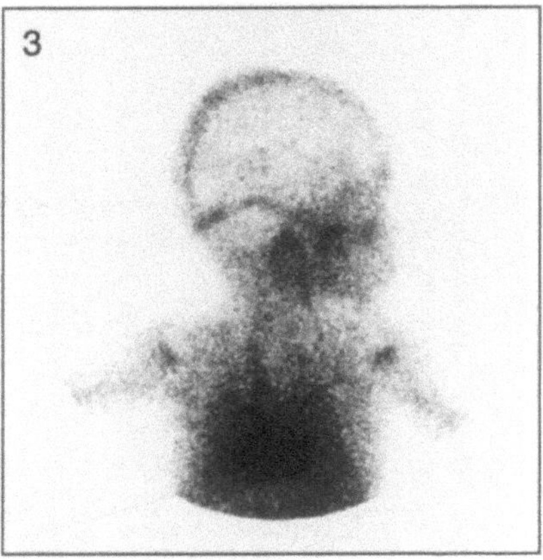

Technical Comments
- Note the clarity of the venous sinuses on all the views
- The "hot" epiphyseal plates at the upper ends of the humeri are also clearly seen, this is normal

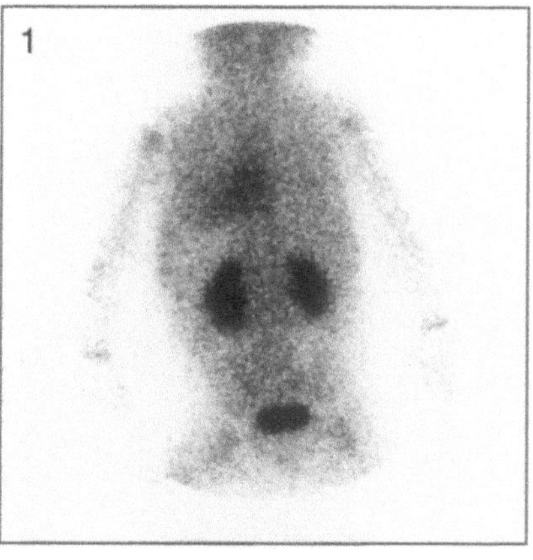

Fig. 1. Posterior view of thorax, spine, pelvis and upper limbs

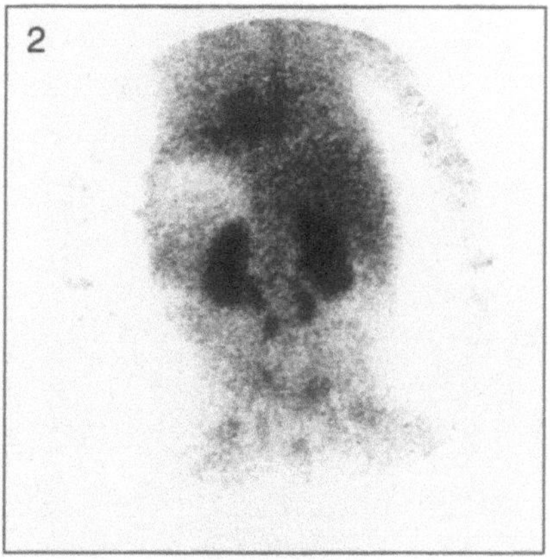

Fig. 2. Posterior view of thorax, spine, pelvis and upper limbs

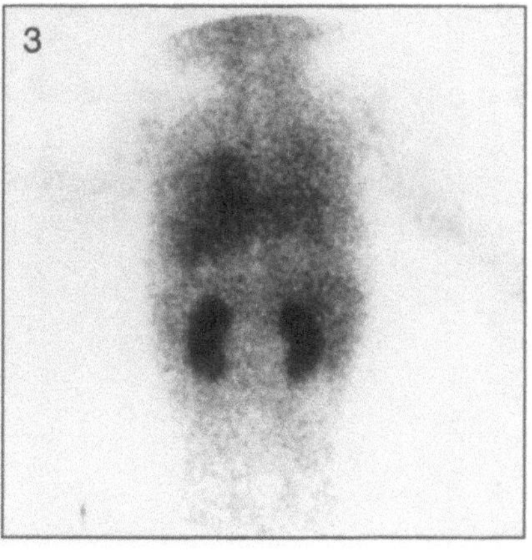

Fig. 3. Posterior view of thorax, spine and pelvis

Technical Comments
- Fig. 3 is an image 1 minute after the injection of tracer since no activity is yet seen in the bladder, while Figs. 1 and 2 are at 4–5 minutes with activity noted in the bladder
- Fig. 2 shows a photon deficient area above the left kidney. This is due to a full stomach

▶ **Potential Pitfall**
- Curvilinear radioactivity between and below the kidneys in Fig. 2 is probably due to the tortuous ureter, this is normal

Fig. 1. Posterior view of pelvis and lower limbs

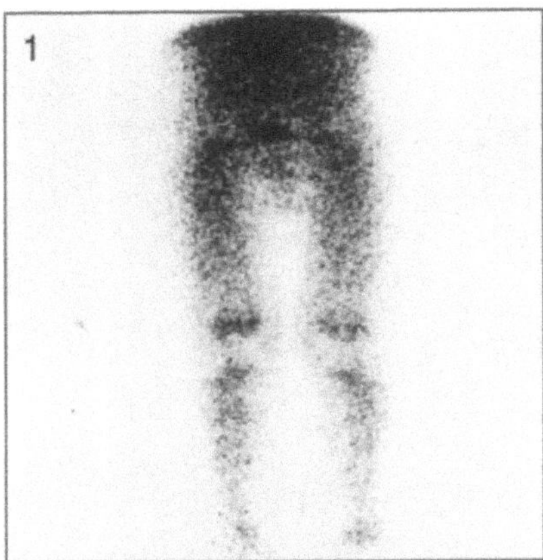

Fig. 2. Posterior view of pelvis and lateral view of lower limbs

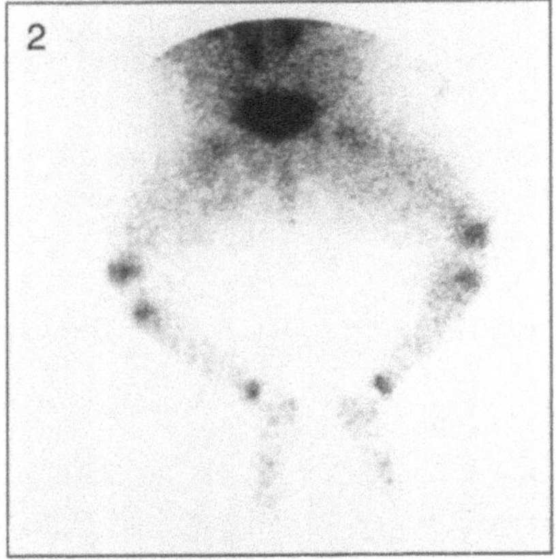

Technical Comments
- Note the normal increased perfusion of the epiphyseal plates of the lower limbs
- The position of the lower limbs in Fig. 1 is ideal for the knees and tibiae while in Fig. 2 the position is ideal for the feet
- Bladder activity is noted in Fig. 2

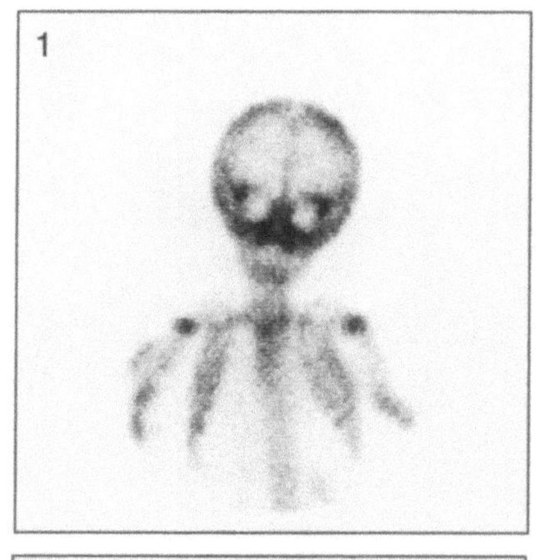

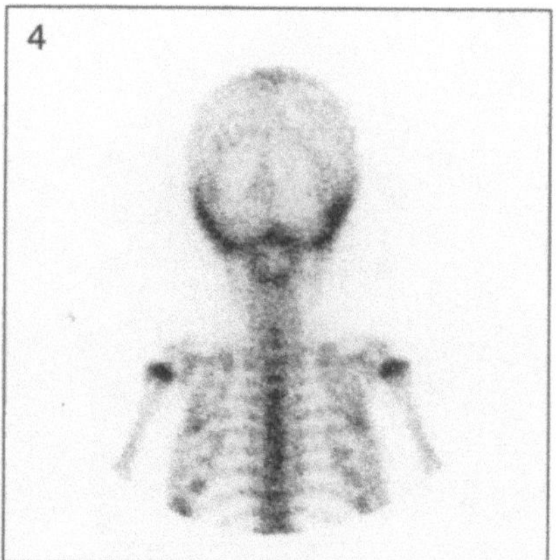

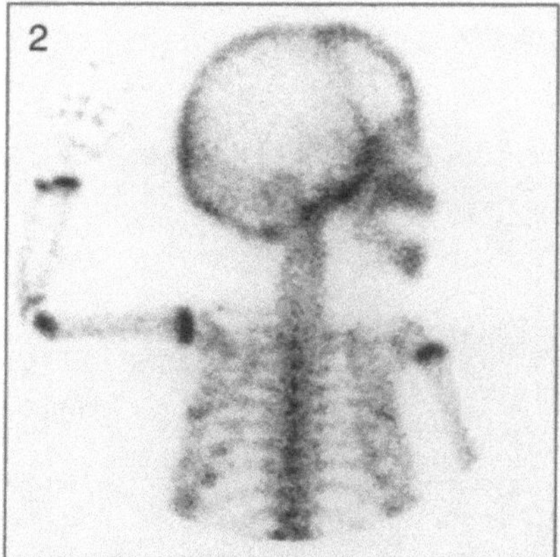

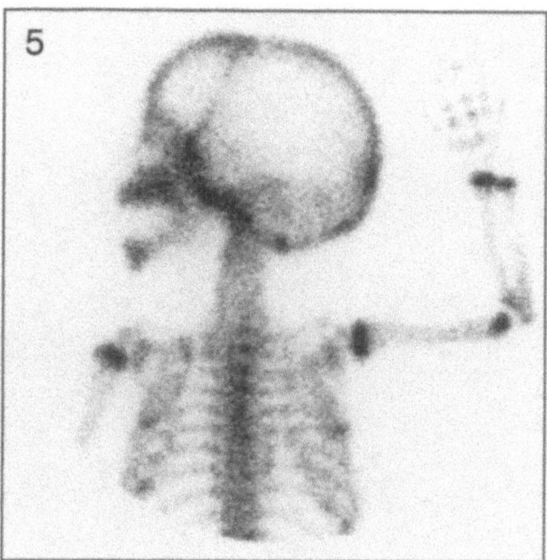

Fig. 1. Anterior view of skull and thorax

Fig. 4. Posterior view of skull and thorax

Fig. 2. Right lateral view of skull, left upper limb and posterior view of thorax

Fig. 5. Left lateral view of skull, right upper limb and posterior view of thorax

Technical Comment

- Note the difference between the coronal sutures on the lateral images in Figs. 2 and 5

► Potential Pitfalls

- Fig. 1 shows asymmetry in the activity of the skull vault. This is due to rotation and should not be mistaken for a subdural haematoma
- Focal increased activity is seen over the ribs laterally of Figs. 2, 4 and 5, this is due to activity from the costo-chondral junctions (see pp. 9 and 10)

Fig. 1. Anterior view of skull and thorax

Fig. 4. Posterior view of skull and thorax

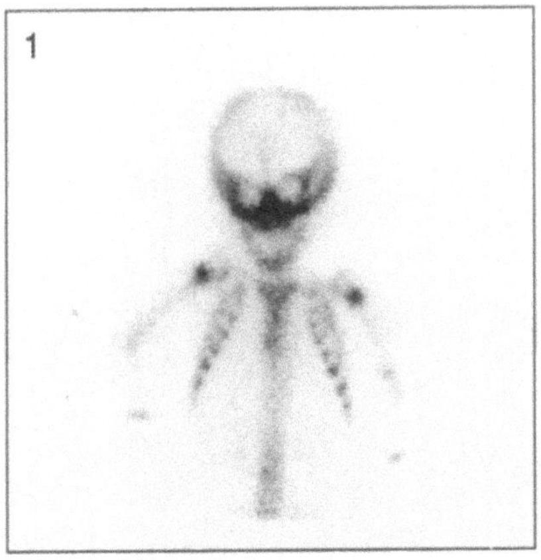

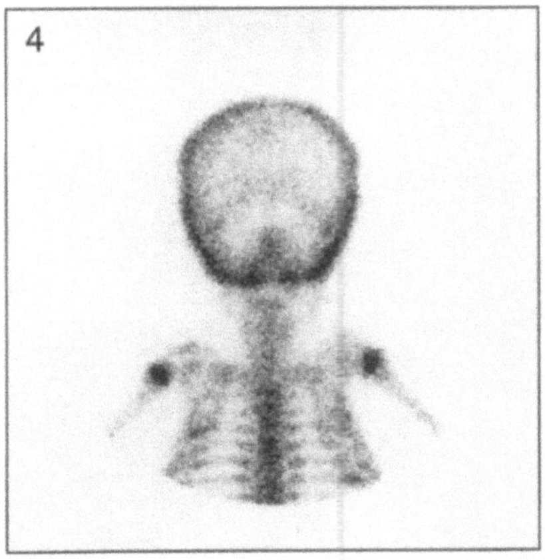

Fig. 2. Right lateral view of skull, right upper limb and right anterior oblique view of thorax

Fig. 5. Left lateral view of skull, left upper limb and anterior view of thorax

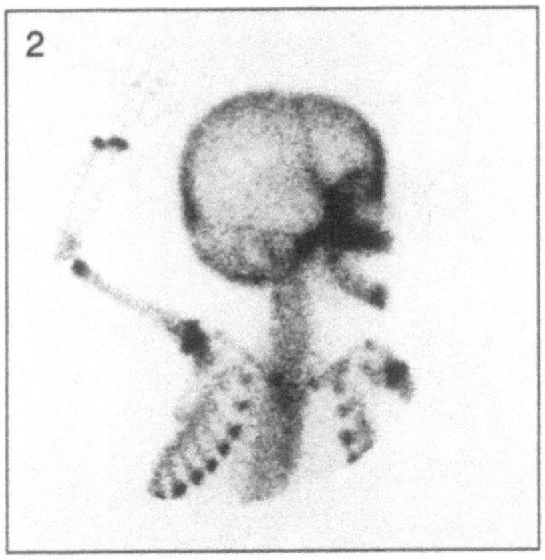

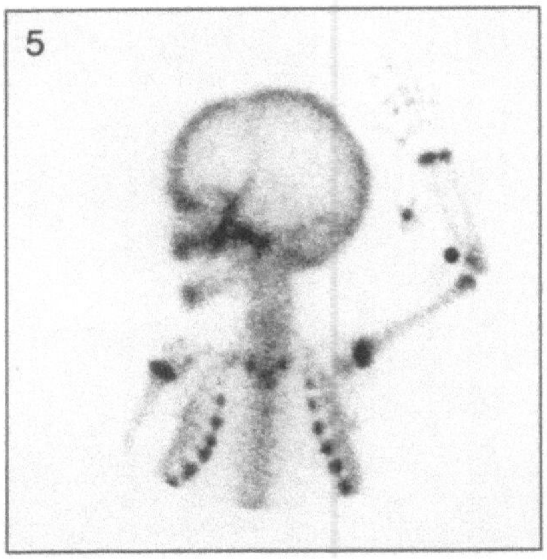

Technical Comments

- Focal increased activity is seen over the ribs laterally in Fig. 4, this is due to activity from the costo-chondral junctions (see p. 21)
- Note extravasation of isotope at the site of injection in the left elbow in Fig. 5

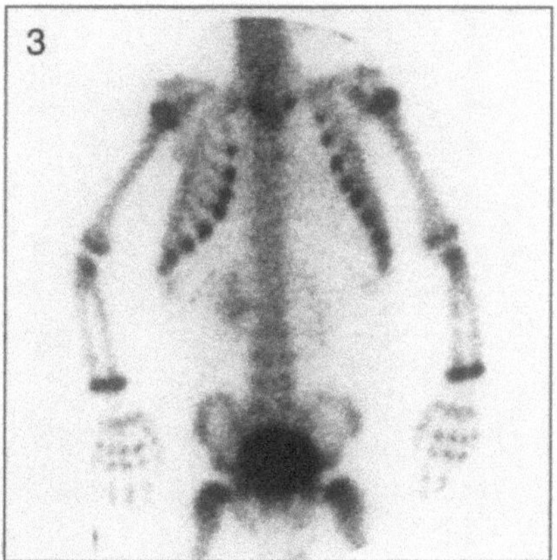

Fig. 1. Left lateral view of skull, anterior view of thorax and both upper limbs

Fig. 2. Anterior view of thorax and spine

Fig. 3. Anterior view of thorax, spine, pelvis and upper limbs

Technical Comments

- Note the position of the upper limbs in Fig. 1 versus Fig. 3, both are adequate at this age
- There is variability in the ossification of the sternum. The entire sternum is clearly seen in Fig. 2. This is not true in Fig. 1 or Fig. 3 (Note also Fig. 1, p. 21; Figs. 1 and 5, p. 22; Figs. 1 and 4, p. 24)
- In Fig. 3 the anterior thorax is obliquely positioned and the bladder is full of activity. These factors make this image difficult to interpret

▶ **Potential Pitfall**

- There is asymmetry between the kidneys in Fig. 3, with the right kidney having more activity than the left, this is a variation of normal

Fig. 1. Right anterior oblique view of thorax

Fig. 4. Left lateral view of skull, left anterior oblique view of thorax and upper limbs

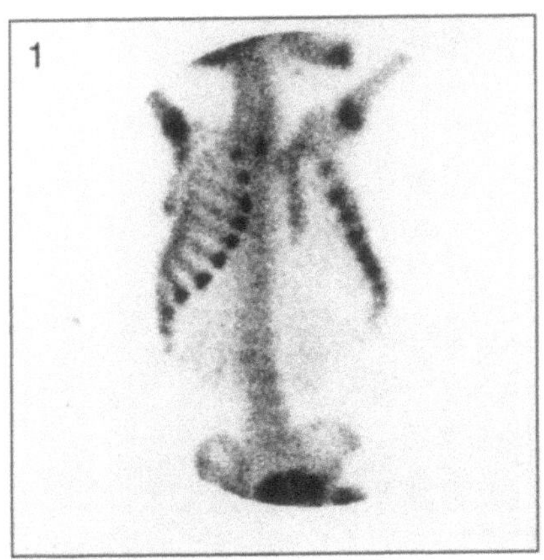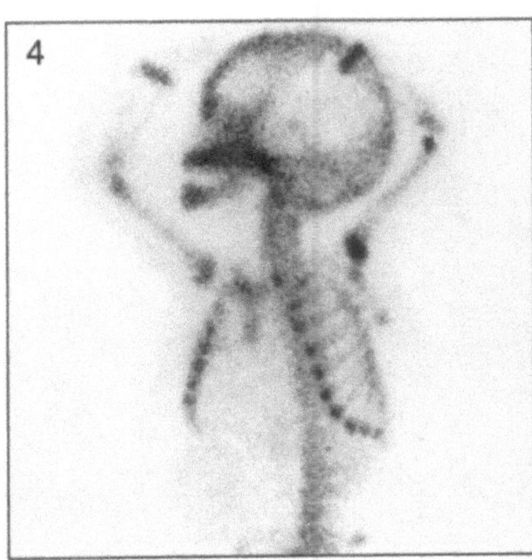

Technical Comment

– Fig. 4 is of poor quality because of the slight movement causing blurring of the upper limbs. The distal left radius and ulna overlie the skull, this is not ideal.

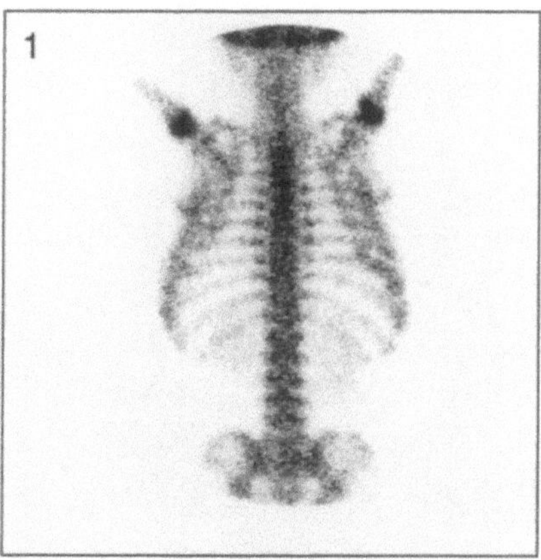

Fig. 1. Posterior view of thorax and spine

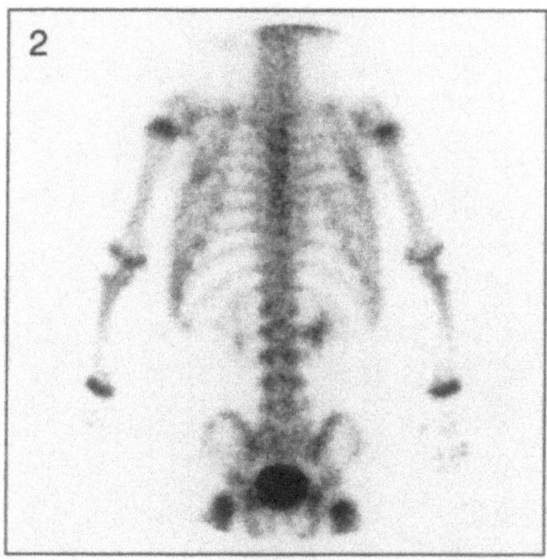

Fig. 2. Posterior view of thorax, spine, pelvis and upper limbs

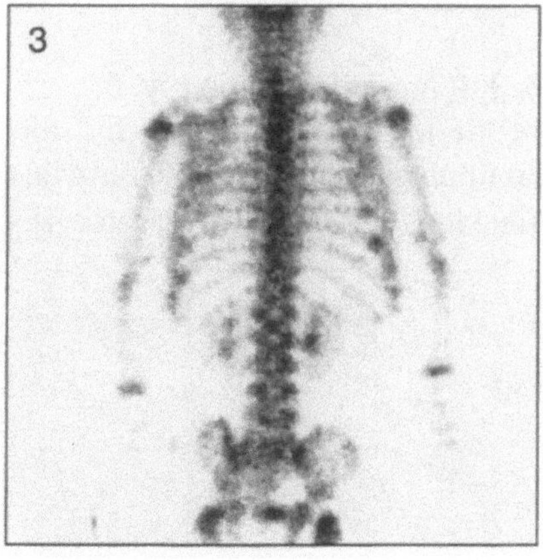

Fig. 3. Posterior view of thorax, spine, pelvis and upper limbs

Technical Comments
- Note the different positioning of the upper limbs, this results in variation of the appearances of the scapulae in Fig. 1 compared with Fig. 2
- The renal activity is different between the two kidneys in Figs. 2 and 3. These appearances are within normal limits

▶ **Potential Pitfall**
- Focal patchy increased activity is noted in the lateral aspect of the ribs in all images. This is due to the normal increased activity in the costochondral junctions "shining through" on this projection

Fig. 1. Anterior view of pelvis
and upper portion of lower limbs

Fig. 4. Posterior view of spine,
pelvis and upper portion of lower
limbs

Fig. 5. Posterior view of pelvis
and lower limbs

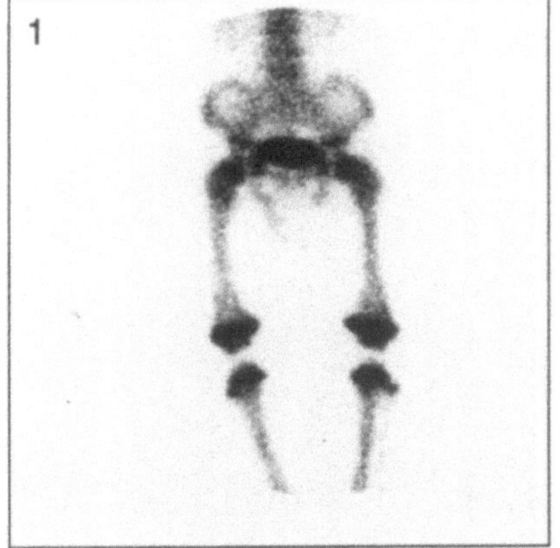

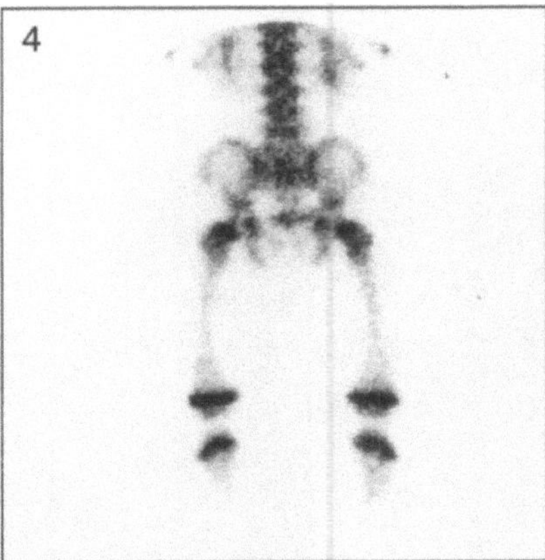

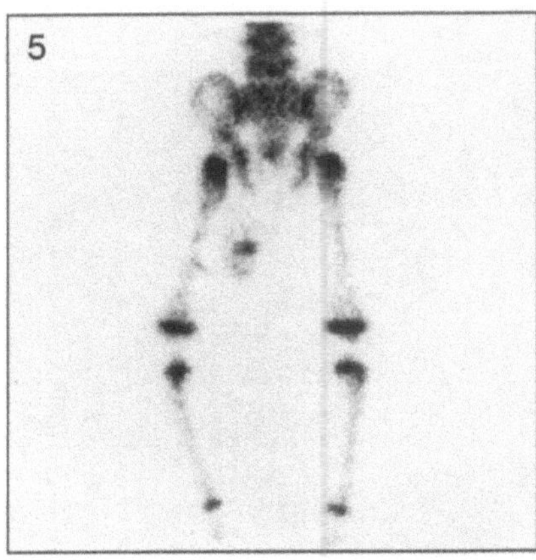

Technical Comments
- Urine contamination medial to the left femur is seen in Fig. 5
- Fig. 1 shows good positioning of the left knee and foot, this allows
  visualization of the left fibula. The fibula is not seen on the right due to
  poor positioning of both the knee and the foot, the same is seen in Fig. 5

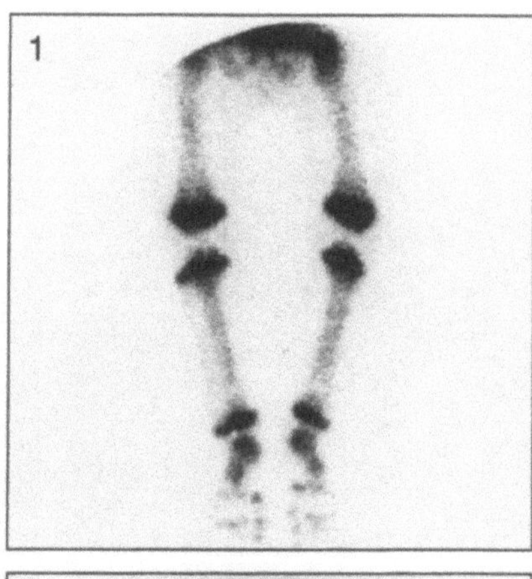

Fig. 1. Anterior view of lower limbs and feet

Fig. 4. Posterior view of lower limbs

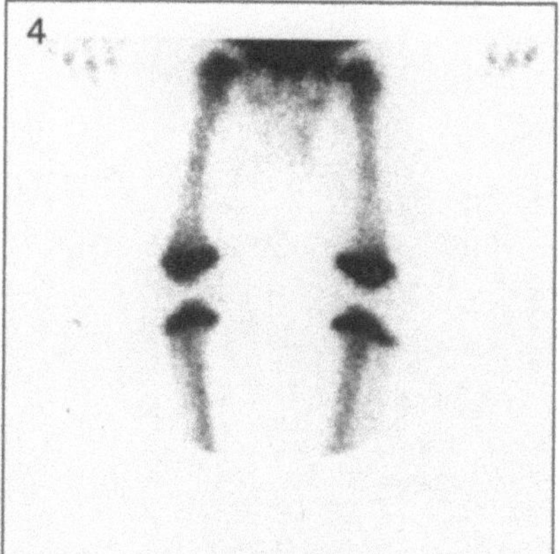

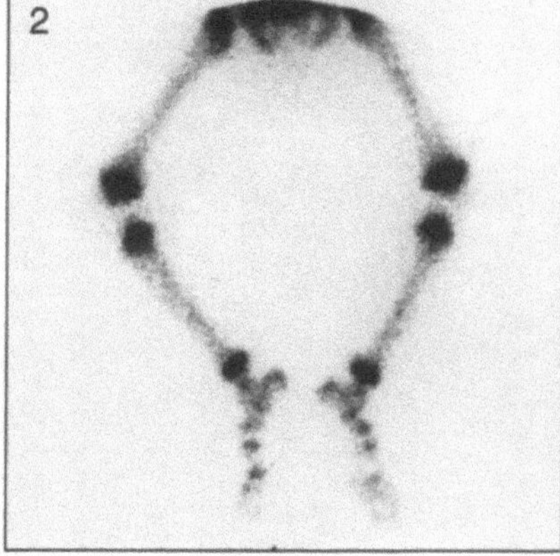

Fig. 2. Lateral view of lower limbs and feet

Fig. 5. Posterior view of lower limbs and feet

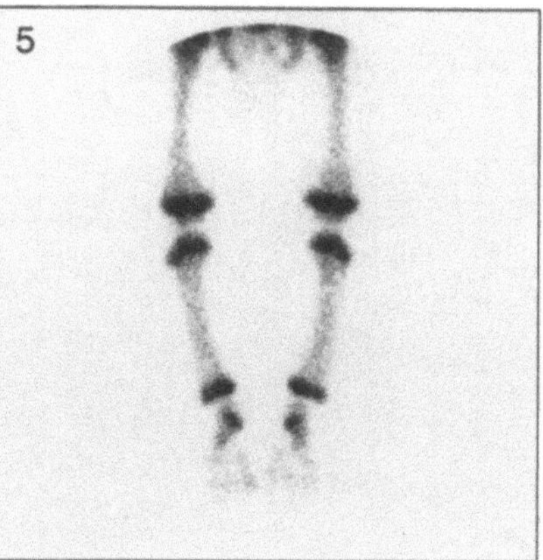

Technical Comments

- Note the shape of the epiphyseal plates around the knees
- Fig. 1 shows good positioning of the right knee and foot, this allows visualization of the right fibula. The fibula is not seen on the left due to poor positioning of both the knee and the foot, the same is seen in Fig. 5
- Fig. 2 is a lateral view of the lower limbs, this is good for the feet but inadequate for the knees

3: Age 1–2 Years

Fig. 1. Posterior view of skull and thorax

Fig. 4. Posterior view of thorax, spine and pelvis

Fig. 2. Posterior view of lower limbs

Fig. 5. Posterior view of lower limbs

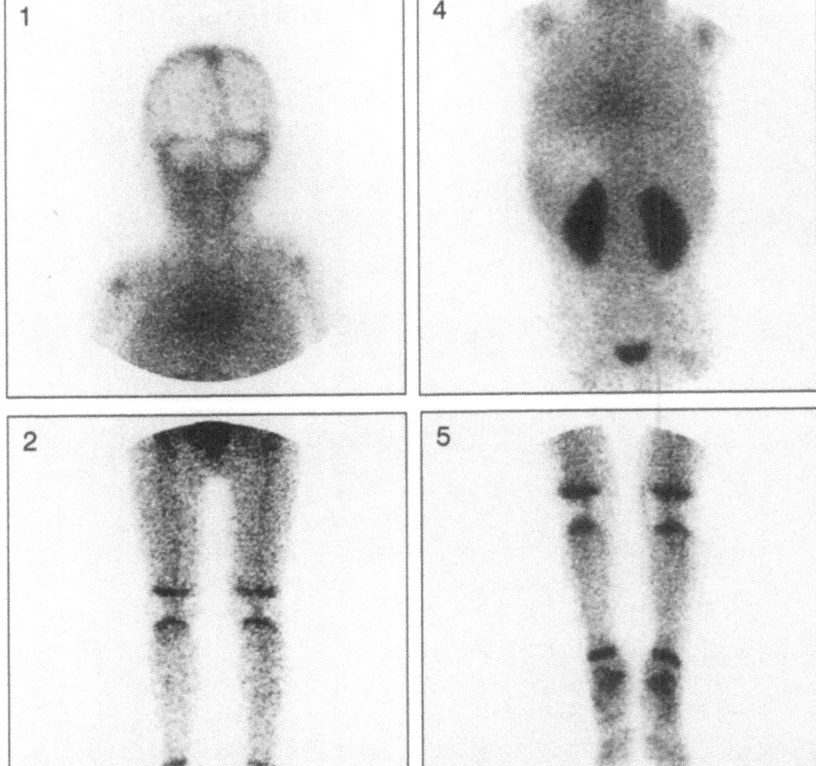

Technical Comments
- Note the photon deficient area above the left kidney in Fig. 4 due to a full stomach
- Slight rotation of the head in Fig. 1 causes an apparent deviation of the sagittal sinus

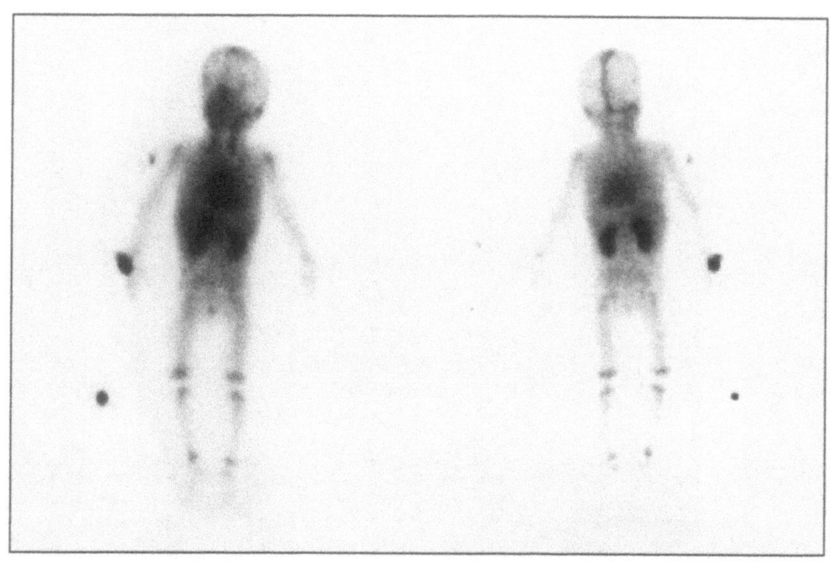

- A double headed whole body gamma camera was used
- Left image is the anterior view
- Right image is the posterior view

Technical Comments
- Note rotation of the head causing apparent deviation of the sagittal sinus
- Note extravasation of isotope at the site of injection in the right hand
- Marker on child's right side

3: Age 1–2 Years

- A single headed whole body gamma camera was used
- Left image is the anterior view
- Right image is the posterior view

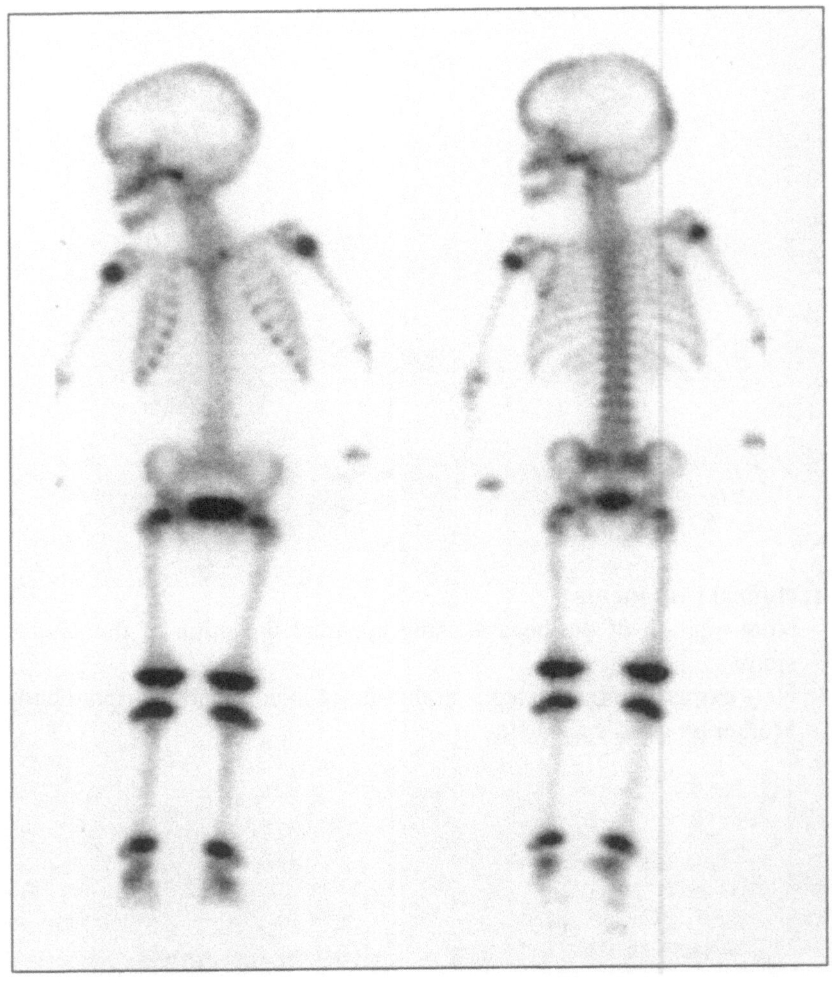

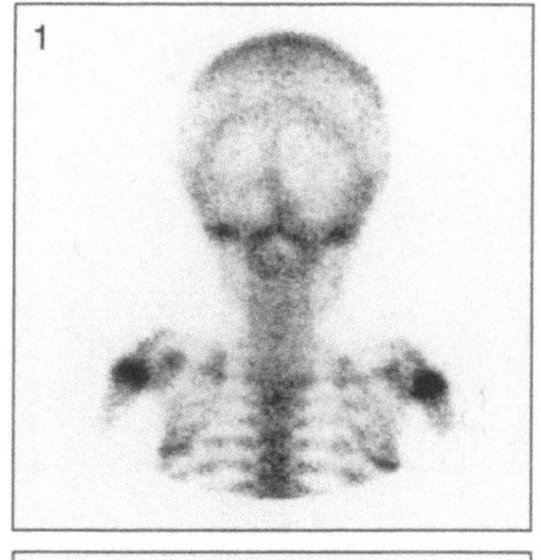

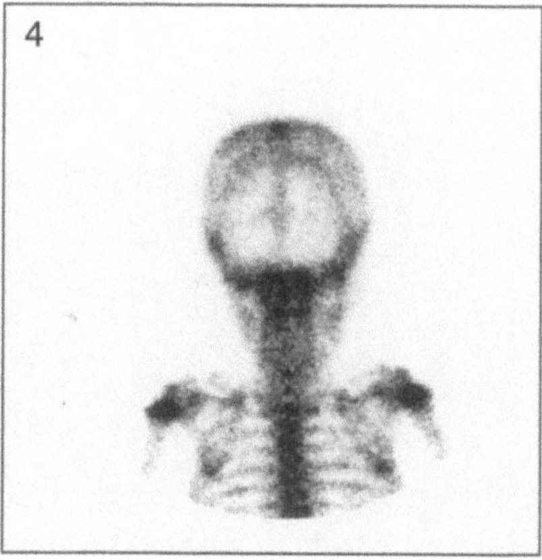

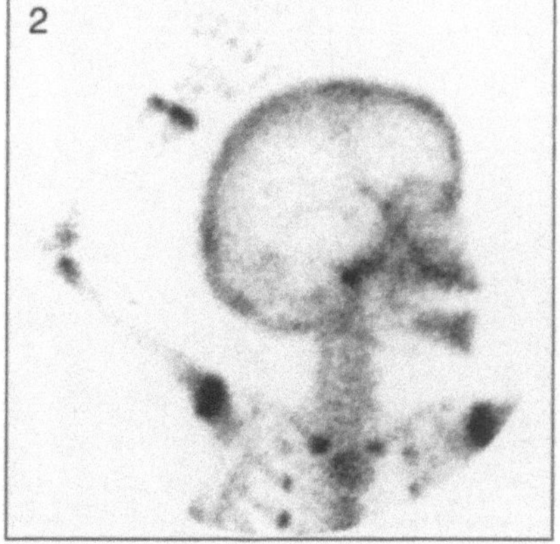

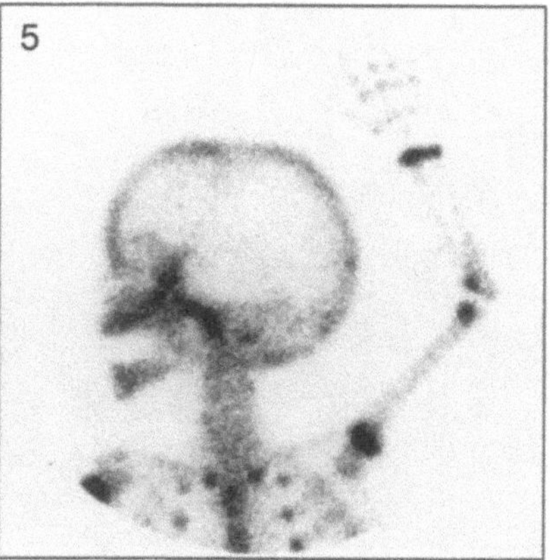

Fig. 1. Posterior view of skull and thorax

Fig. 4. Posterior view of skull with chin down and thorax

Fig. 2. Right lateral view of skull and right upper limb

Fig. 5. Left lateral view of skull and left upper limb

Technical Comments
- Note the different appearances of the skull in the posterior projection dependent on the degree of flexion of the head (see p. 34)
- Note the clear separation of the distal radius and ulna in Figs. 2 and 5
- The anterior ribs and sterno-clavicular joints are well seen Figs. 2 and 5
- The lateral views of the skull (Figs. 2 and 5) were taken anteriorly

Fig. 1. Posterior view of skull
and thorax

Fig. 4. Posterior view of skull
and thorax

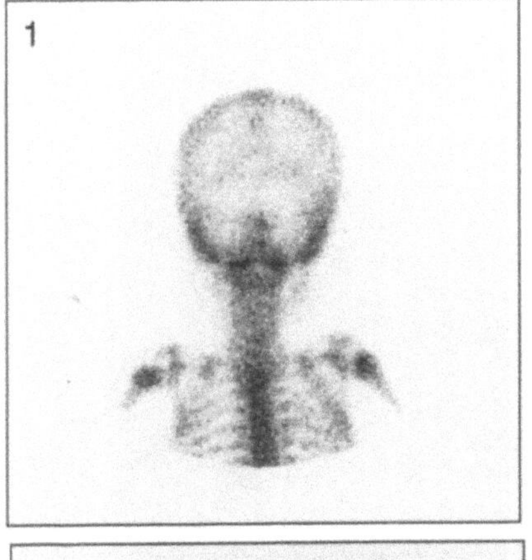

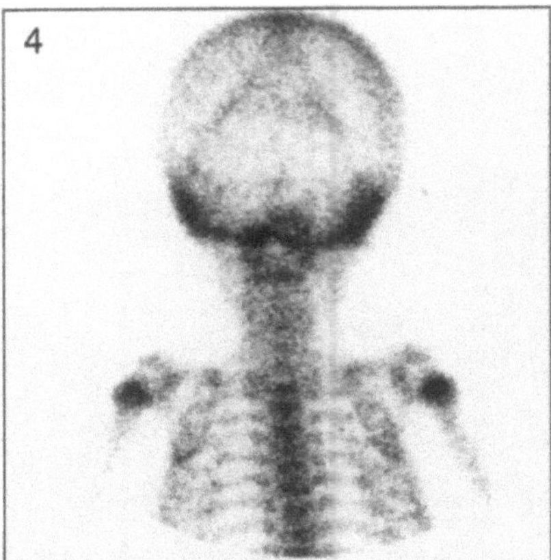

Fig. 2. Right lateral view of skull
and right upper limb

Fig. 5. Left lateral view of skull
and left upper limb

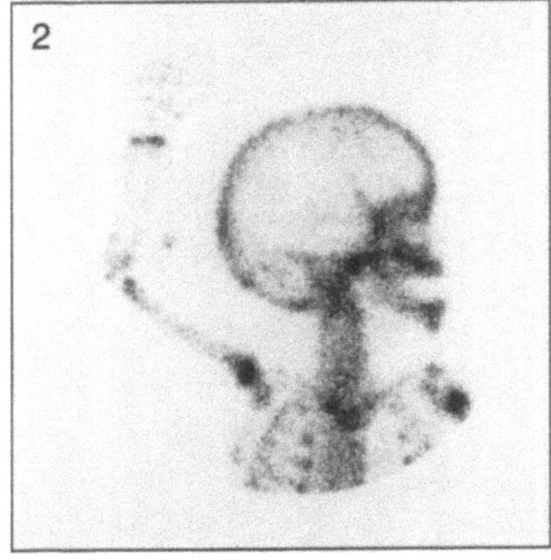

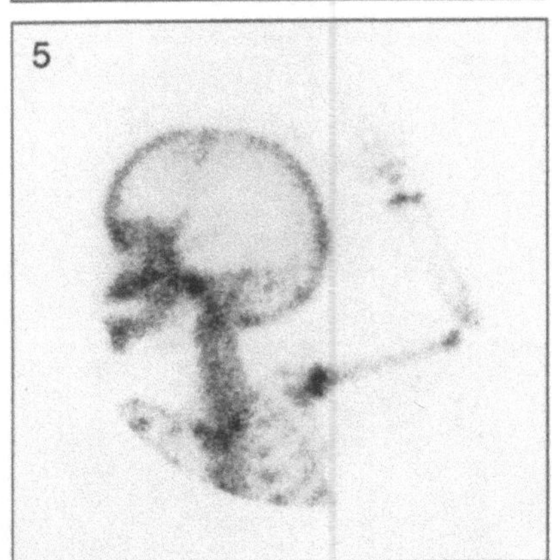

Technical Comments
- Note the different appearances of the skull in the posterior projection
  dependent on the degree of flexion of the head (see p. 33)
- The anterior ribs and sterno-clavicular joints are well seen in Figs. 2
  and 5
- The lateral views of the skull (Figs. 2 and 5) were taken anteriorly
- Note extravasation of isotope at the site of injection in the right elbow
  in Fig. 2

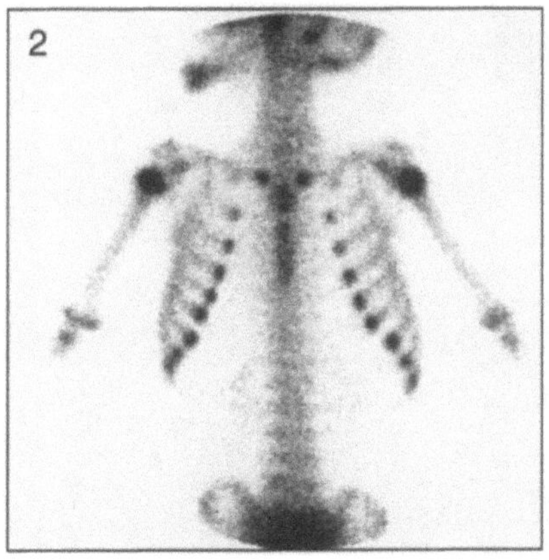

Fig. 1. Anterior view of thorax and spine

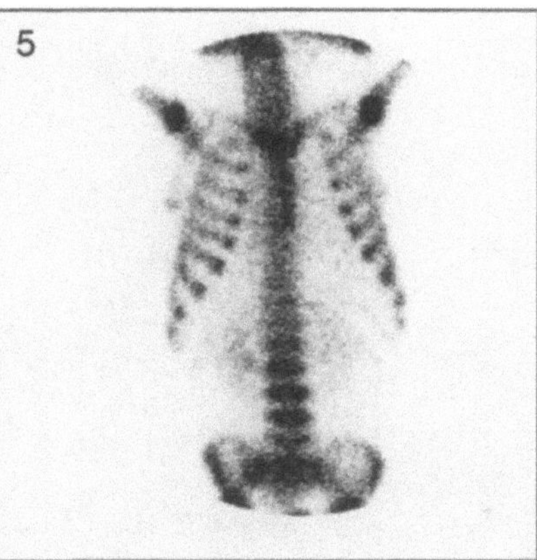

Fig. 2. Anterior view of thorax and spine

Fig. 5. Anterior view of thorax and spine

Technical Comments

- The variability in sternal ossification is clearly seen
- The lumbar vertebrae are very clearly seen in Figs. 1 and 5. This appearance on the anterior thoracic view of the lumbar spine will be seen on other anterior thoracic images
- The different positions of the upper limbs in the three illustrations result in the scapulae having a different appearance on the illustrations

Fig. 4. Posterior view of thorax and spine

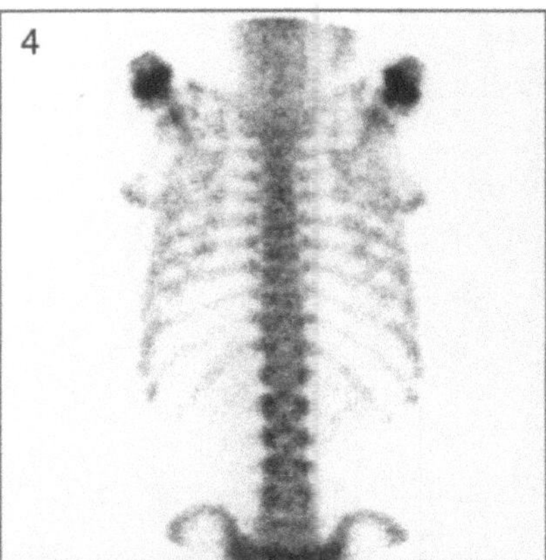

Fig. 2. Anterior view of thorax and spine

Fig. 5. Posterior view of thorax and spine

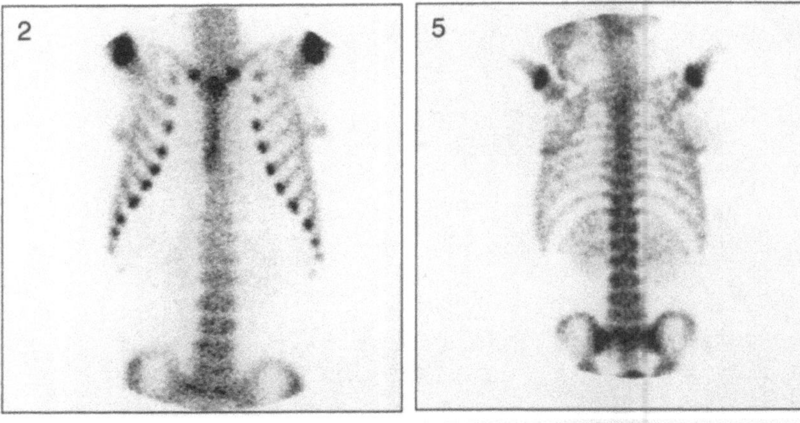

Fig. 6. Posterior view of thorax and spine

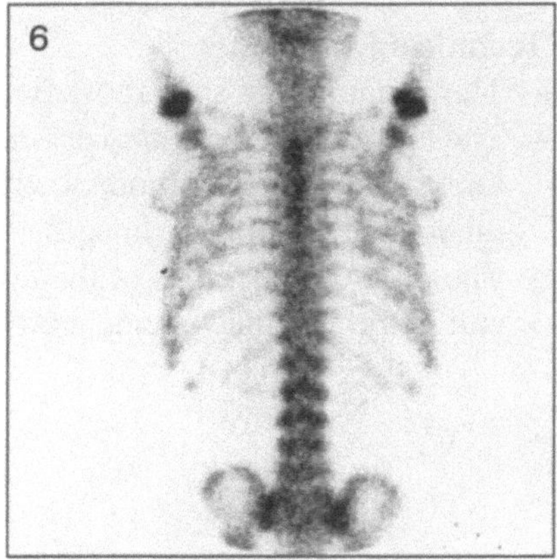

▶ **Potential Pitfall**
- Increased activity in the lateral aspect of the ribs posteriorly best seen in Figs. 4 and 6 are due to the normal increased activity in the costo-chondral junction clearly seen in Fig. 2

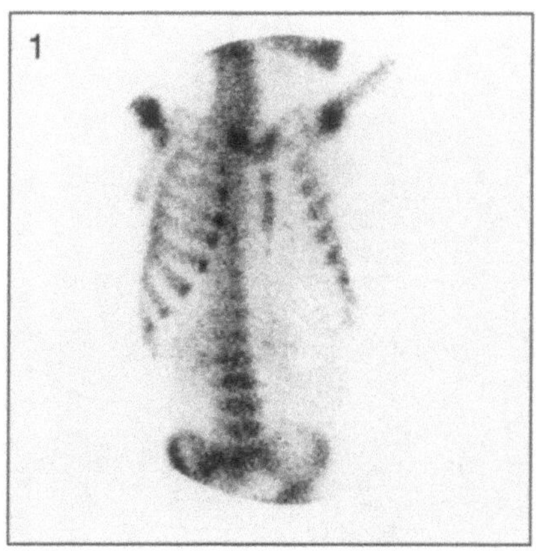

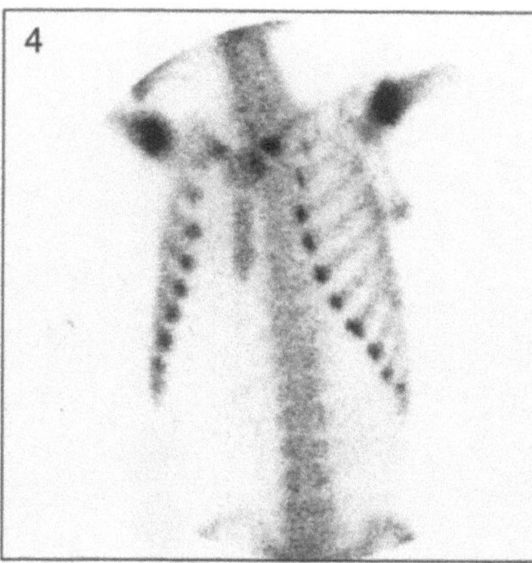

Fig. 1. Right anterior oblique view of thorax

Fig. 4. Left anterior oblique view of thorax

Technical Comment
– Note the variability in sternal ossification

Fig. 1. Anterior view of pelvis and femora

Fig. 4. Posterior view of pelvis and femora

Fig. 2. Anterior view of pelvis and femora

Fig. 5. Posterior view of pelvis and femora

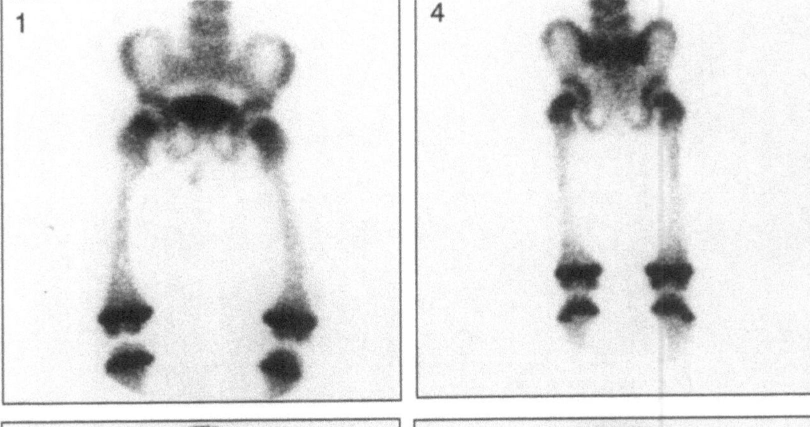

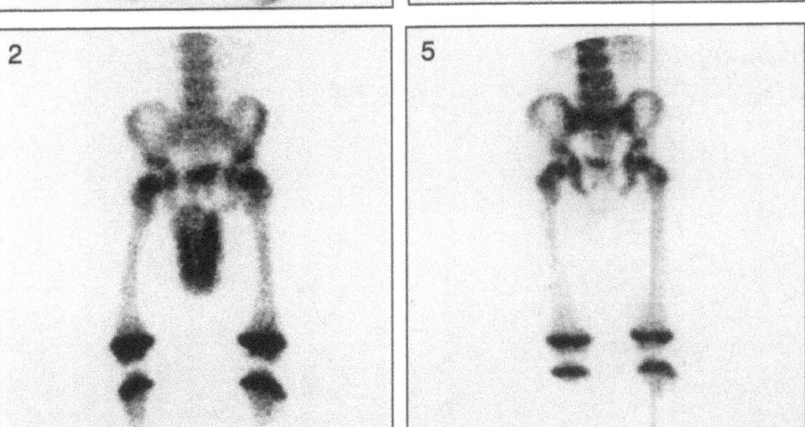

Technical Comment
– Urine contamination below the pelvis is seen in Figs. 1 and 2

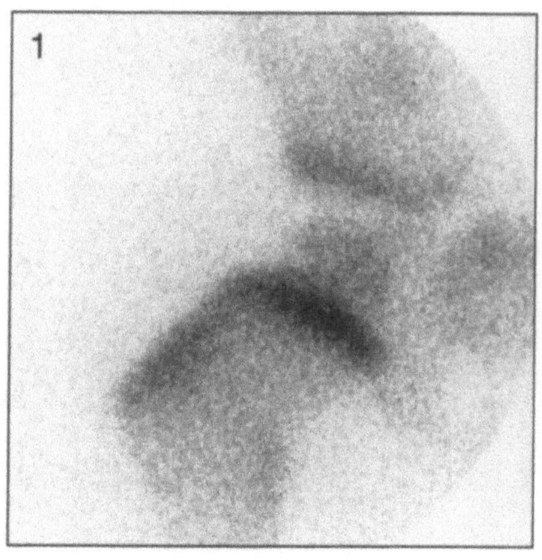 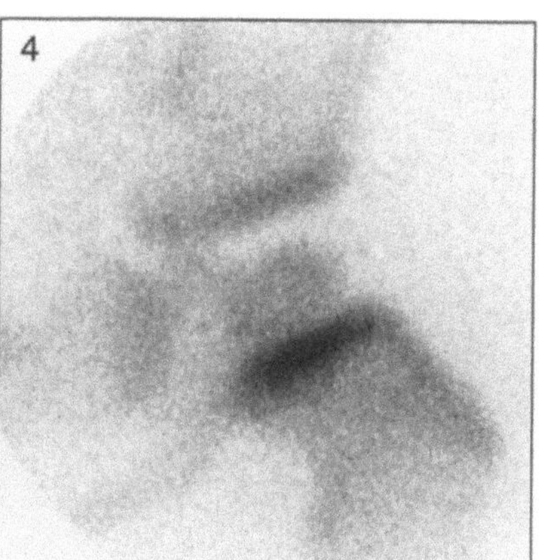

Fig. 1. Pinhole view of right hip

Fig. 4. Pinhole view of left hip

Figs. 1. Posterior view of lower limbs and feet

Figs. 4. Anterior view of lower limbs and feet

Figs. 2. Lateral view of lower limbs and feet

Figs. 5. Lateral view of lower limbs and feet

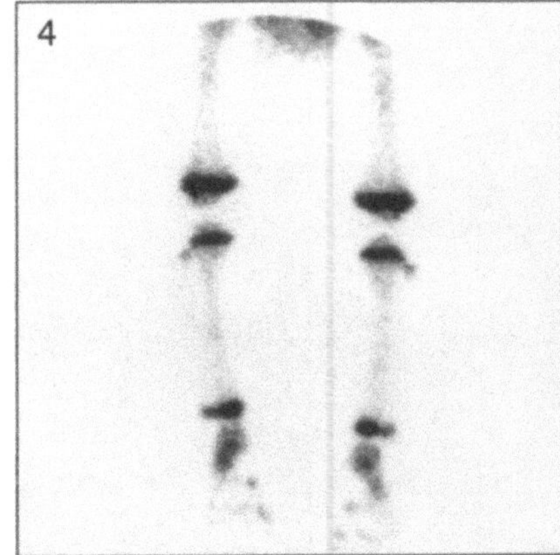

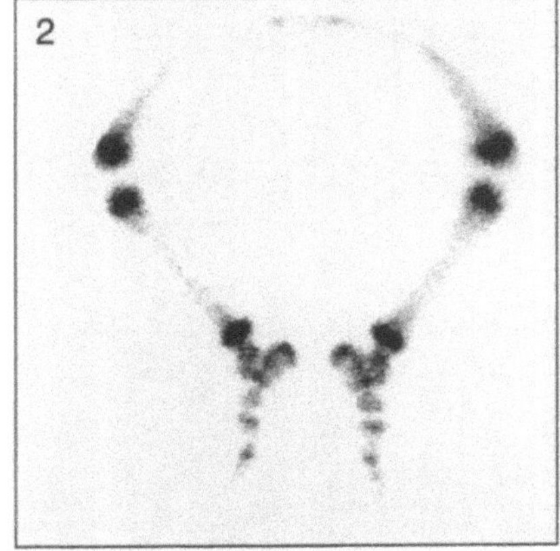

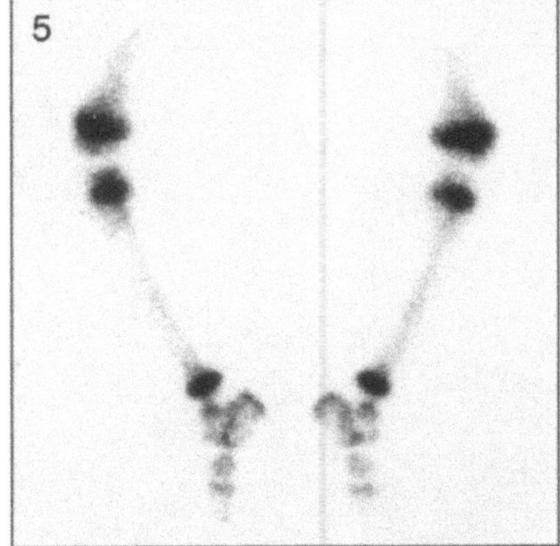

Technical Comments

- Figs. 1 and 4 are good positioning for the knees, tibia and fibula. The fibula is clearly seen separate from the tibia
- Figs. 2 and 5 are the position for imaging the feet. This position is inadequate for the knees since there is overlapping of the epiphyseal plates of the tibia and fibula

4: Age 2–3 Years

4: Age 2–3 Years

Fig. 1. Posterior view of skull
and upper limbs

Fig. 4. Left lateral view of skull

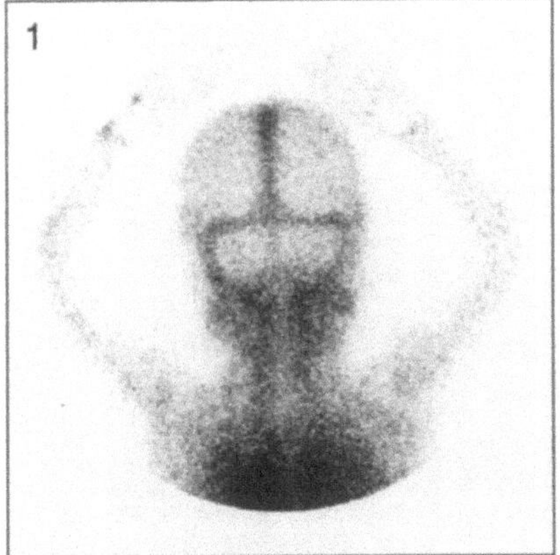

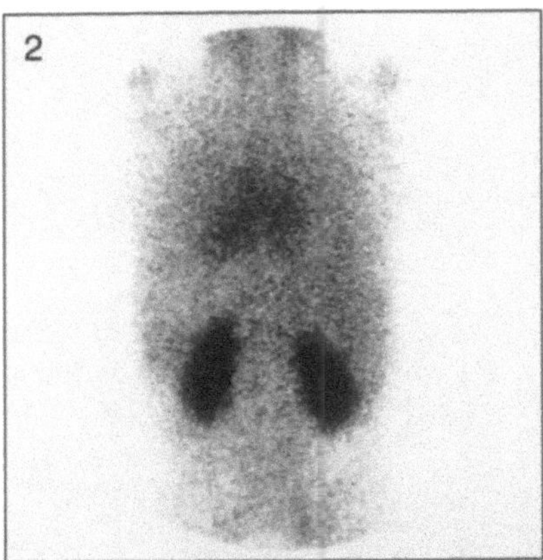

Fig. 2. Posterior view of thorax
and spine

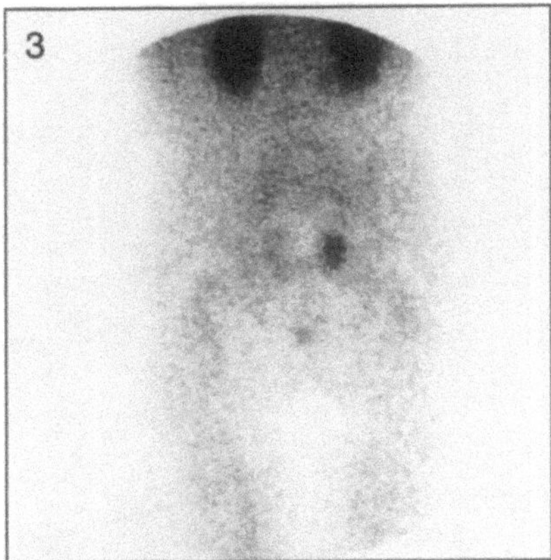

Fig. 3. Posterior view of spine,
pelvis and lower limbs

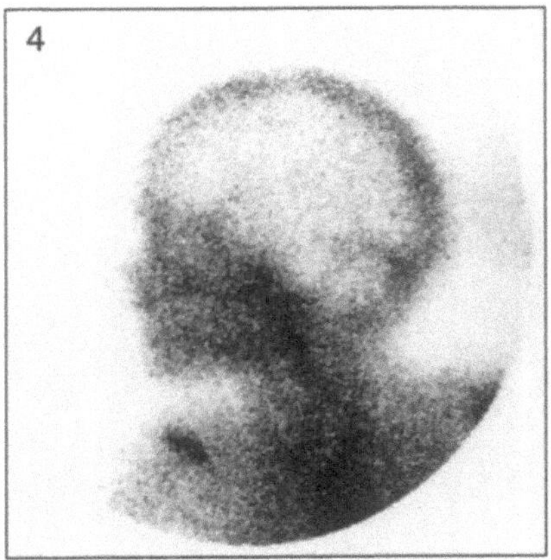

Technical Comments
- Note the venous sinuses in Fig. 1
- Focal accumulation of isotope in the mid portion of the image in Fig. 3, this is due to isotope in the bladder
- Note extravasation of isotope at the site of injection in the left hand in Fig. 1

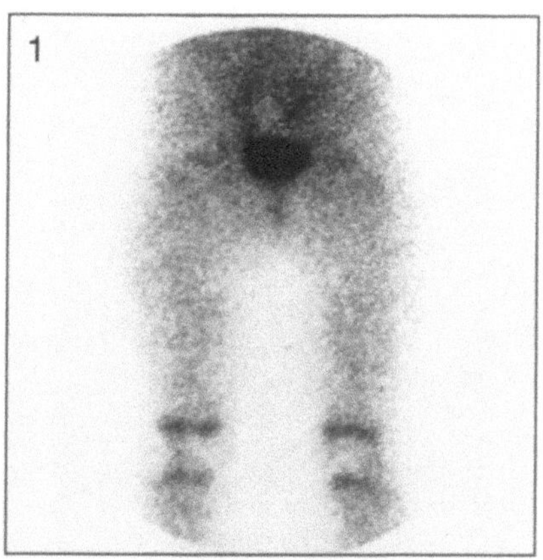

Fig. 1. Posterior view of pelvis and upper part of lower limbs

Fig. 2. Posterior view of lower part of lower limbs

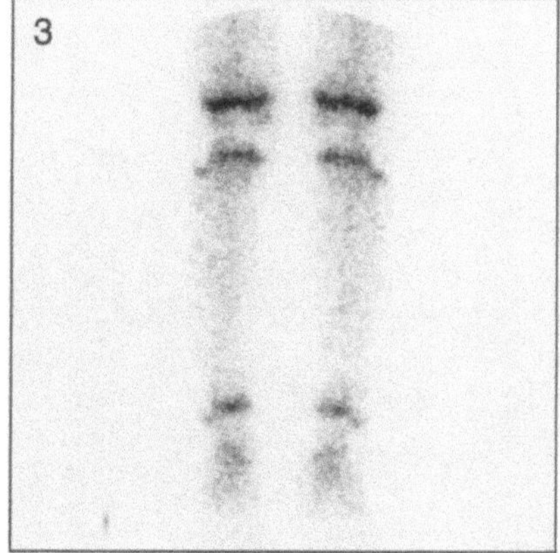

Fig. 3. Posterior view of lower part of lower limbs

Technical Comments
- Note the vascularity of the epiphyseal plates showing up as areas of increased uptake of tracer
- Bladder activity in Fig. 1 is noted

– A double headed whole body
  gamma camera was used
– Left image is the anterior view
– Right image is the posterior
  view

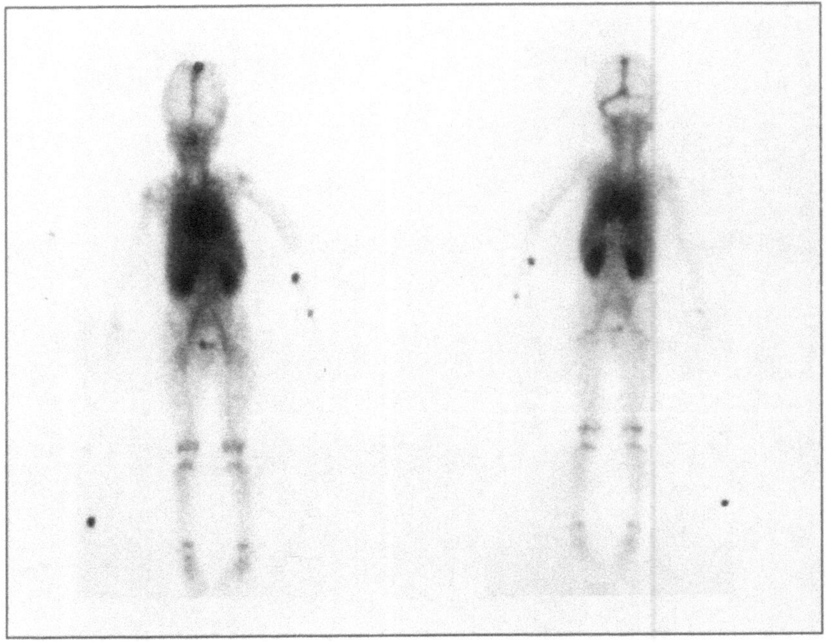

Technical Comments
– Note rotation of the head causing apparent deviation of the sagittal
  sinus
– Note extravasation of isotope at the site of injection in the left hand
– Marker on child's right side

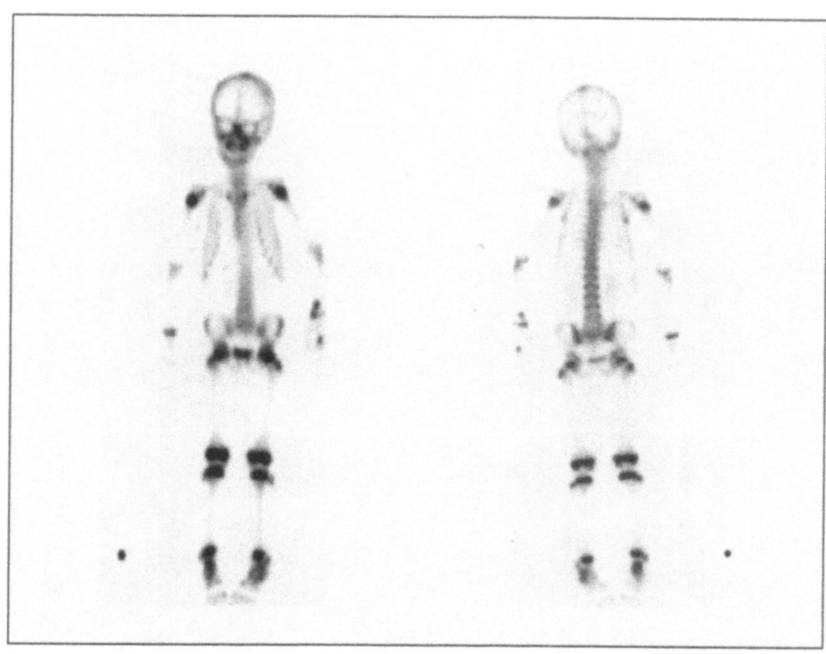

- A double headed whole body gamma camera was used
- Left image is the anterior view
- Right image is the posterior view

Technical Comments
- Note rotation of the head causing asymmetry
- Note extravasation of isotope at the site of injection in the left hand
- Marker on child's right side

– A double headed whole body
  gamma camera was used
– Left image is the anterior view
– Right image is the posterior
  view

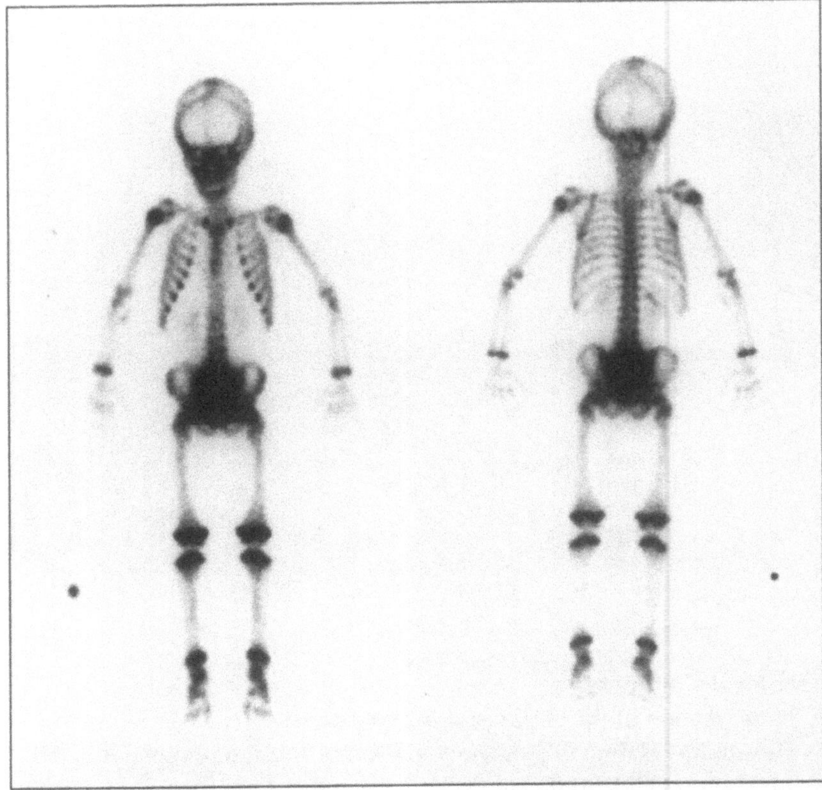

Technical Comments
– Note the full bladder, not an ideal situation
– Marker on child's right side

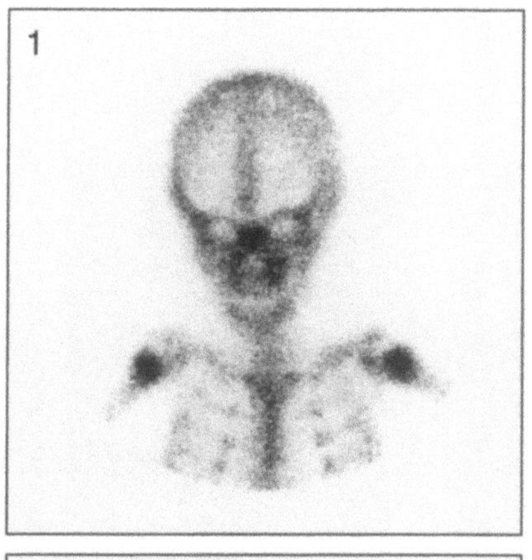

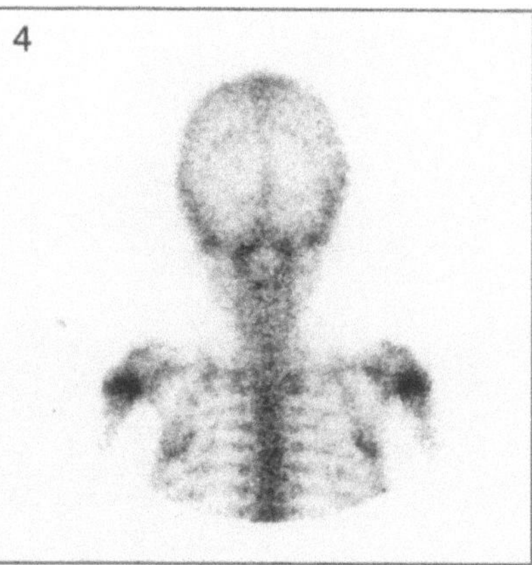

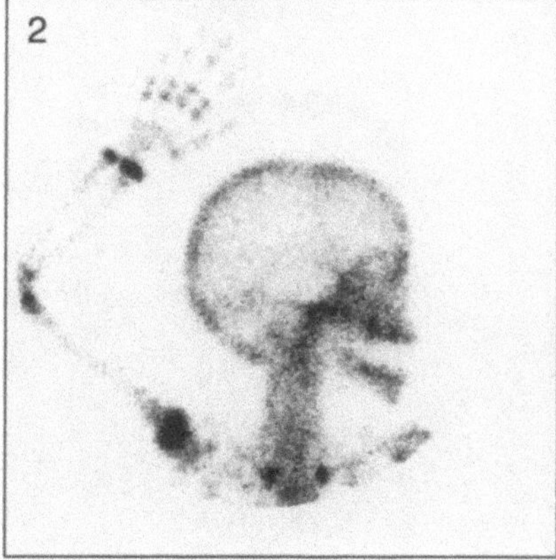

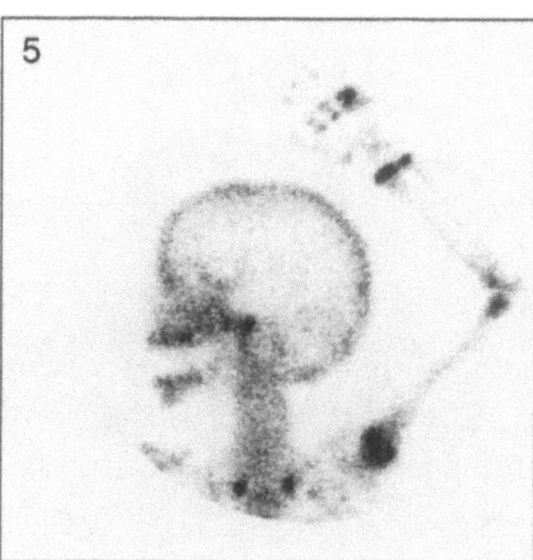

Fig. 1. Anterior view of skull and thorax

Fig. 4. Posterior view of skull and thorax

Fig. 2. Right lateral view of skull and right upper limb

Fig. 5. Left lateral view of skull and left upper limb

Technical Comments

- Focal increased activity is noted in the 4th finger of the left hand due to extravasation of isotope at the site of injection in Fig. 5
- The lateral views of the skull (Figs. 2 and 5) were taken anteriorly

Fig. 4. Posterior view of skull
and thorax

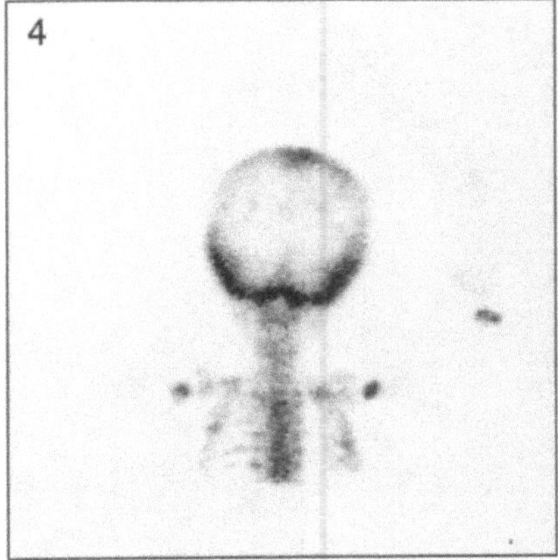

Fig. 2. Right lateral view of skull
and right upper limb

Fig. 5. Left lateral view of skull
and left upper limb

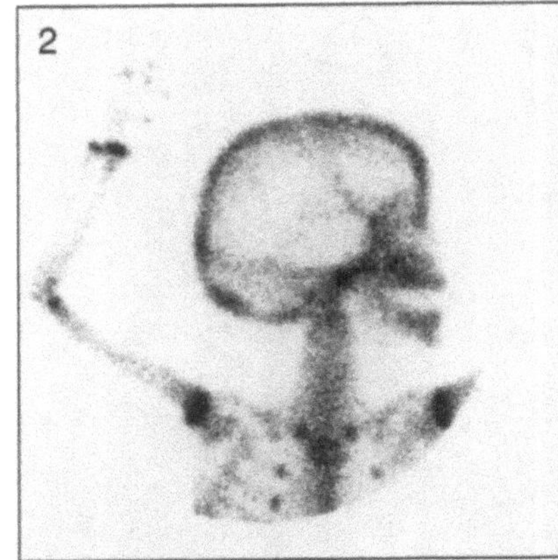

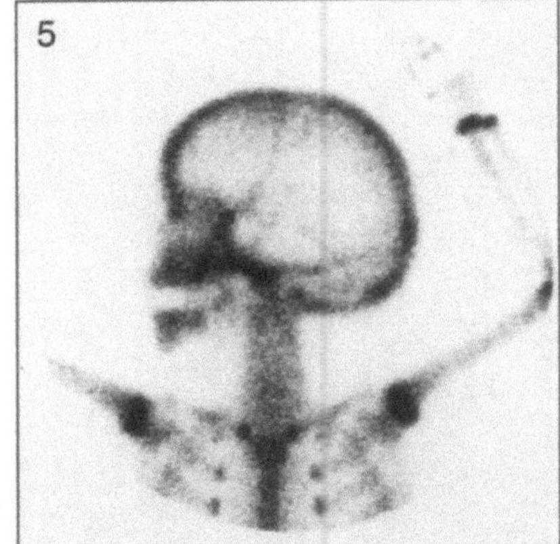

Technical Comments
– Note the rather poor positioning of the hands in all three figures com-
  pared to the images on p. 47
– The lateral views of the skull (Figs. 2 and 5) were taken anteriorly

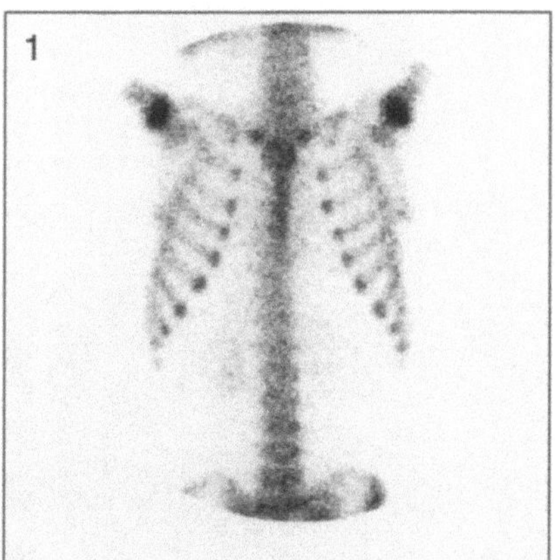

Fig. 1. Anterior view of thorax and spine

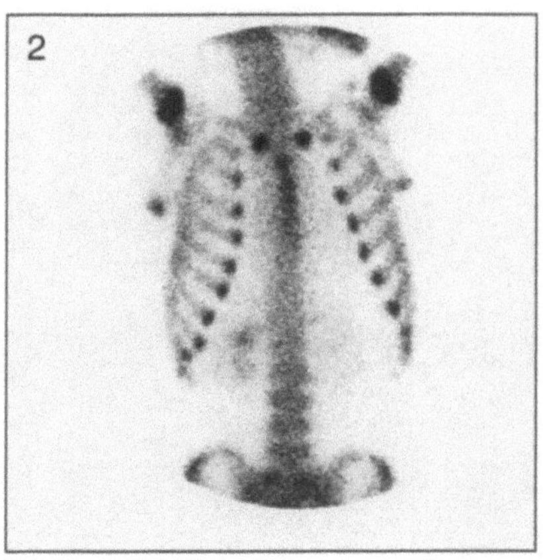
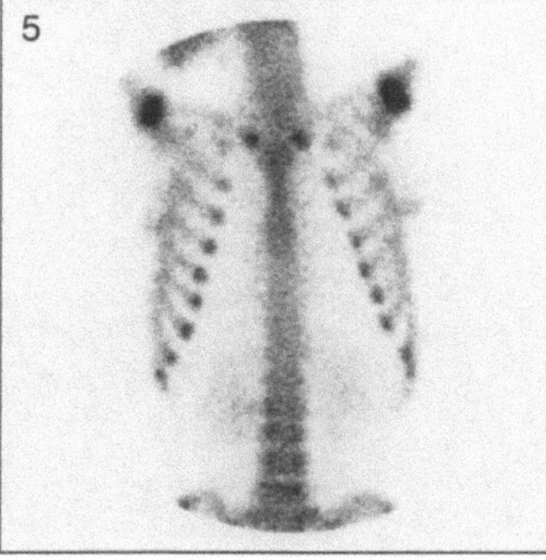

Fig. 2. Anterior view of thorax and spine

Fig. 5. Anterior view of thorax and spine

Technical Comments
- Note the right kidney in Fig. 2, this is within normal limits
- Note the clarity with which the lower lumbar spine is seen on all three images

Fig.1. Posterior view of thorax
and spine

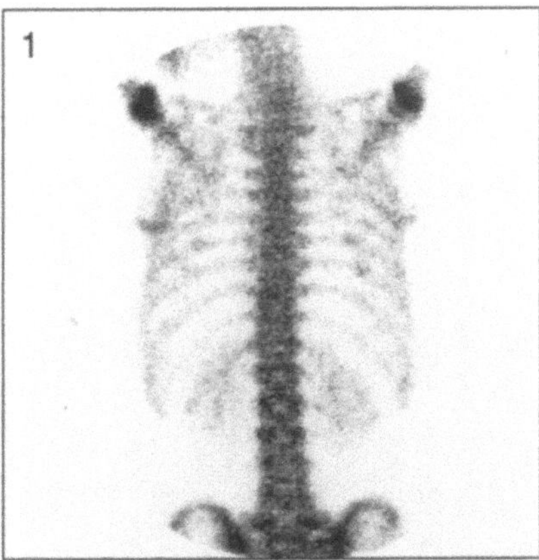

Fig. 2. Posterior view of thorax
and spine

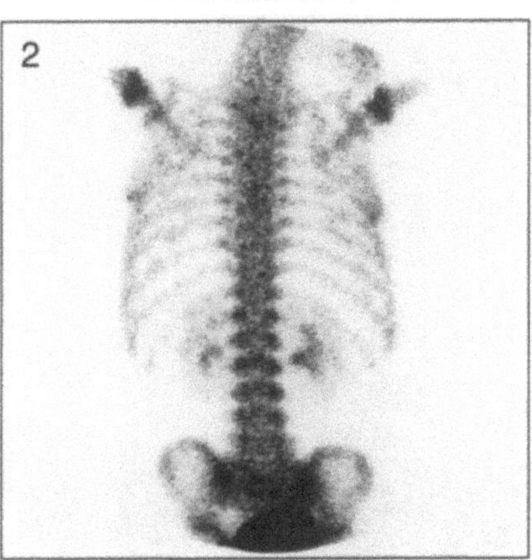

▶ **Potential Pitfall**
- The focal increased activity noted in the mid portion of the posterior
  ribs best seen in Fig.1 is due to activity from the costo-chondral junctions and should not be mistaken for a rib fracture. This is also seen in
  Fig. 2

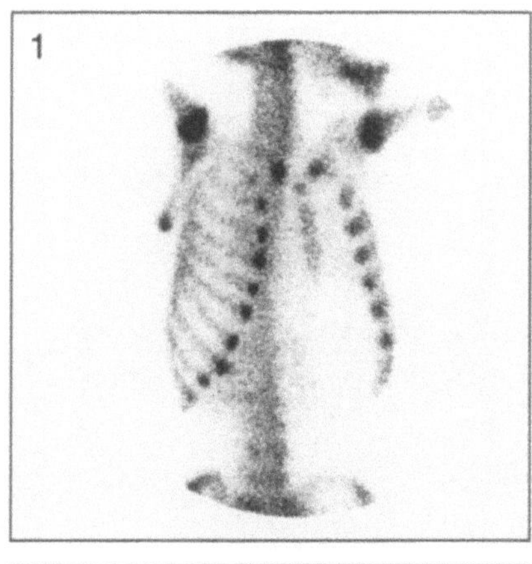

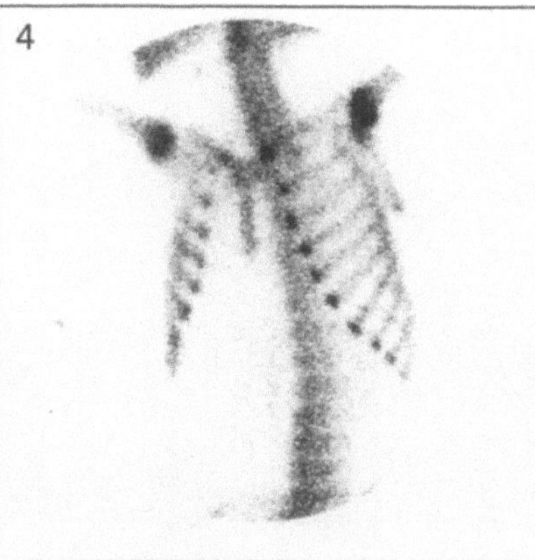

Fig. 1. Right anterior oblique view of thorax

Fig. 4. Left anterior oblique view of thorax

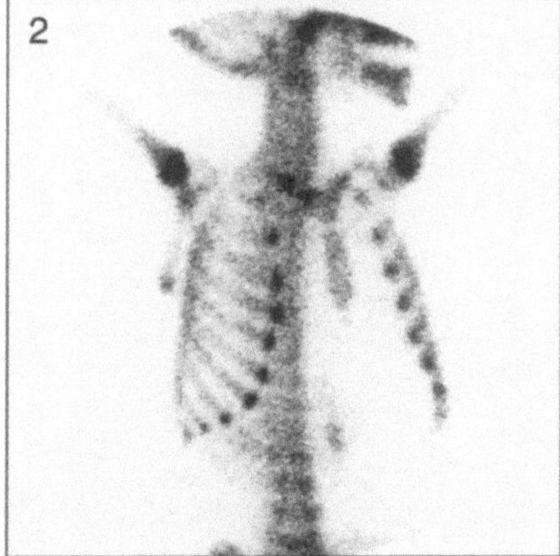

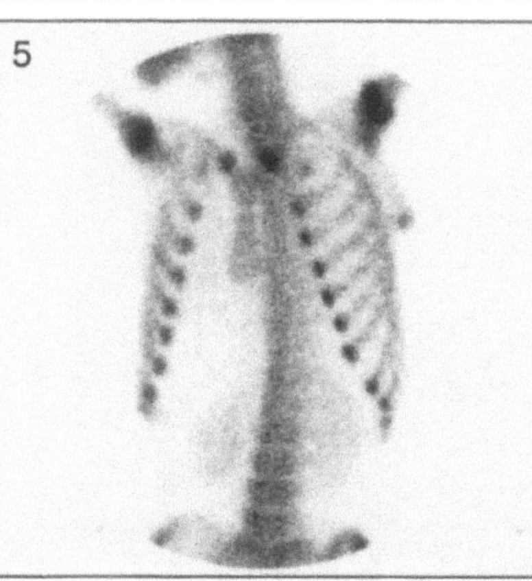

Fig. 2. Right anterior oblique view of thorax

Fig. 5. Left anterior oblique view of thorax

Technical Comments

- The sternum continues to show its variation but all sternal ossification centres are now present
- In Fig. 2 a small amount of activity is noted in the kidney to the left of the midline, this is normal

Fig. 1. Anterior view of pelvis and femora

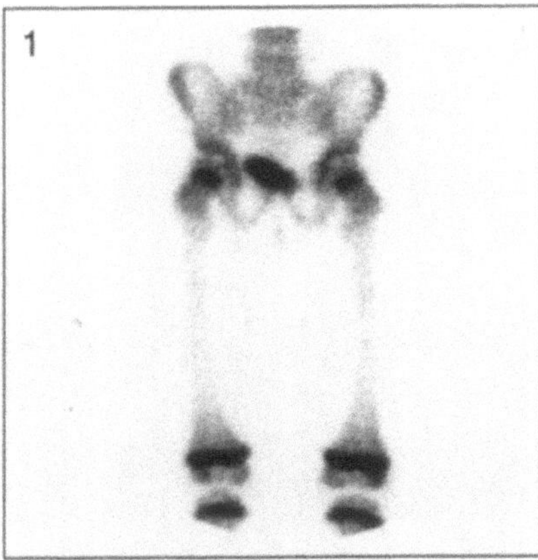

Fig. 2. Anterior view of spine, pelvis and femora

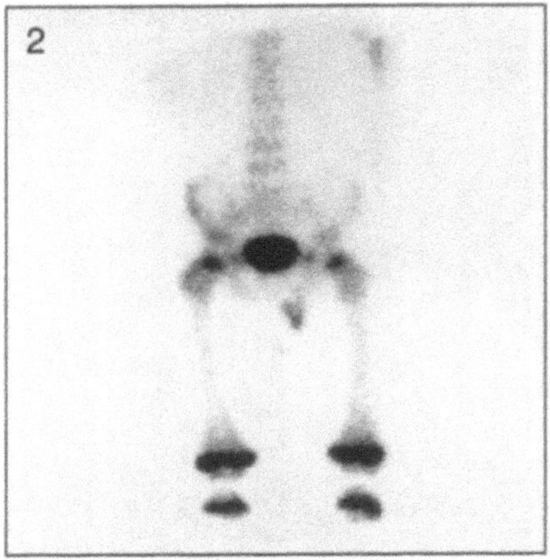

Technical Comment

– Urine contamination below the pelvis is seen in Fig. 2

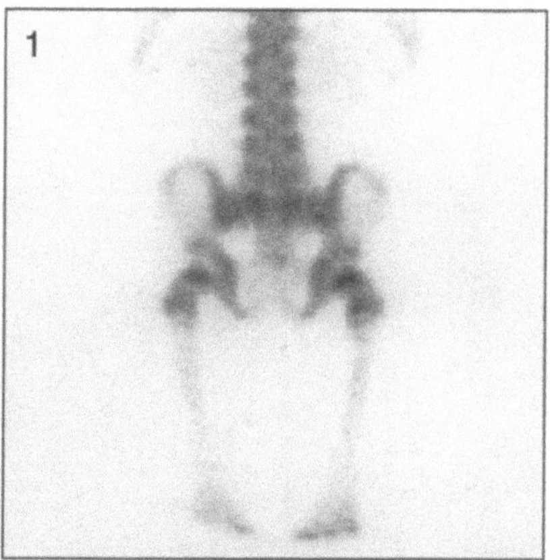

Fig. 1. Posterior view of spine, pelvis and femora

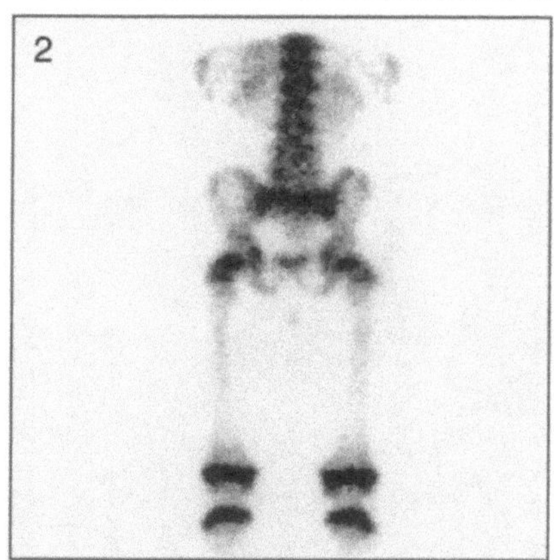

Fig. 2. Posterior view of spine, pelvis and femora

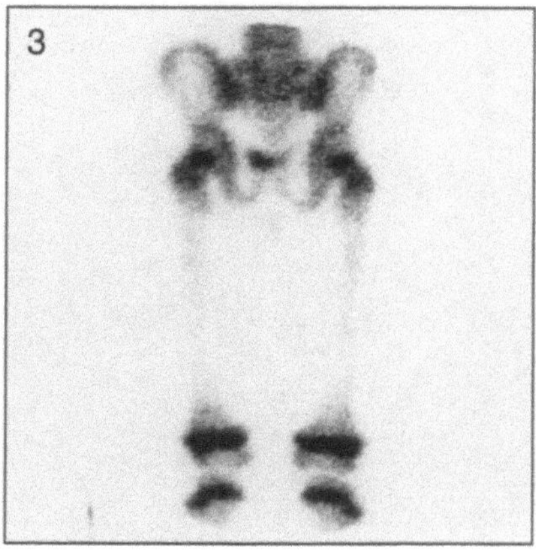

Fig. 3. Posterior view of pelvis and femora

Technical Comment

– Note that the bladder is virtually empty on all three images, an ideal situation

Fig. 1. Pinhole view of right hip

Fig. 4. Pinhole view of left hip

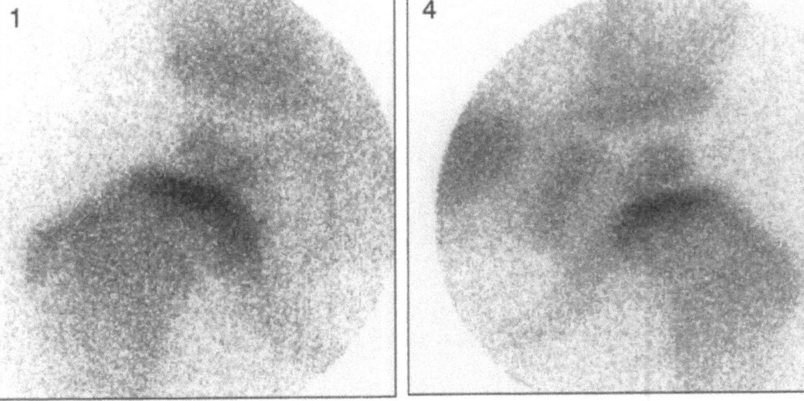

Fig. 2. Pinhole view of right hip

Fig. 5. Pinhole view of left hip

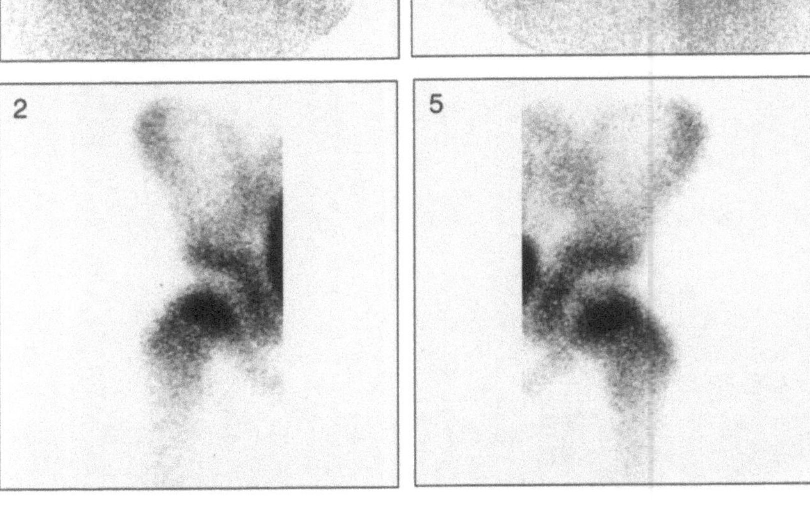

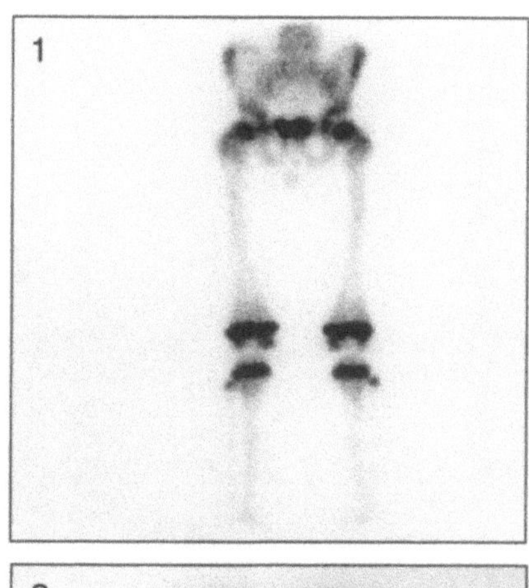

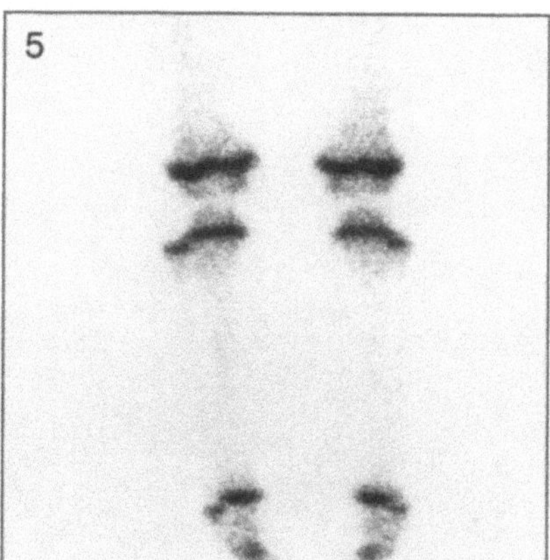

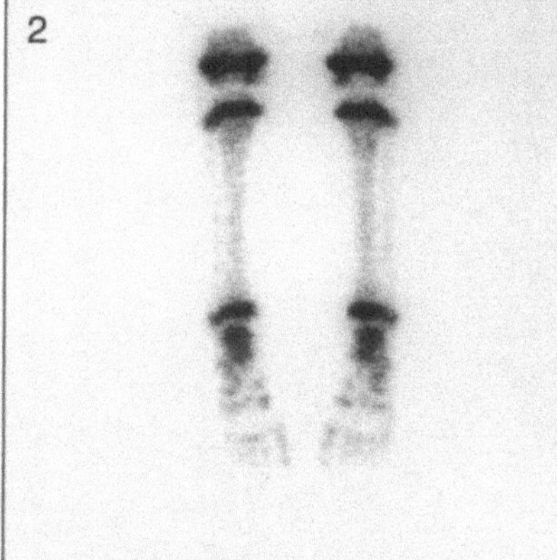

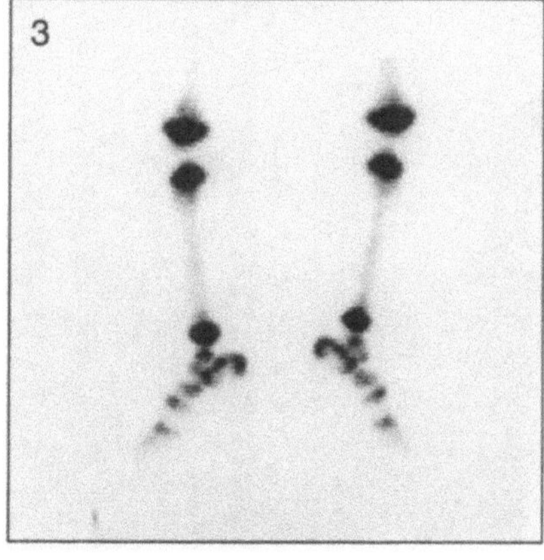

Fig. 1. Anterior view of pelvis and lower limbs

Fig. 4. Posterior view of lower limbs

Fig. 2. Anterior view of tibia, fibula and feet

Fig. 5. Posterior view of tibia and fibula

Fig. 3. Lateral view of lower limbs and feet

Technical Comments

- Note the clarity of the fibula separate from the tibia in Figs. 1, 2, 4 and 5. This is due to the "radiographic neutral position" of the feet
- Lateral views of the feet (Fig. 3) are inadequate for the evaluation of the knees

5: Age 3–4 Years

Fig. 1. Posterior view of skull
and thorax

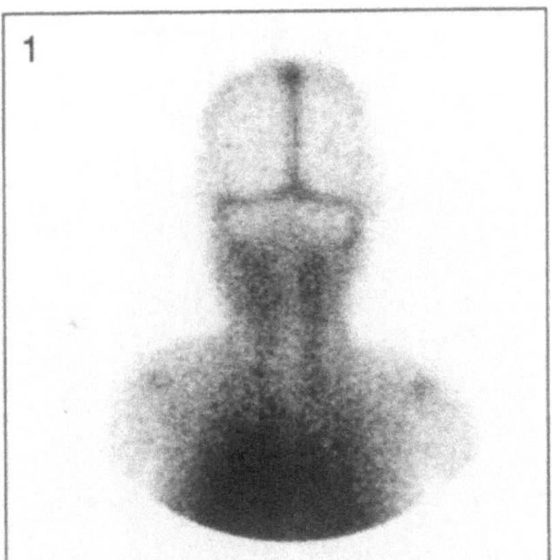

Fig. 2. Right lateral view of skull
and anterior view of thorax

Fig. 5. Left lateral view of skull
and anterior view of thorax

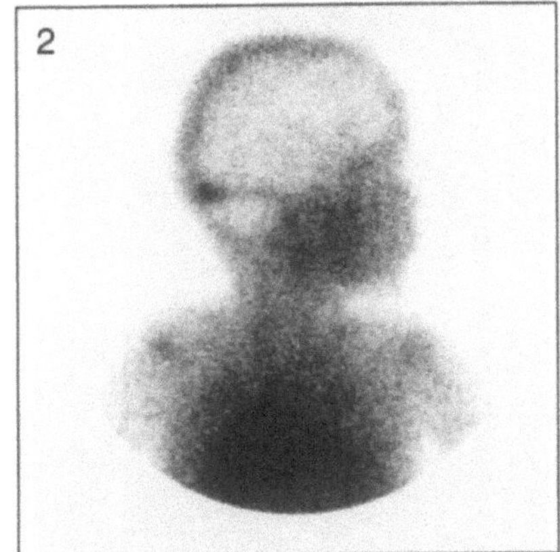
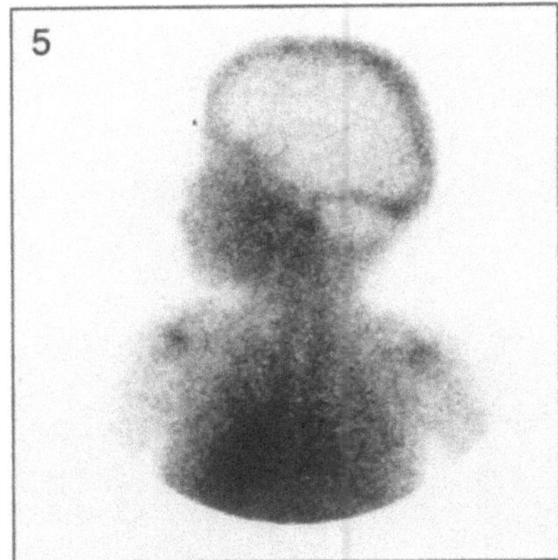

Technical Comment
– Note the high activity in the venous sinuses on all images

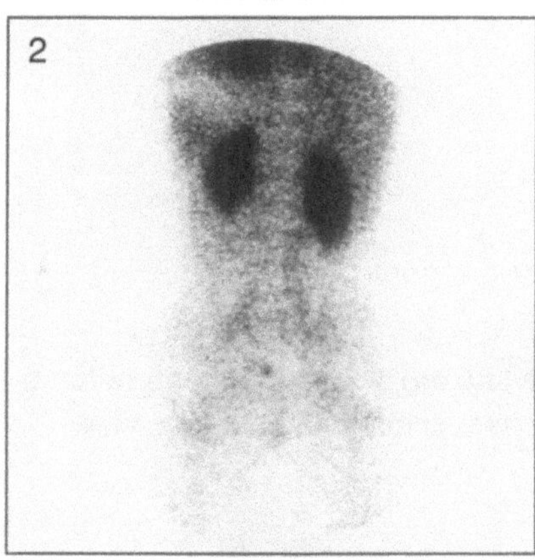

Fig. 1. Posterior view of thorax
and spine

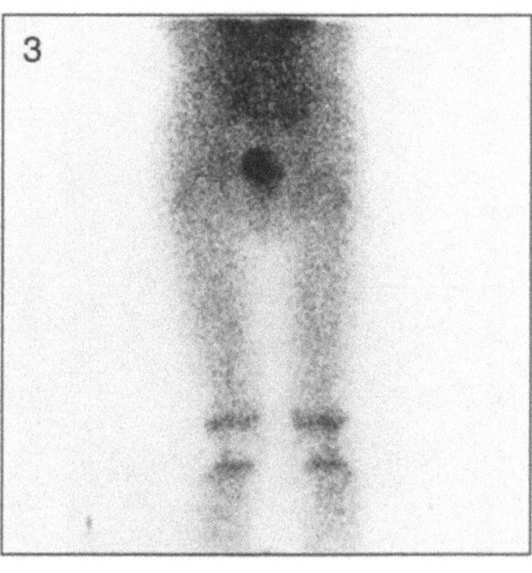

Fig. 2. Posterior view of spine
and pelvis

Fig. 3. Posterior view of pelvis
and lower limbs

Technical Comment
– Note activity in the bladder in Fig. 3

- A double headed whole body
  gamma camera was used
- Left image is the anterior view
- Right image is the posterior
  view

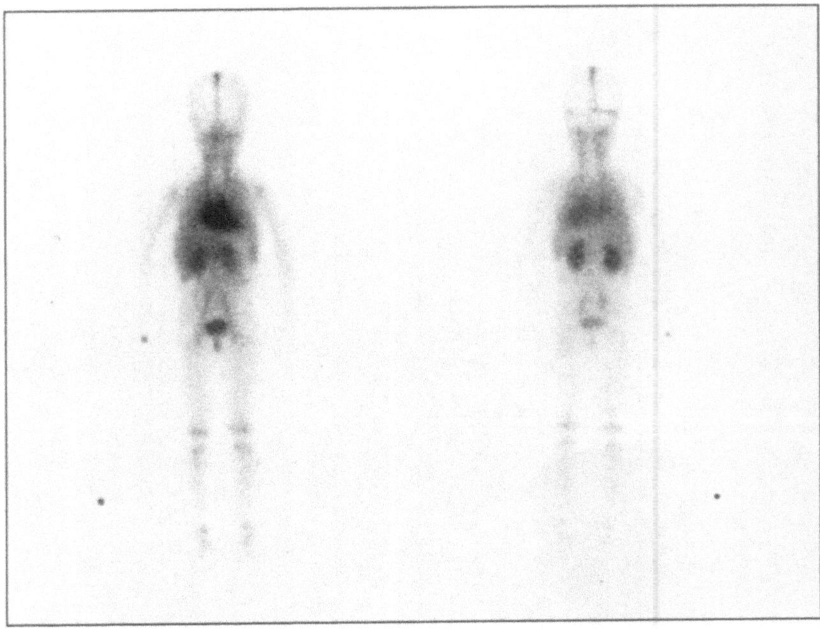

Technical Comments
- Note extravasation of isotope at the site of injection in the right hand
- Urine contamination below the pelvis is seen on the anterior view
- Marker on child's right side

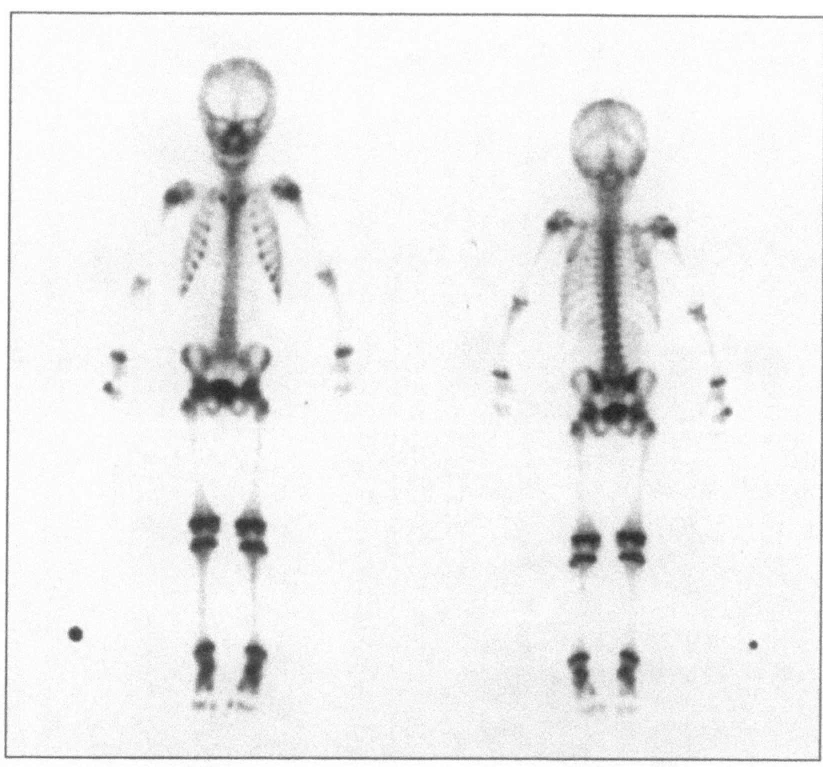

- A double headed whole body gamma camera was used
- Left image is the anterior view
- Right image is the posterior view

Technical Comments

- The left foot is well positioned, with the toes turned inward allowing good visualization of the fibula
- Note extravasation of isotope at the site of injection in the right hand
- Marker on child's right side

Fig. 1. Anterior view of skull

Fig. 4. Posterior view of skull and thorax

Fig. 2. Right lateral view of skull and right upper limb

Fig. 5. Left lateral view of skull and left upper limb

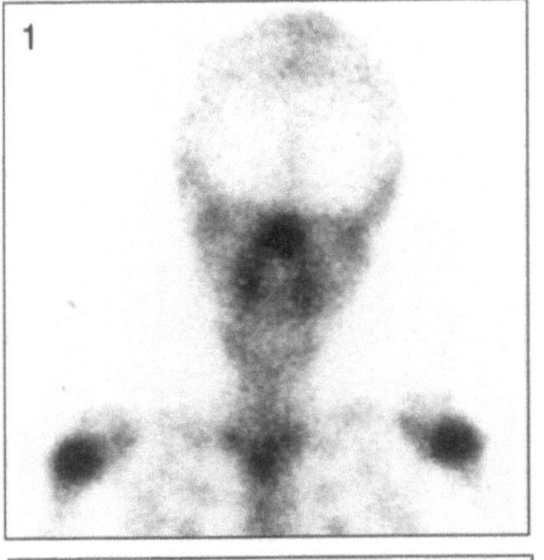

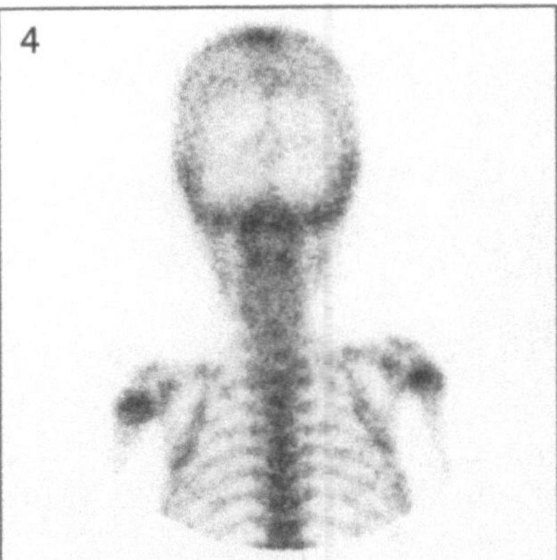

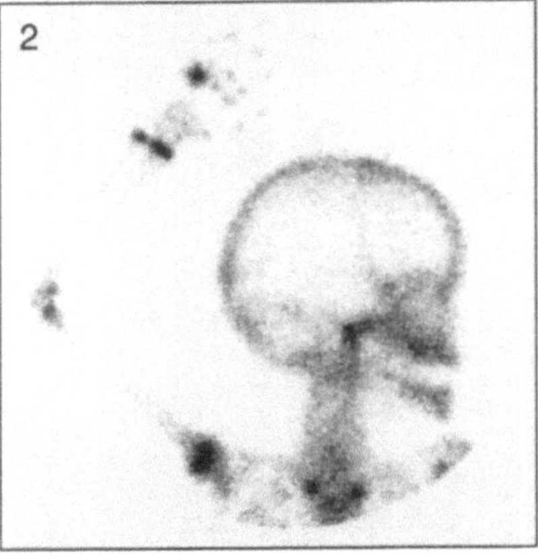

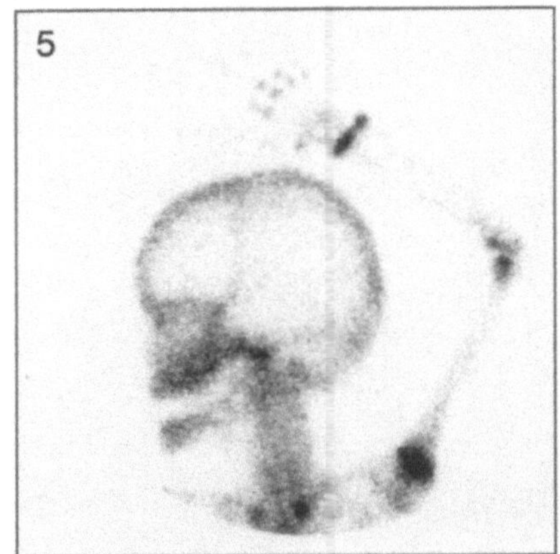

Technical Comments

- Note the clear visualization of the coronal sutures on both lateral skull images, compare these to the poor visualization of the sutures seen on the next page (Figs. 2 and 5, p. 63)
- Note extravasation of isotope at the site of injection in the right hand in Fig. 2
- The lateral views of the skull (Figs. 2 and 5) were taken anteriorly

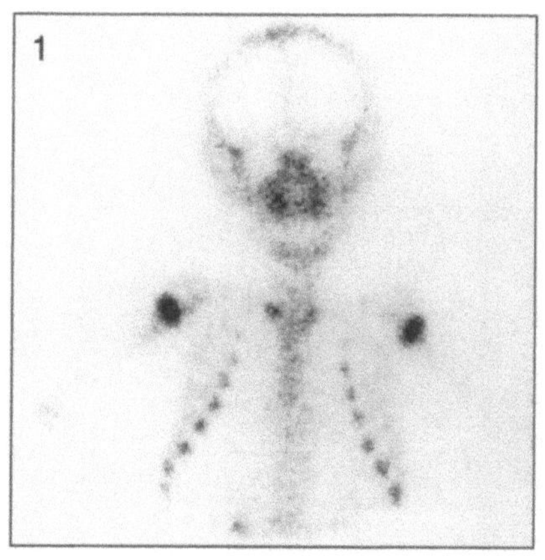

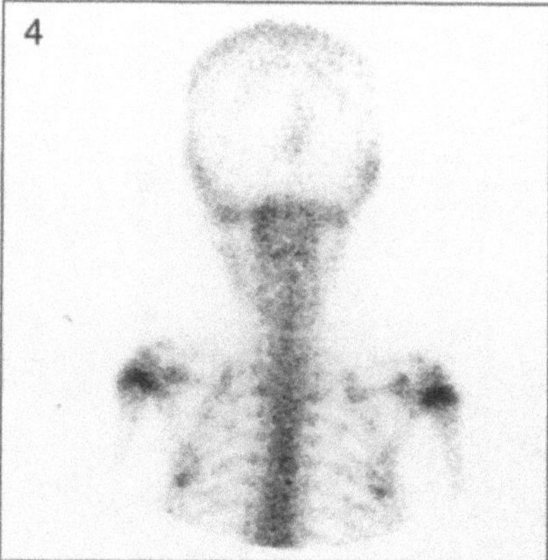

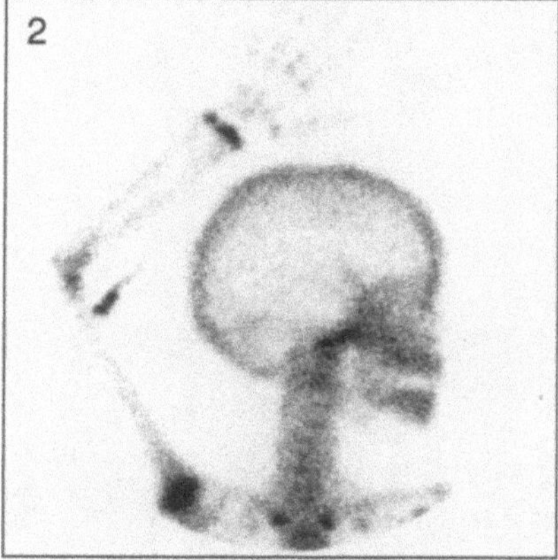

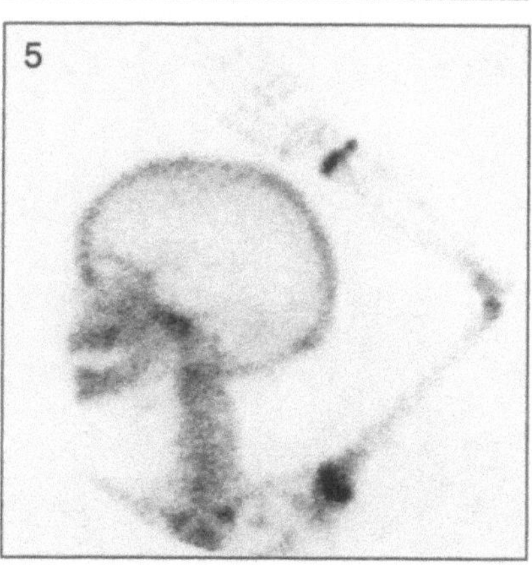

Fig. 1. Anterior view of skull and·thorax

Fig. 4. Posterior view of skull and thorax

Fig. 2. Right lateral view of skull and right upper limb

Fig. 5. Left lateral view of skull and left upper limb

Technical Comments

- Note the poor visualization of the coronal sutures on both lateral skull images, compare these to the clarity of the sutures seen on the previous page (Figs. 2 and 5, p. 62)
- Note extravasation of isotope at the site of injection in the right elbow in Fig. 2
- The lateral views of the skull (Figs. 2 and 5) were taken anteriorly

Fig.1. Anterior view of thorax and spine

Fig. 4. Left lateral view of skull and anterior view of thorax

Fig. 2. Anterior view of thorax and spine

Fig. 5. Anterior view of thorax and spine

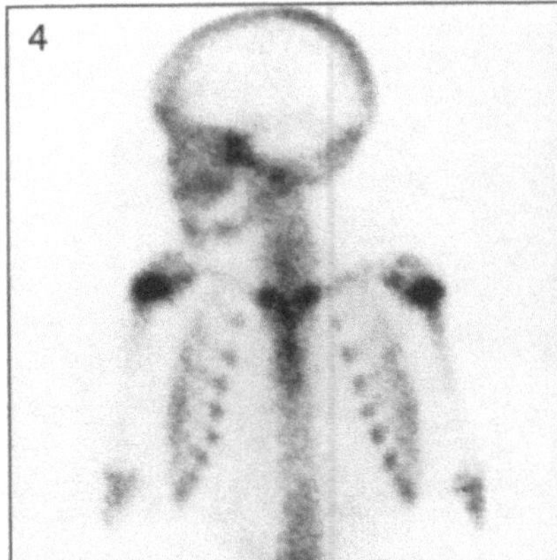

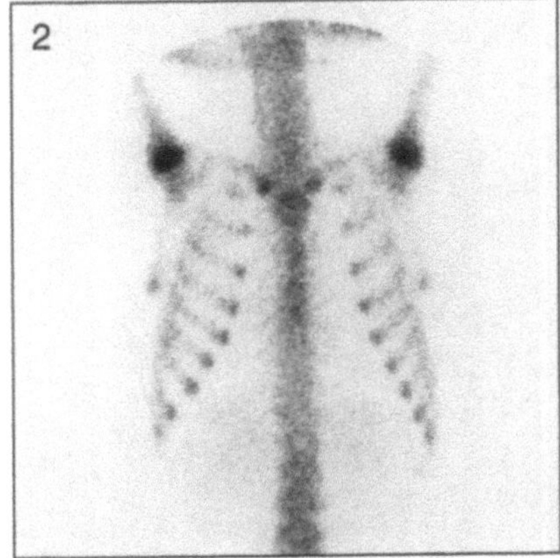

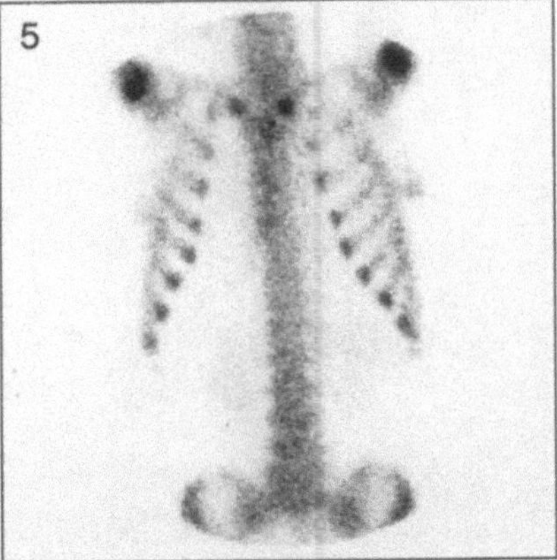

Technical Comments

- Fig. 1 shows activity in the thyroid gland and also in the stomach due to free 99m-Tc-pertechnetate. This is rather unusual but does occasionally occur
- Fig. 1 shows the lower lumbar spine to good advantage
- Note variation in sternal ossification

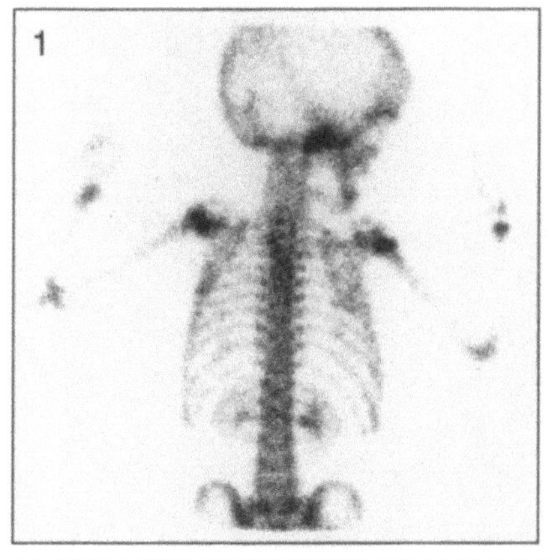

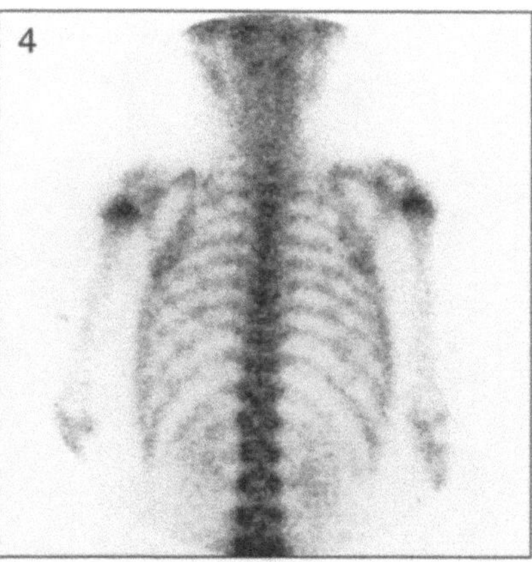

Fig. 1. Right lateral view of skull, posterior view of thorax, spine and upper limbs

Fig. 4. Posterior view of thorax and spine

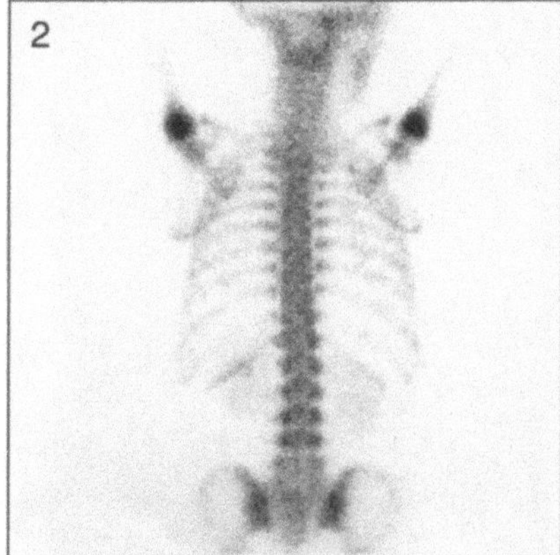

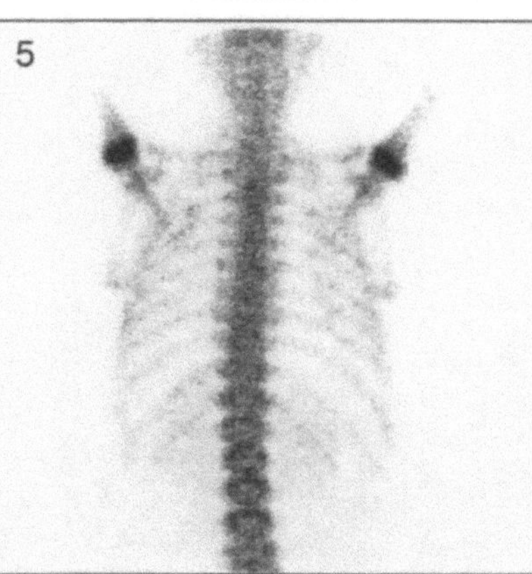

Fig. 2. Posterior view of thorax and spine

Fig. 5. Posterior view of thorax and spine

Technical Comments
- With the upper limbs elevated in Figs. 2 and 5, the scapula clears the thorax and is well seen
- Poor positioning of the hands is noted in Fig. 1

Fig.1. Left anterior oblique view of thorax

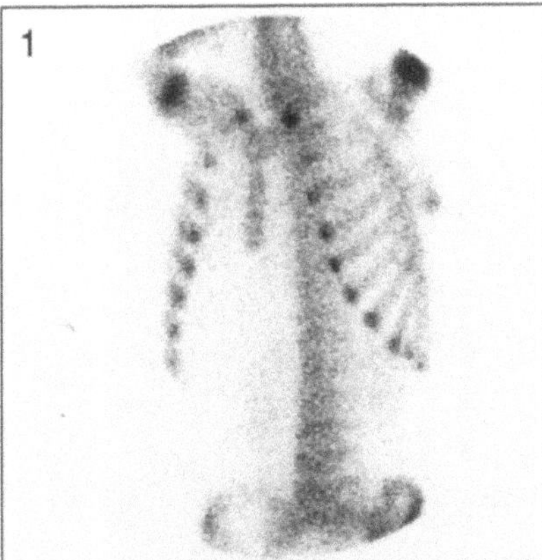

Fig.2. Left anterior oblique view of thorax

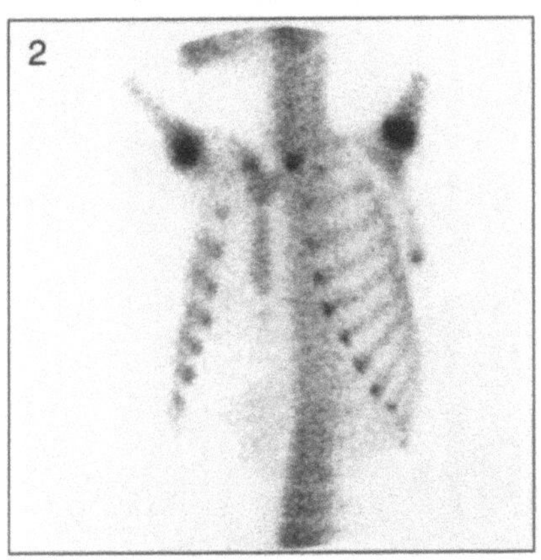

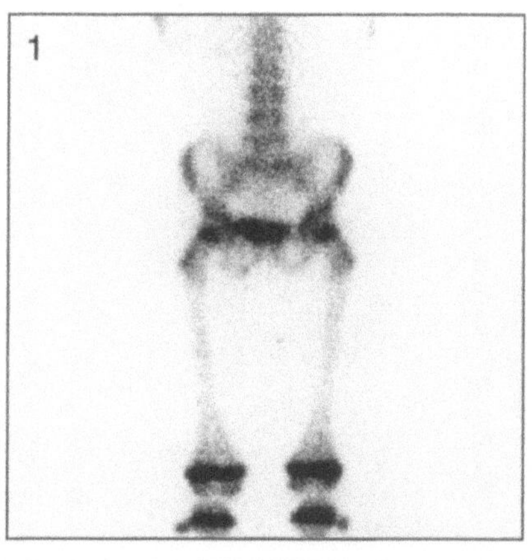

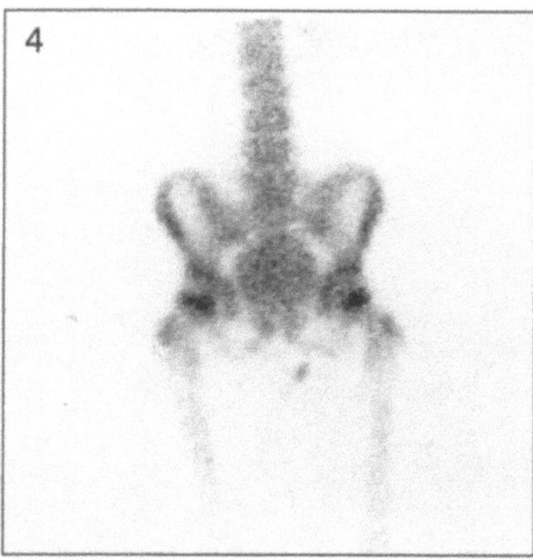

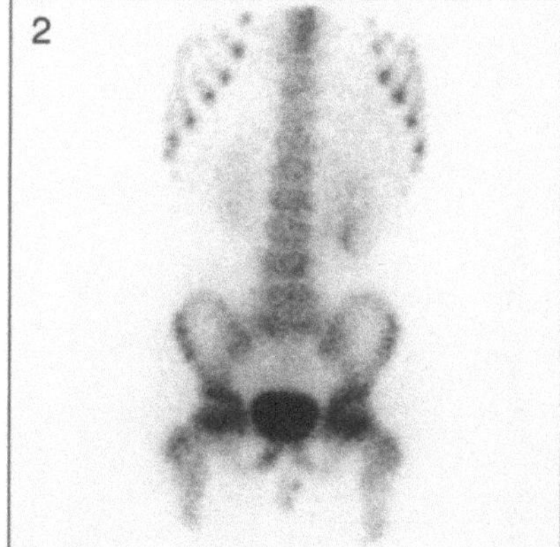

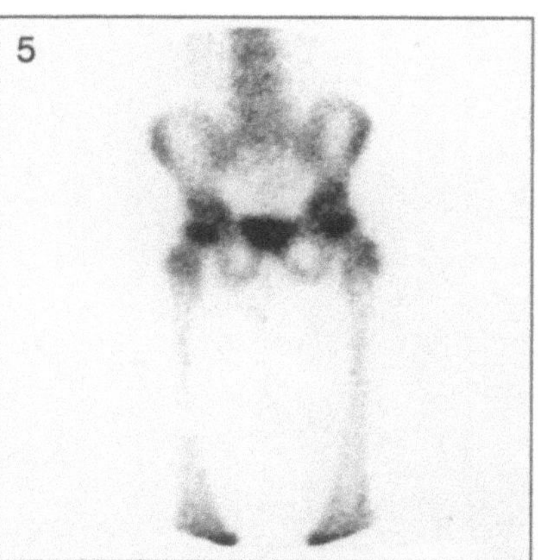

Fig. 1. Anterior view of spine, pelvis and femora

Fig. 4. Anterior view of spine, pelvis and femora

Fig. 2. Anterior view of spine and pelvis

Fig. 5. Anterior view of pelvis and femora

Technical Comments
- Note the clarity of the lower lumbar spine on all images
- The bladder in Fig. 4 is of relatively large volume but has little activity suggesting that the child is well hydrated. The other three images show high activity and low volume bladders. This is **not** recommended
- Urine contamination below the pelvis is seen in Figs. 2, 4 and 5

Fig. 1. Posterior view of spine, pelvis and femora

Fig. 4. Posterior view of spine, pelvis and femora

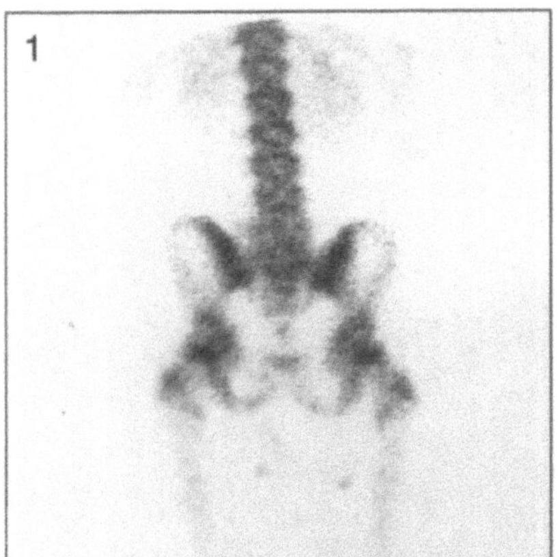
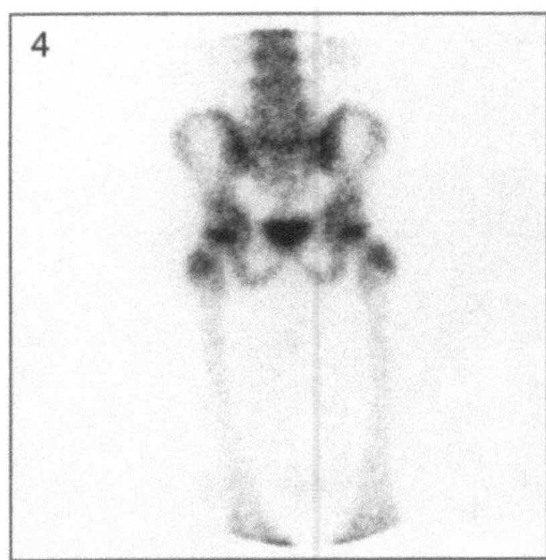

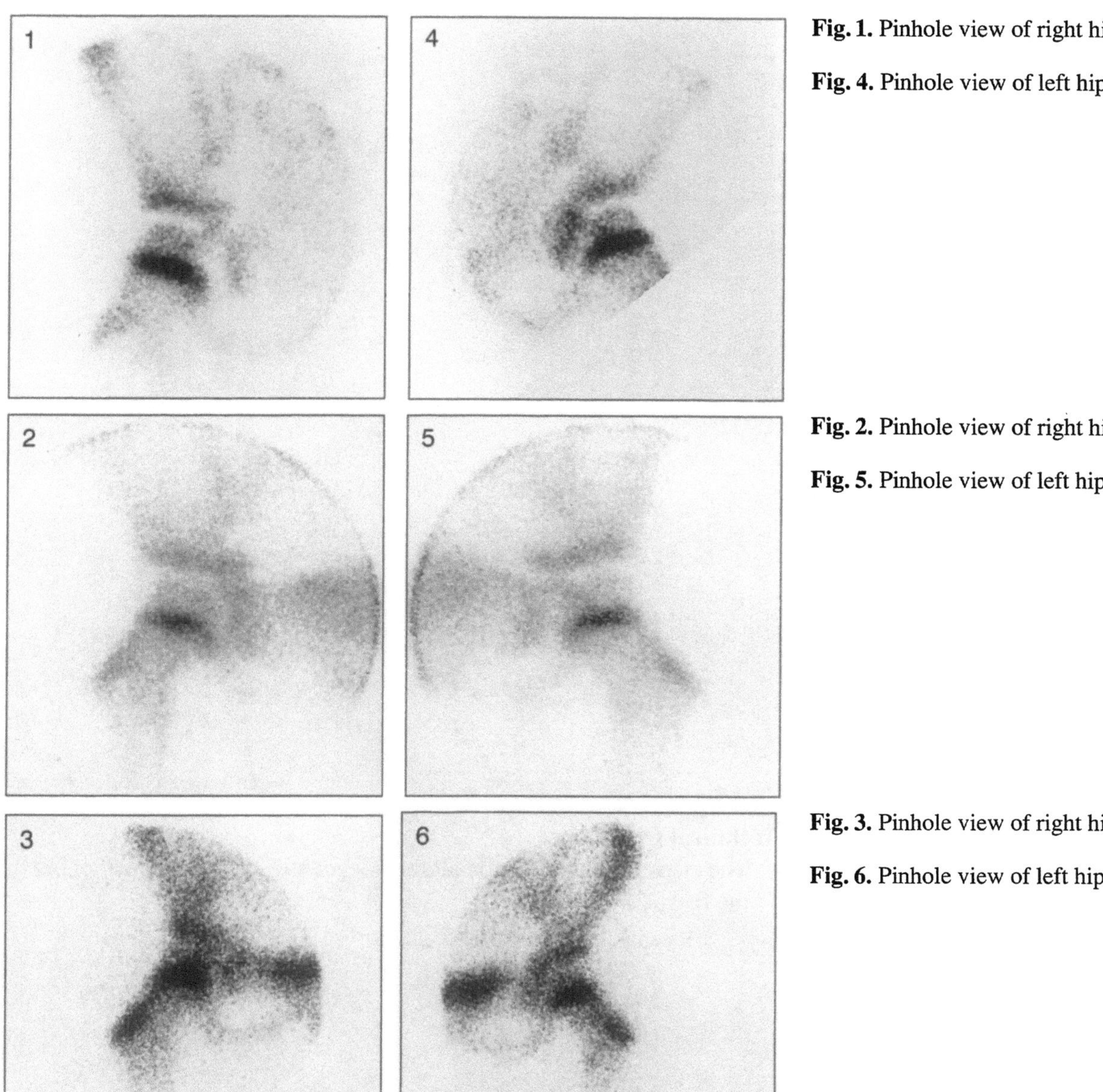

Fig. 1. Pinhole view of right hip

Fig. 4. Pinhole view of left hip

Fig. 2. Pinhole view of right hip

Fig. 5. Pinhole view of left hip

Fig. 3. Pinhole view of right hip

Fig. 6. Pinhole view of left hip

Technical Comments
- Different size of pinhole inserts give different degrees of magnification
- The position of the hips in the bottom series (Figs. 3 and 6) is not as good as in the top two. This is related to the positioning of the knees and the lack of inturning of the feet. "The radiographic neutral position" of the feet and knees is essential when doing pinhole views of the hips

Fig. 1. Posterior view of femora and knees

Fig. 4. Posterior view of lower limbs

Fig. 2. Posterior view of knees, tibia, fibula and ankles

Fig. 5. Posterior view of lower limbs

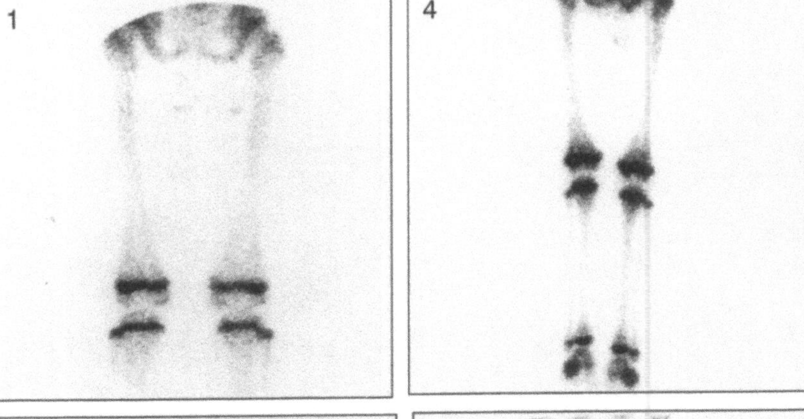

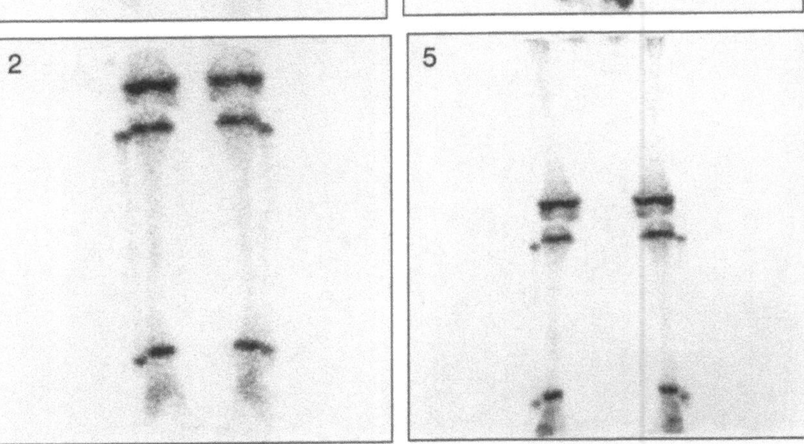

Technical Comment
– The clarity of the fibula on all the images suggests good positioning of the feet

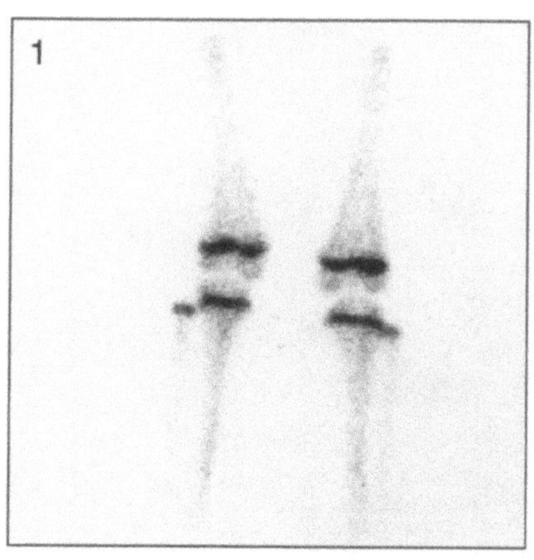

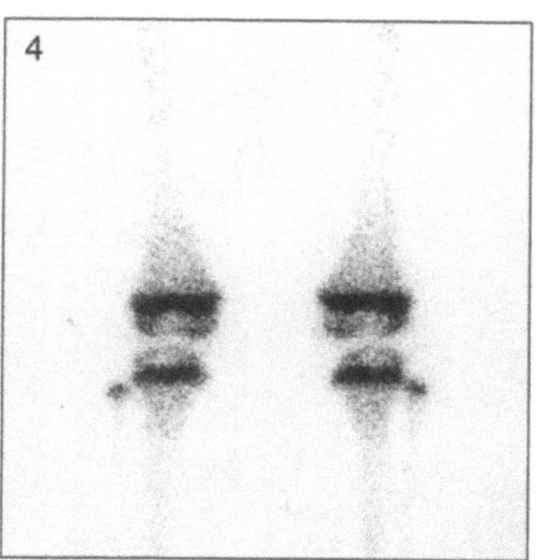

Fig. 1. Anterior view of knees

Fig. 4. Posterior view of knees

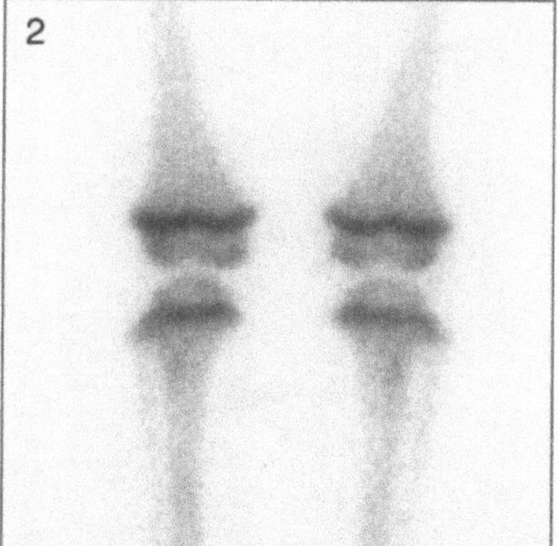

Fig. 2. Anterior view of knees

Fig. 1. Lateral view of right knee and foot

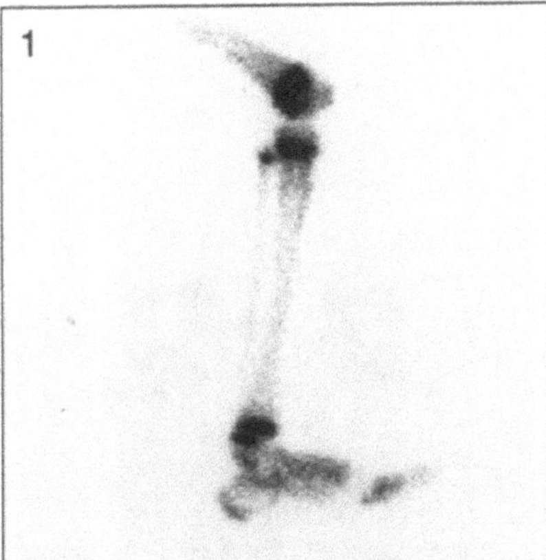

Fig. 2. Anterior pinhole view of right knee

Fig. 5. Anterior pinhole view of left knee

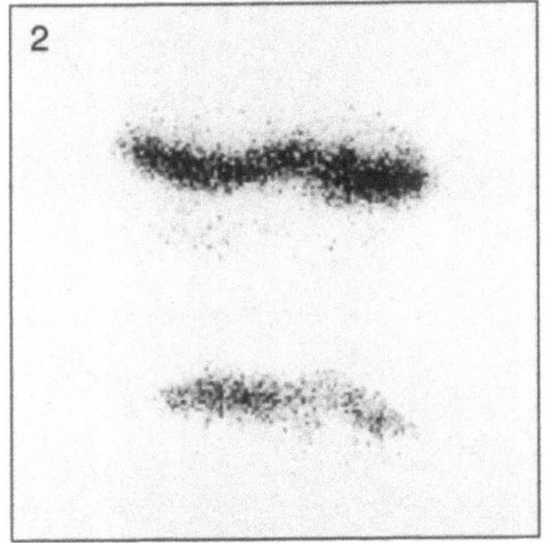

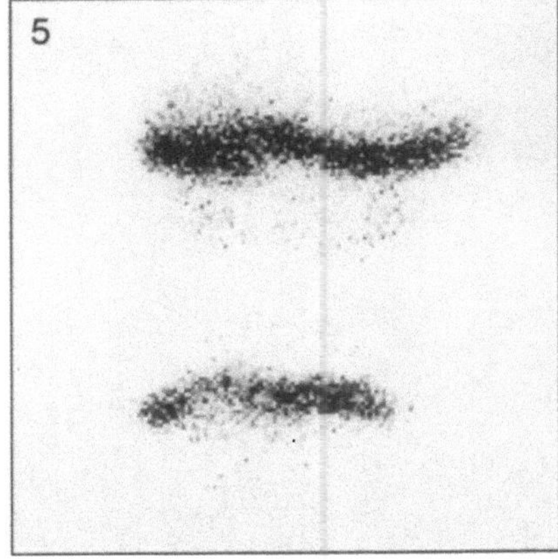

Technical Comment

– The pinhole projections are rather infrequently required clinically

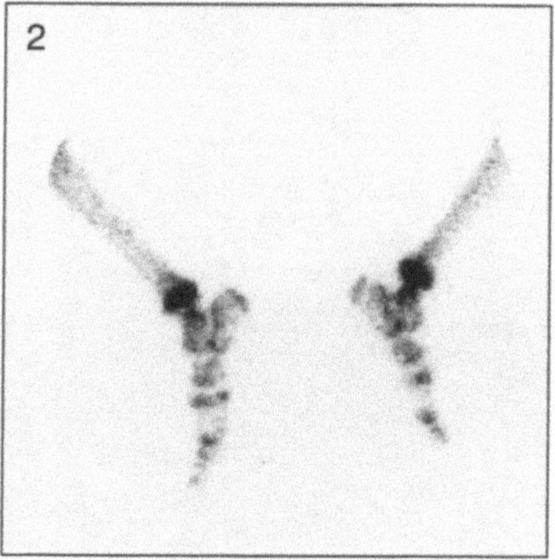

Fig. 1. Posterior view of tibia, fibula and ankles

Fig. 2. Lateral view of feet

6: Age 4–5 Years

6: Age 4–5 Years

Fig. 1. Posterior view of skull

Fig. 4. Posterior view of thorax and spine

Fig. 2. Anterior view of spine and pelvis

Fig. 5. Posterior view of spine and pelvis

Fig. 3. Anterior view of femora and knees

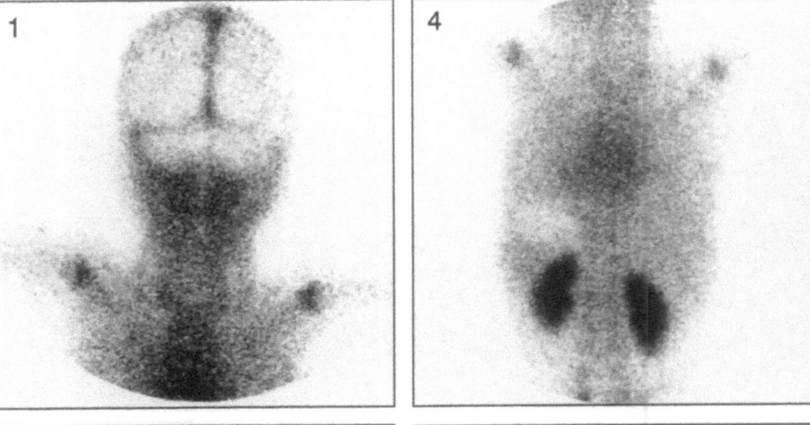

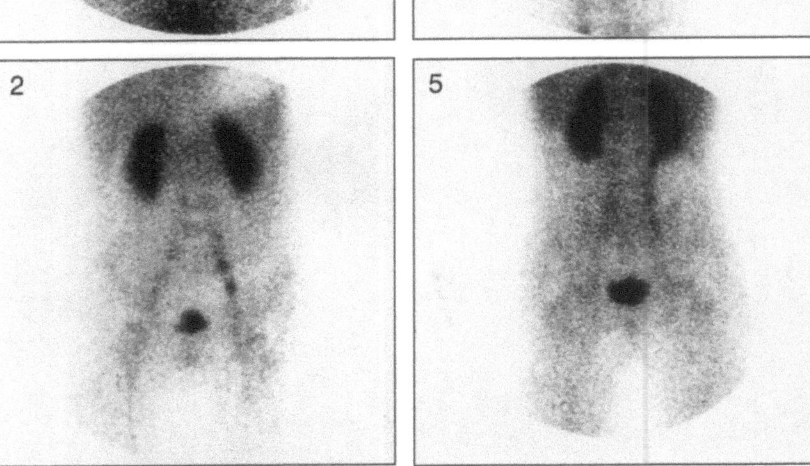

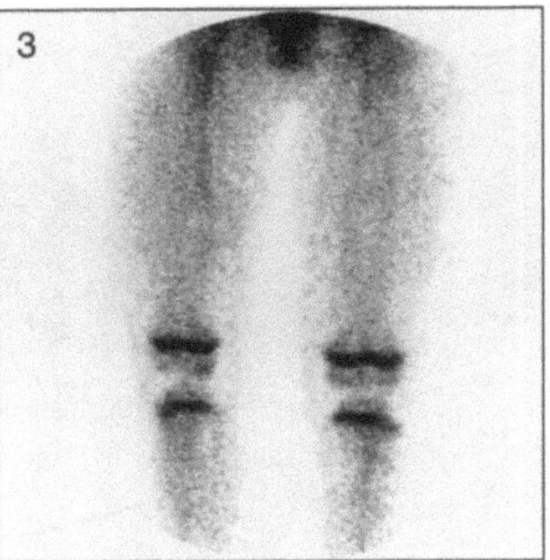

Technical Comments
– The photon deficient area above the left kidney in Fig. 4 represents the stomach full of food. This is also seen in Fig. 2 immediately above the left kidney
– Fig. 2 shows focal accumulation of isotope above the bladder on the left, this may be in the ureter

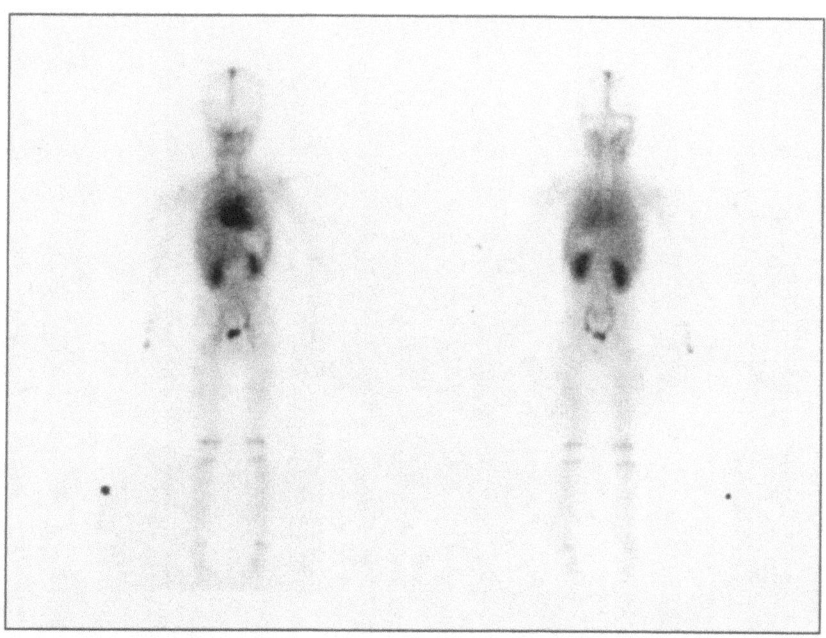

- A double headed whole body gamma camera was used
- Left image is the anterior view
- Right image is the posterior view

Technical Comments
- Isotope is seen in both ureters
- The full stomach is seen as a photon deficient area on the anterior view above the left kidney
- Note extravasation of isotope at the site of injection in the right hand
- Marker on child's right side

- A double headed whole body gamma camera was used
- Left image is the anterior view
- Right image is the posterior view

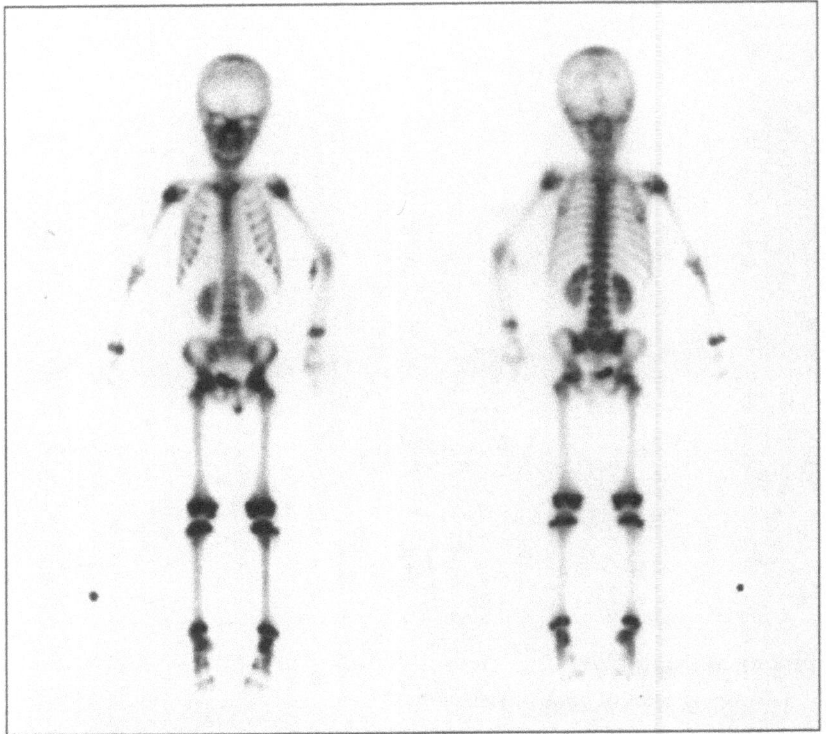

Technical Comments

- The left foot is better positioned than the right, with the toes turned inward allowing good visualization of the fibula
- Diffuse parenchymal renal activity is seen, no pathological cause was evident
- Note extravasation of isotope at the site of injection in the left elbow
- Urine contamination below the pelvis is seen on the anterior view
- Marker on child's right side

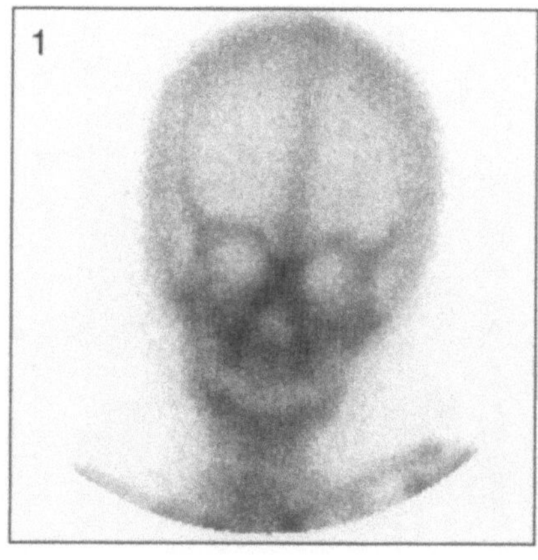

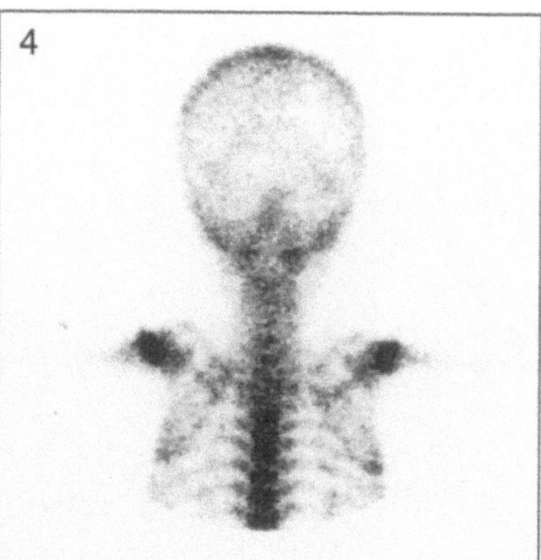

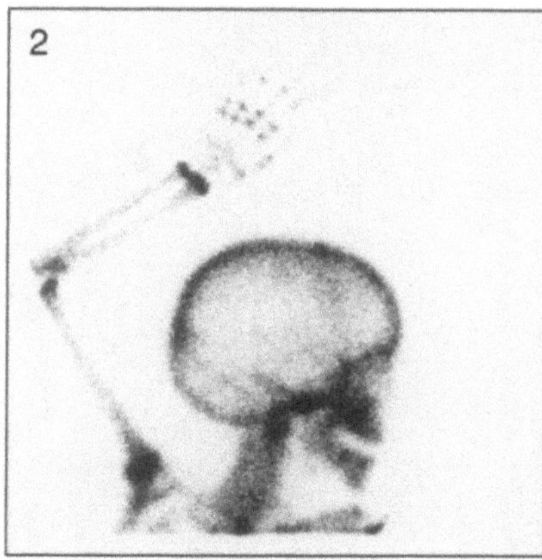

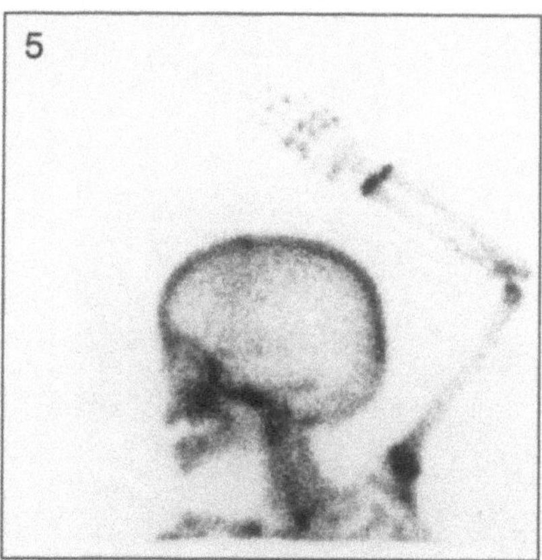

Fig. 1. Anterior view of skull

Fig. 4. Posterior view of skull and thorax

Fig. 2. Right lateral view of skull and right upper limb

Fig. 5. Left lateral view of skull and left upper limb

Technical Comments

- Note the poor positioning of the left hand in Fig. 5, the thumb cannot be seen
- The coronal sutures are indistinct in Figs. 2 and 5, this is a variation of normality
- The lateral views of the skull (Figs. 2 and 5) were taken anteriorly

Fig. 4. Posterior view of skull
and thorax

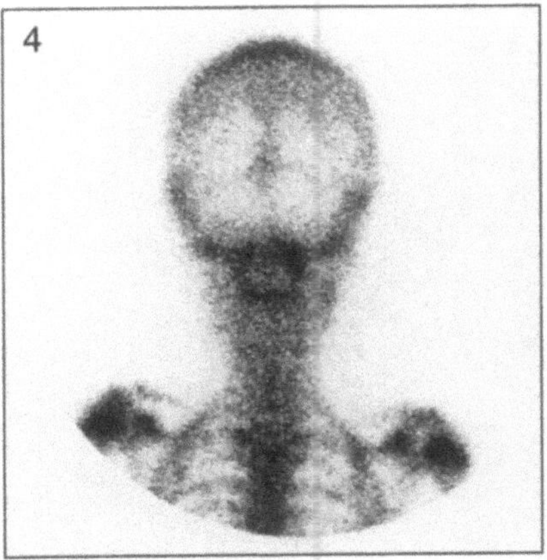

Fig. 2. Right lateral view of skull
and right upper limb

Fig. 5. Left lateral view of skull
and left upper limb

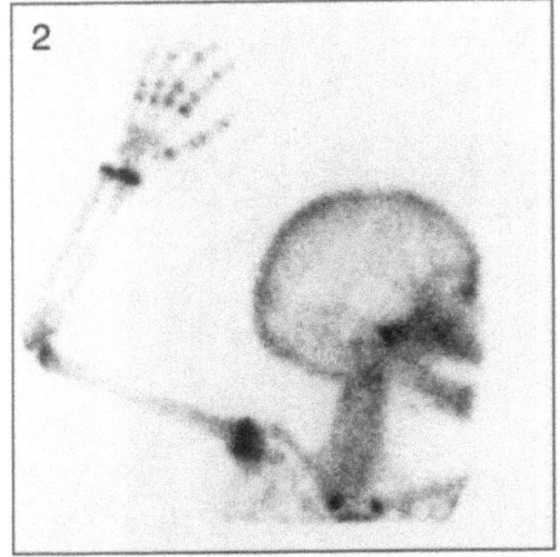

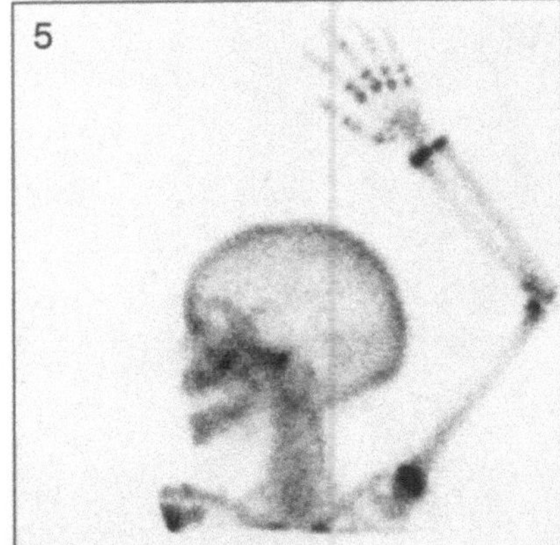

Fig. 3. Left lateral view of skull
and both upper limbs

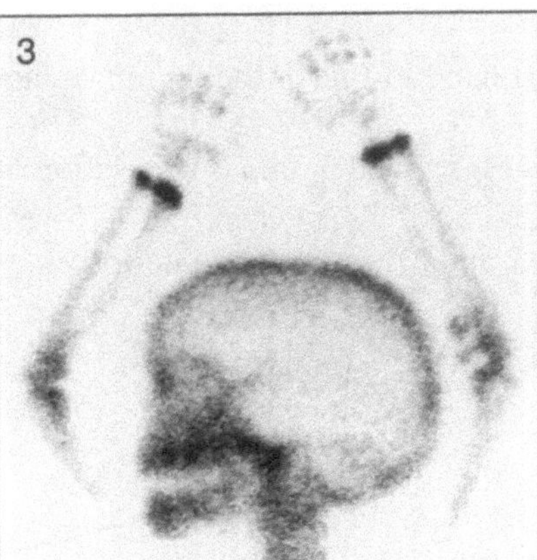

Technical Comments

– Fig. 3 shows the difficulty of attempting to image both hands simultaneously. Imaging of the right upper limb with the right lateral skull and the left upper limb with the left lateral skull is recommended
– The lateral views of the skull (Figs. 2 and 5) were taken anteriorly

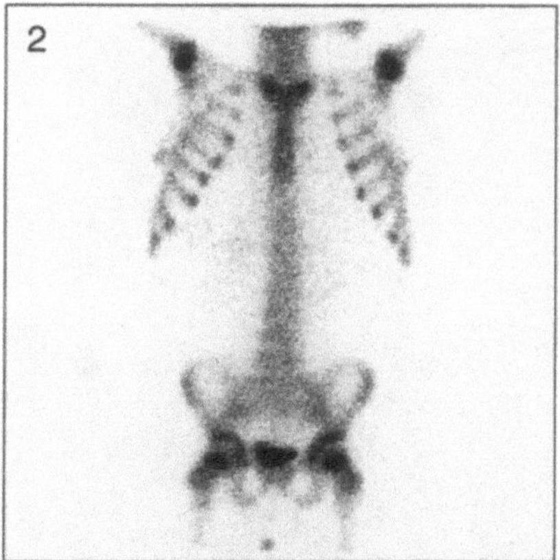

Fig. 1. Right lateral view of skull and anterior view of thorax

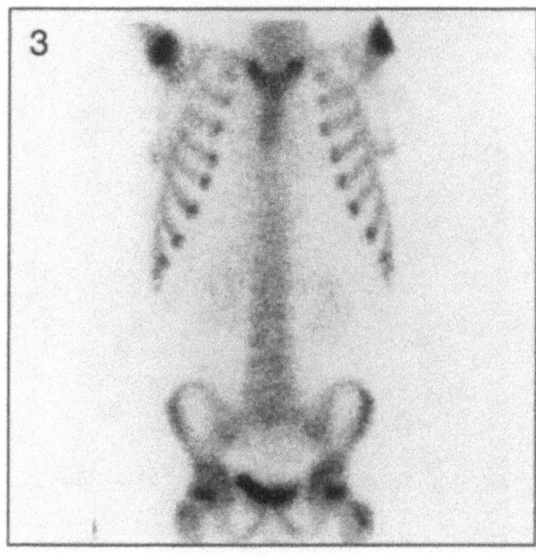

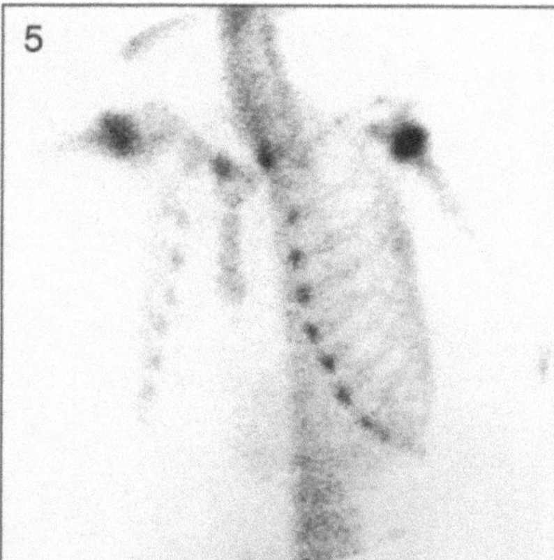

Fig. 2. Anterior view of thorax, spine and pelvis

Fig. 5. Left anterior oblique view of thorax

Fig. 3. Anterior view of thorax, spine and pelvis

Technical Comment
– Urine contamination below the pelvis is seen in Fig. 2

Fig. 1. Anterior view of thorax, spine, upper limbs and pelvis

Fig. 4. Posterior view of thorax, spine and pelvis

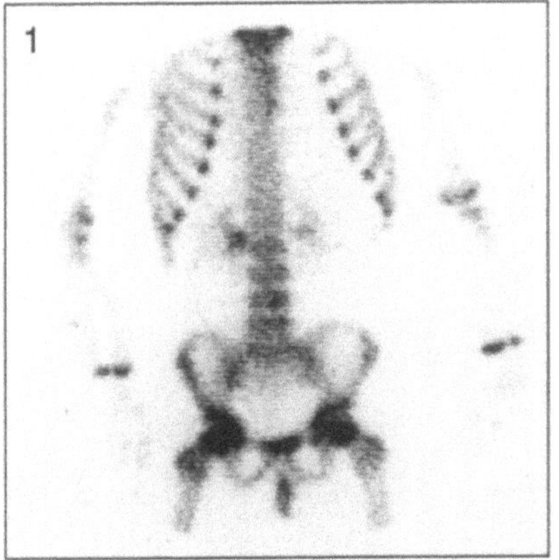

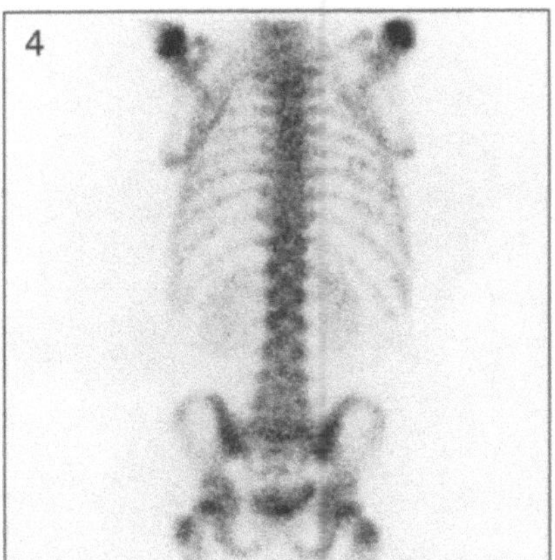

Fig. 5. Posterior view of thorax and spine

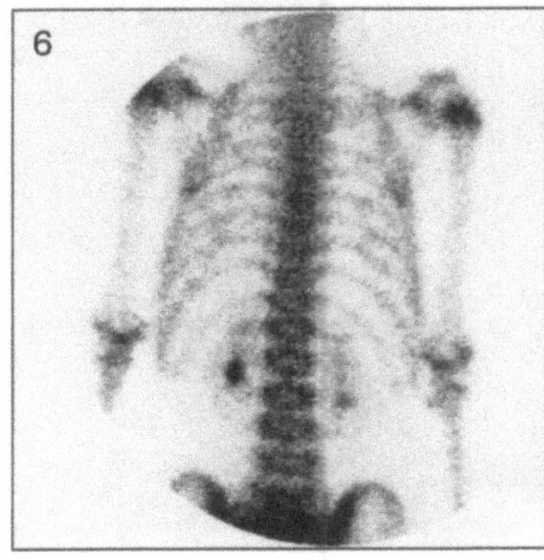

Fig. 6. Posterior view of thorax and spine

Technical Comments

– Different positions of the upper limbs have resulted in the different appearances of the scapulae in Figs. 4 and 5
– Urine contamination below the pelvis is seen in Fig. 1
– Asymmetrical renal pelvic activity is noted in the right renal pelvis in Fig. 1 and in the left in Fig. 6
– Note extravasation of isotope at the site of injection in the right elbow in Fig. 5

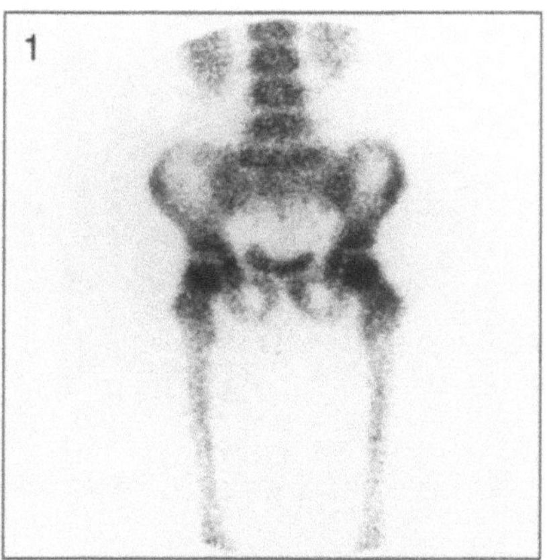

Fig. 1. Anterior view of spine, pelvis and femora

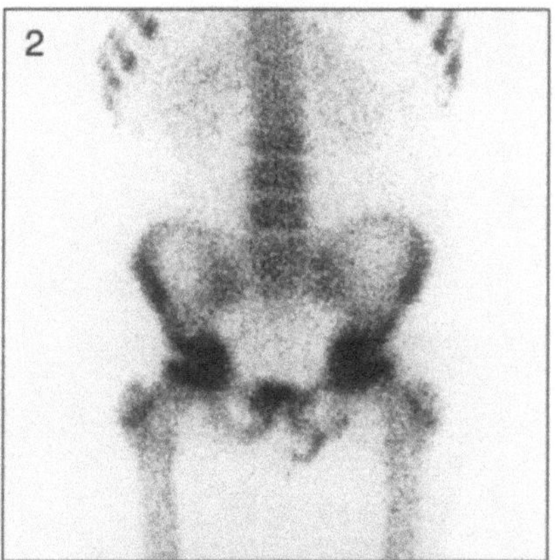

Fig. 2. Anterior view of spine, pelvis and femora

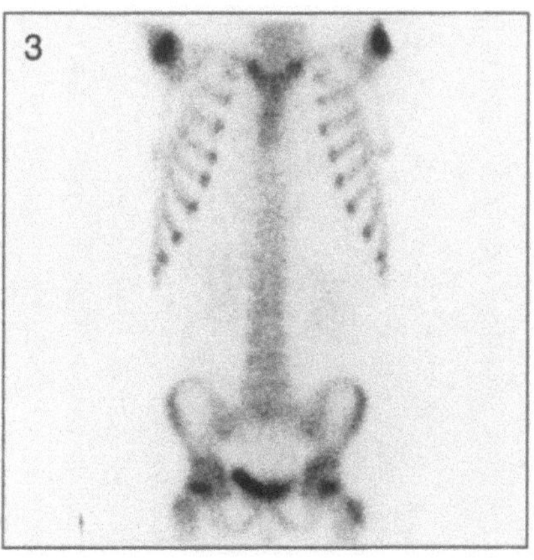

Fig. 3. Anterior view of thorax, spine and pelvis

Technical Comments

– Urine contamination over the right inferior pubic ramus is seen in Fig. 2
– Note the clarity of the lower lumbar spine on these views

Fig. 1. Posterior view of spine, pelvis and femora

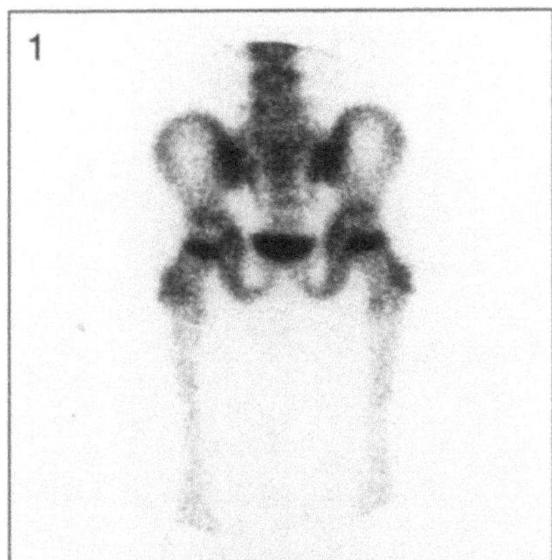

Fig. 2. Posterior view of spine, pelvis and femora

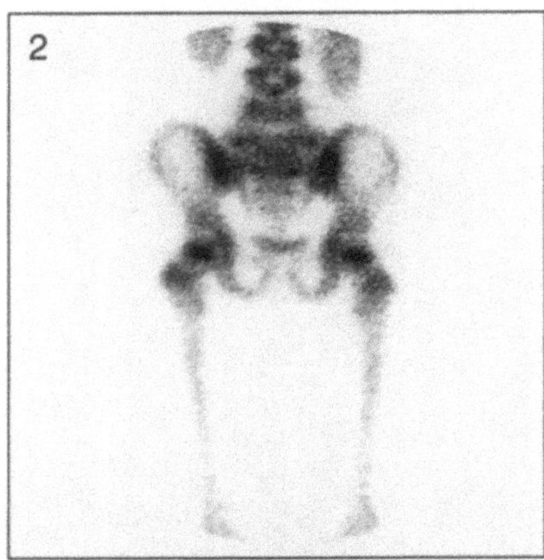

Fig. 3. Posterior view of spine, pelvis and femora

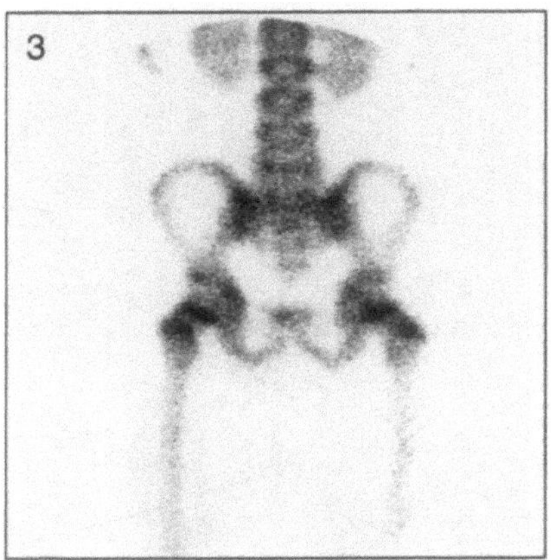

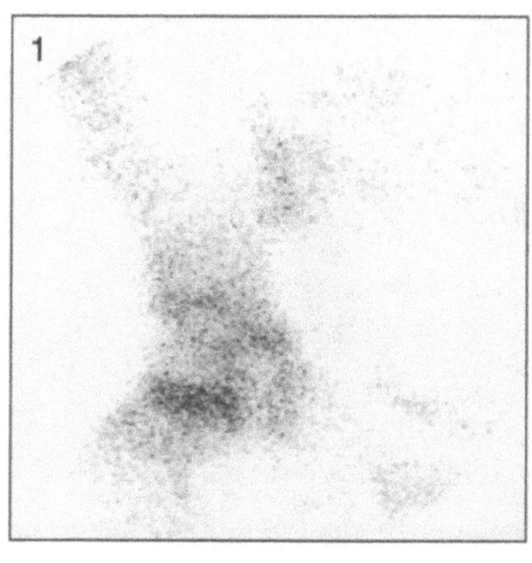

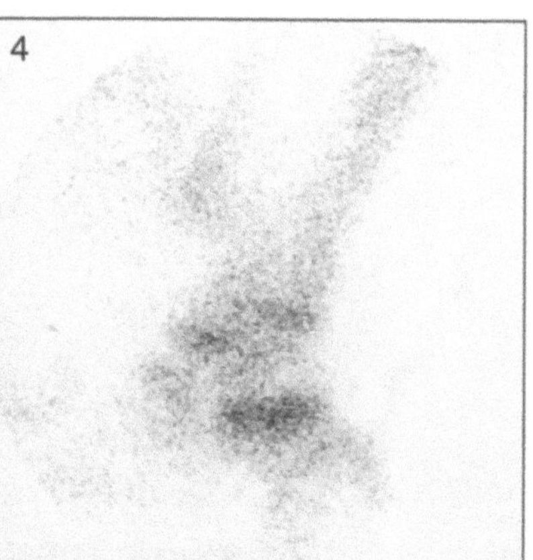

Fig. 1. Pinhole view of right hip

Fig. 4. Pinhole view of left hip

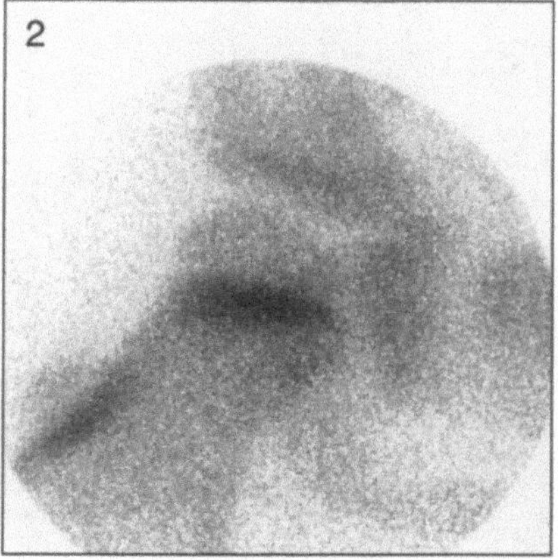

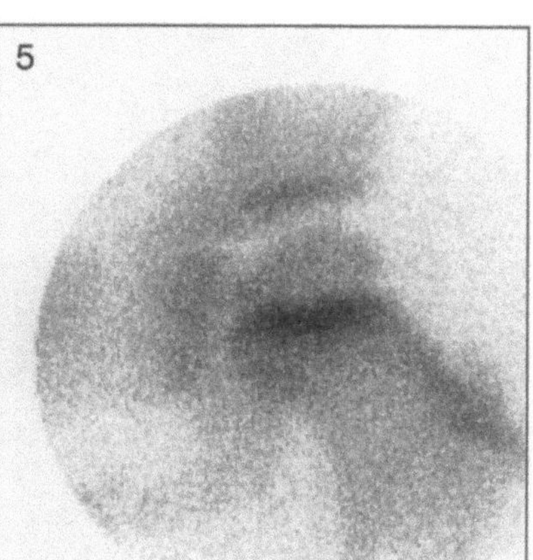

Fig. 2. Pinhole view of right hip

Fig. 5. Pinhole view of left hip

Fig. 1. Posterior view of pelvis
and femora

Fig. 4. Posterior view of lower
limbs

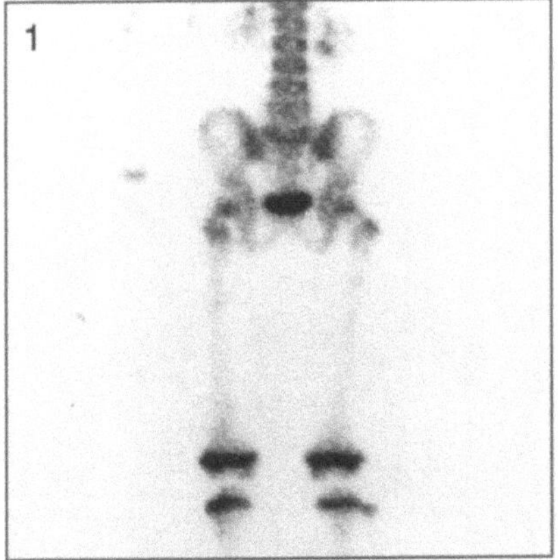

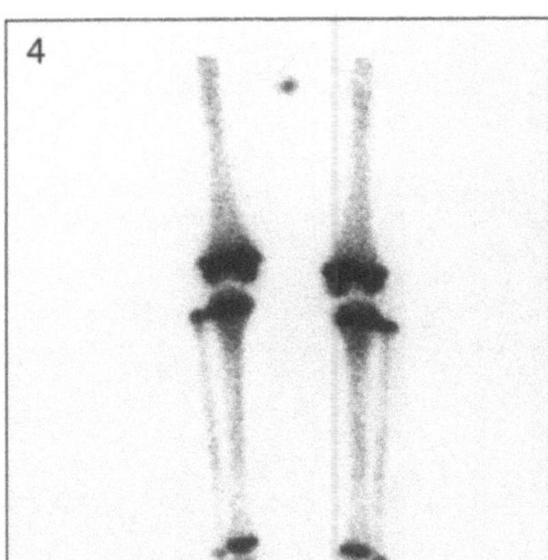

Fig. 2. Posterior view of lower
limbs

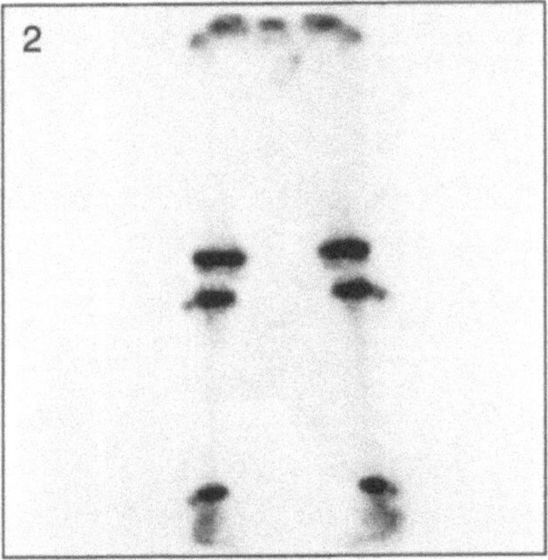

Technical Comments

- Urine contamination is seen at the upper edge of Fig. 4
- Fig. 1 shows the difference in the activity between the two femoral necks. The left leg is poorly positioned as evidenced by lack of visualization of the head of the fibula. This has resulted in the greater trochanter rotating and being seen to overlie the femoral neck. The right side is well positioned.

▶ **Potential Pitfall**

- The lack of clarity of the epiphyseal plates around the knees in Fig. 4 is due to overexposure during the acquisition

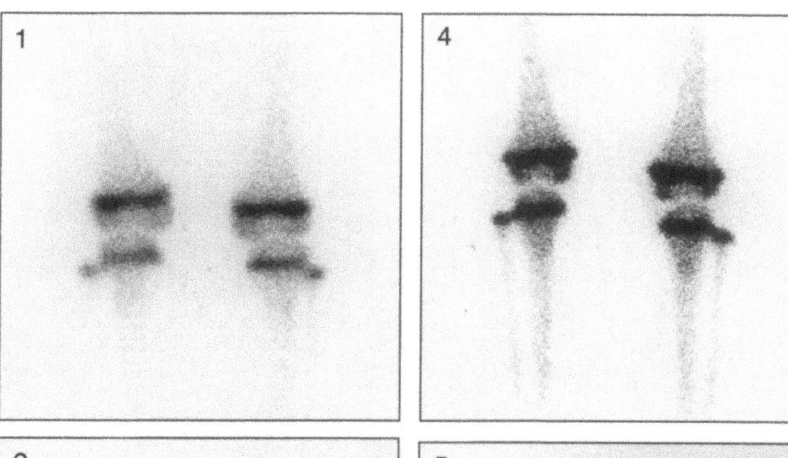

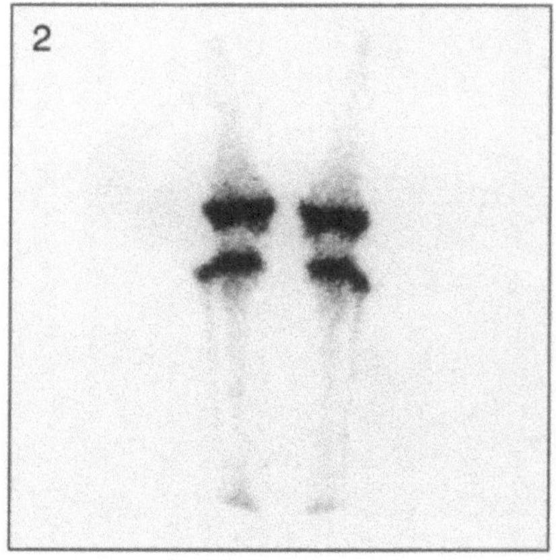

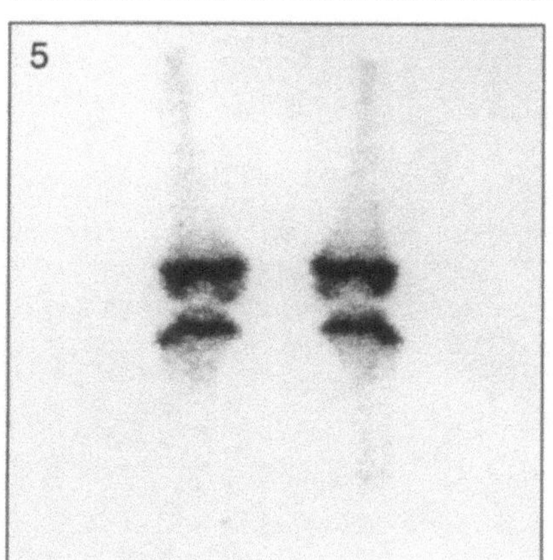

Fig. 1. Posterior view of knees

Fig. 4. Posterior view of knees

Fig. 2. Posterior view of knees

Fig. 5. Posterior view of knees

Technical Comment

– Note the clarity of the heads of the fibulae on all the images. Note the clarity of epiphyseal plates with their well defined margins

Fig. 1. Posterior view of tibia, fibula and ankles

Fig. 4. Anterior view of both feet

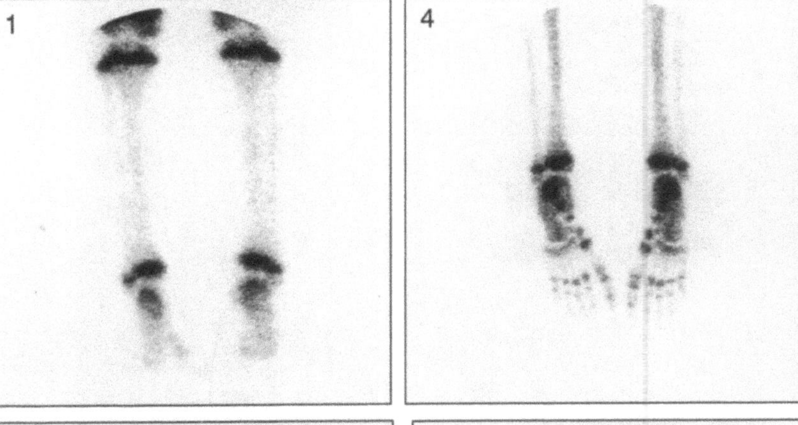

Fig. 2. Lateral view of left lower limb

Fig. 5. Lateral view of right lower limb

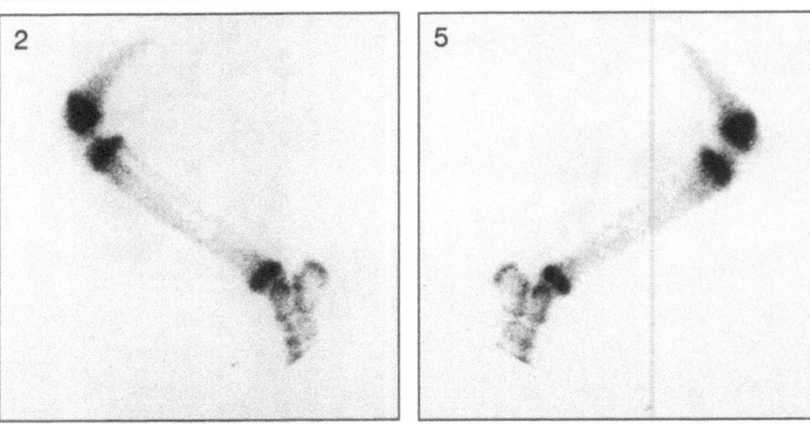

Fig. 3. Lateral view of both feet

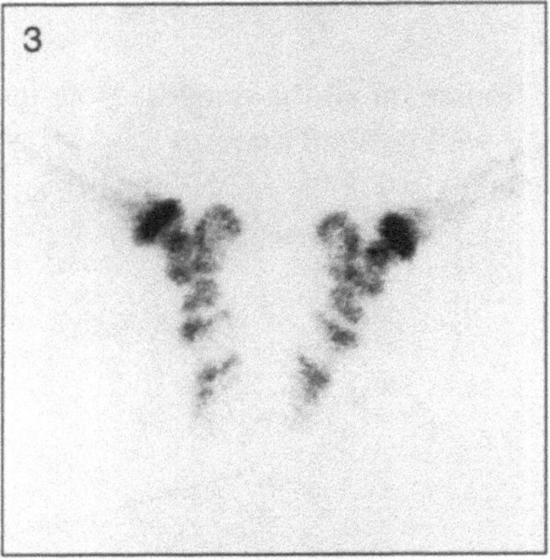

Technical Comment

– The lateral views (Figs. 2 and 5) are of little value for the knees

7: Age 5–6 Years

7: Age 5–6 Years

Fig. 1. Posterior view of thorax and spine

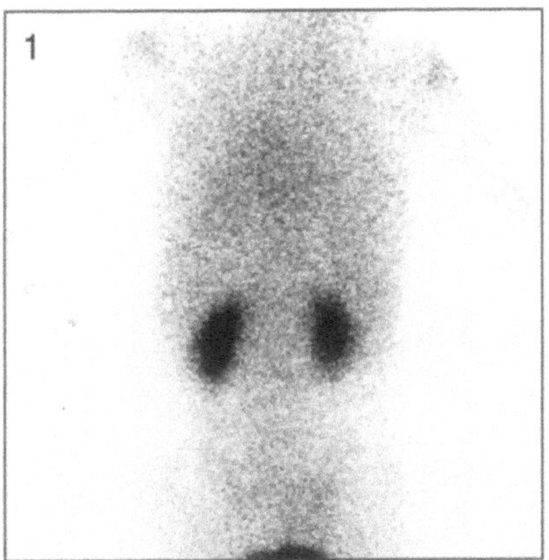

Fig. 2. Posterior view of spine and pelvis

Fig. 5. Anterior view of pelvis

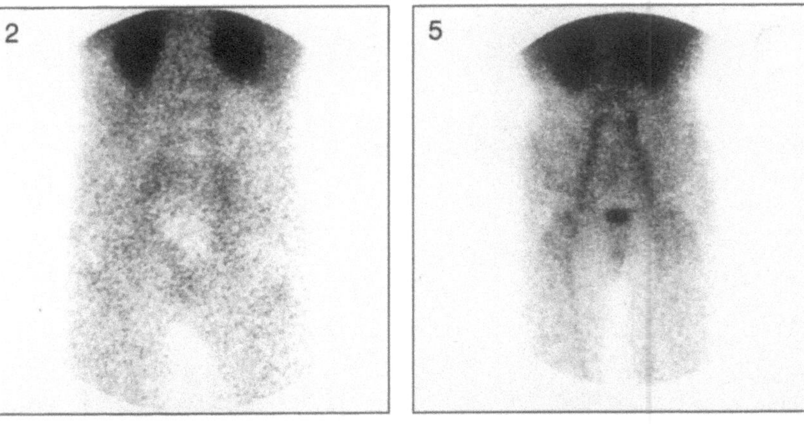

Fig. 3. Posterior view of femora and knees

Fig. 6. Anterior view of lower limbs

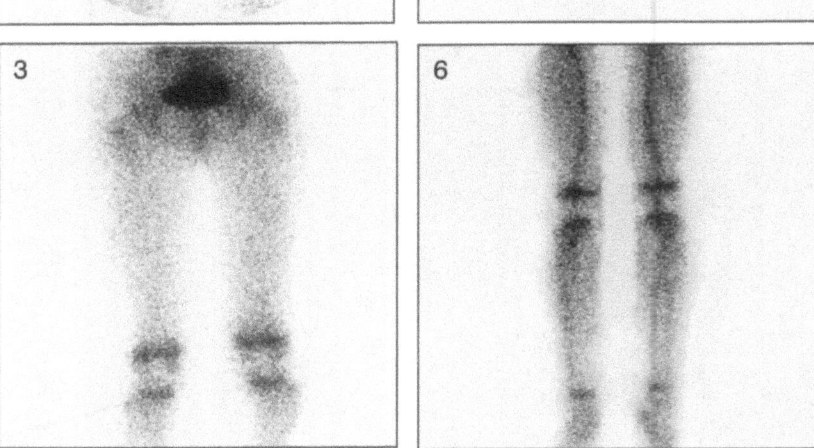

Technical Comments
- The full bladder is seen as a photon deficient area in Fig. 2
- The large vessels are clearly seen on the anterior views in Figs. 5 and 6

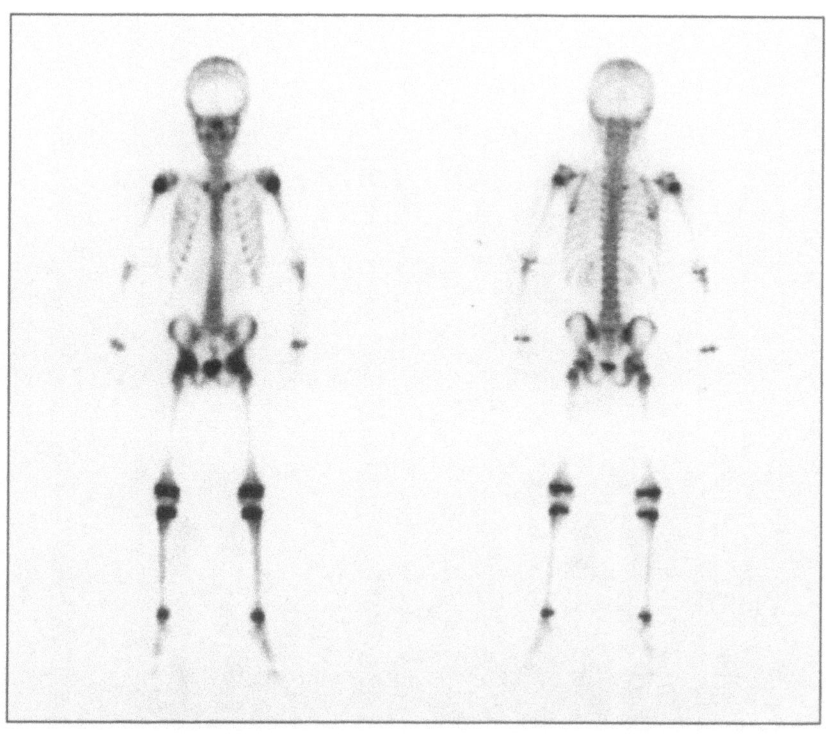

- A double headed whole body gamma camera was used
- Left image is the anterior view
- Right image is the posterior view

Technical Comment
- The feet are poorly positioned with out-turning of the toes bilaterally

- A double headed whole body
  gamma camera was used
- Left image is the anterior view
- Right image is the posterior
  view

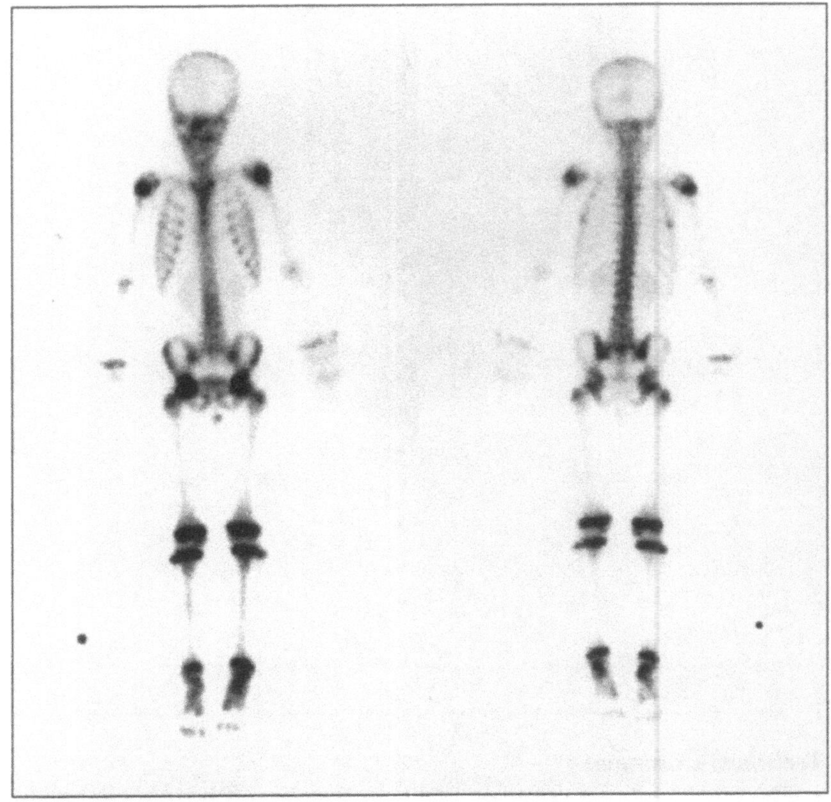

Technical Comments

- Both feet are well positioned with the toes turned inward allowing good
  visualization of the fibula
- Slight movement of the upper limbs has resulted in blurring of the images of the upper limbs
- Urine contamination below the pelvis is seen on the anterior view
- Marker on child's right side

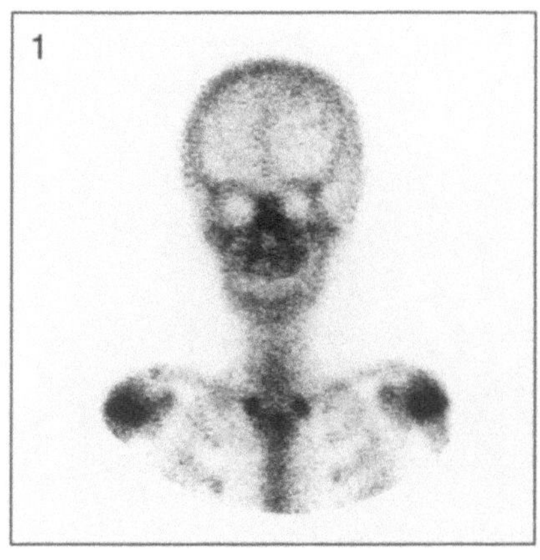

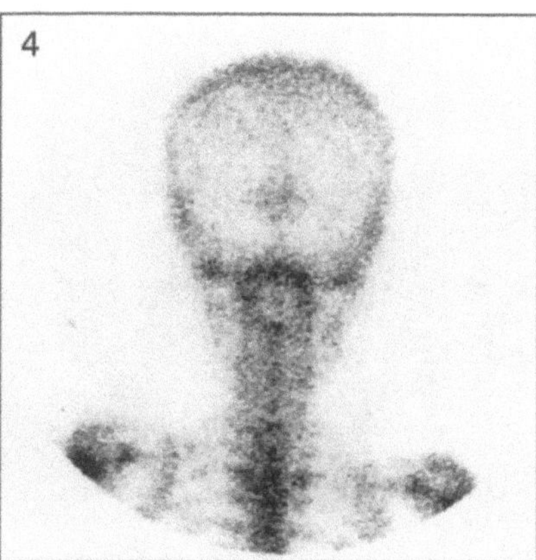

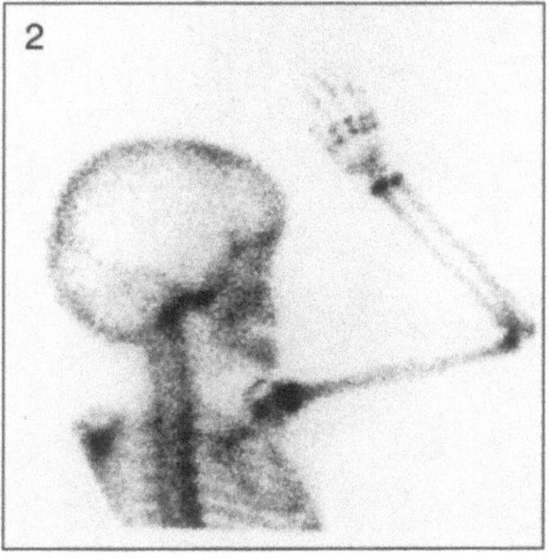

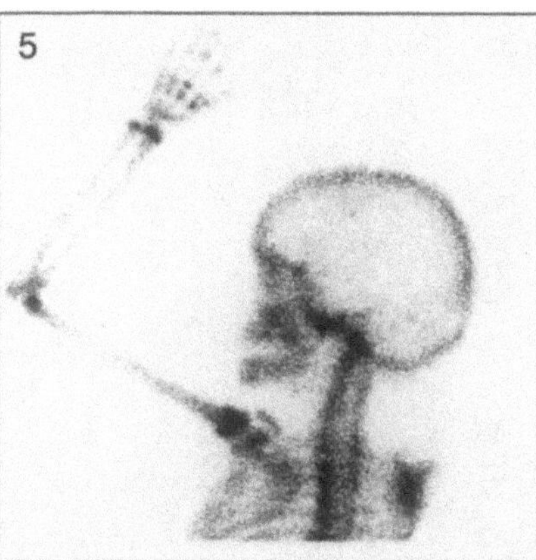

Fig. 1. Anterior view of skull and thorax

Fig. 4. Posterior view of skull and spine

Fig. 2. Right lateral view of skull, right upper limb and posterior view of thorax

Fig. 5. Left lateral view of skull, left upper limb and posterior view of thorax

Technical Comments

- Note the increased activity in the region of the maxillary antra in Fig. 1. This is within normal limits
- The lateral views of the skull (Figs. 2 and 5) were taken posteriorly

Fig. 1. Right lateral view of skull, both upper limbs and posterior view of thorax

Fig. 4. Left lateral view of skull and both upper limbs

Fig. 2. Right lateral view of skull and right upper limb

Fig. 5. Left lateral view of skull and left upper limb

Fig. 3. Anterior view of both hands

Fig. 6. Anterior view of both hands

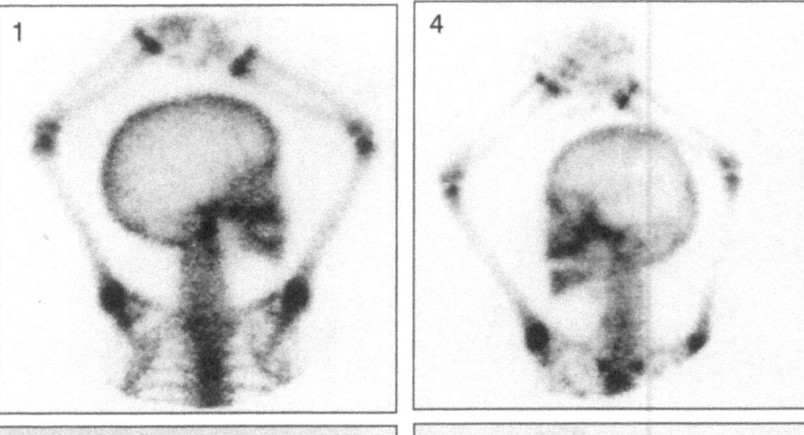

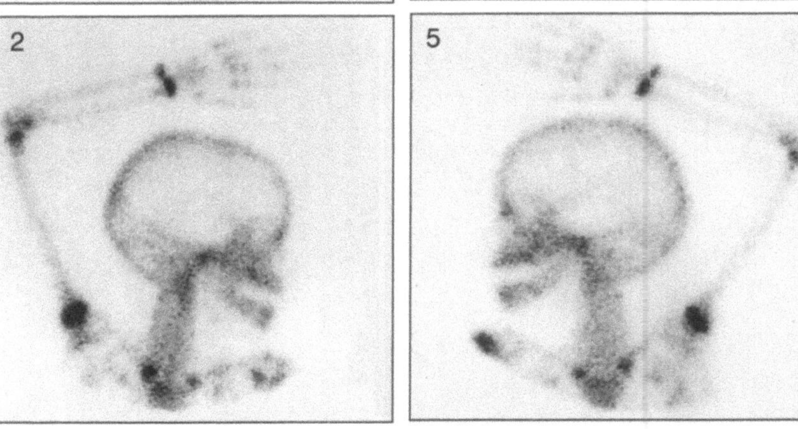

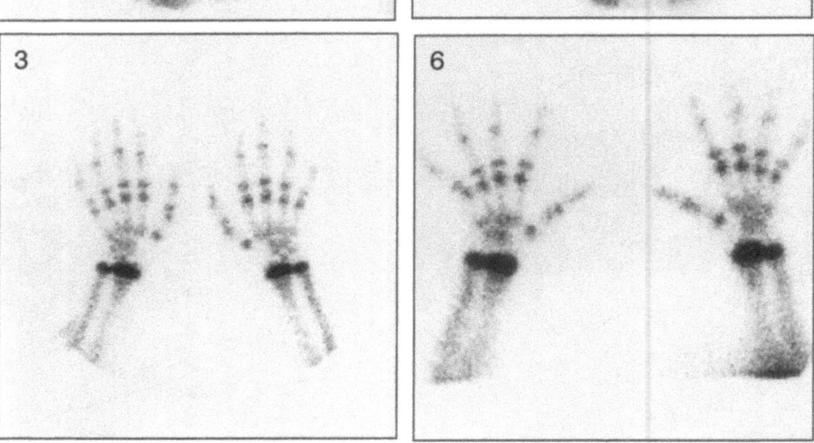

Technical Comments

– Note the difficulty of imaging both hands simultaneously in Figs. 1 and 4
– The lateral views of the skull (Figs. 2 and 5) were taken anteriorly

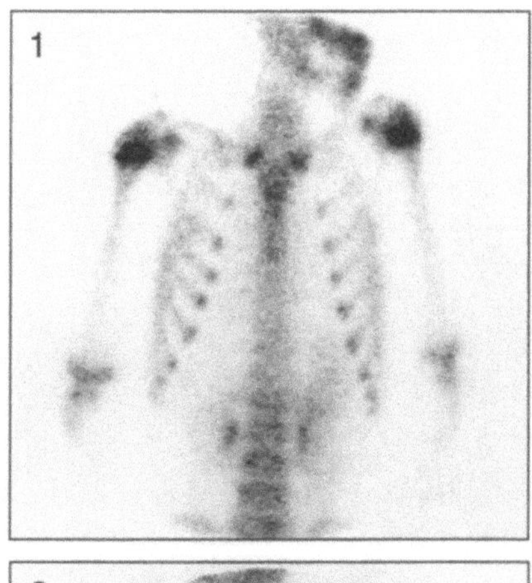

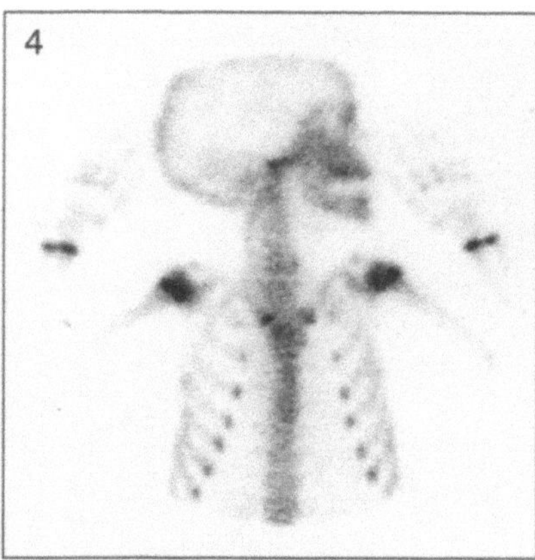

Fig. 1. Anterior view of thorax and spine

Fig. 4. Right lateral view of skull and anterior view of thorax

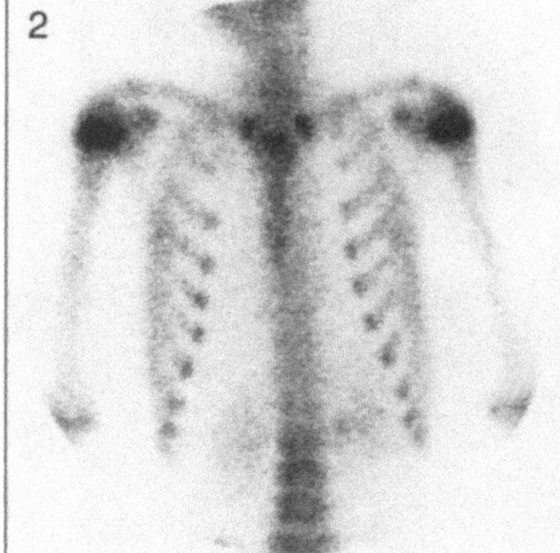

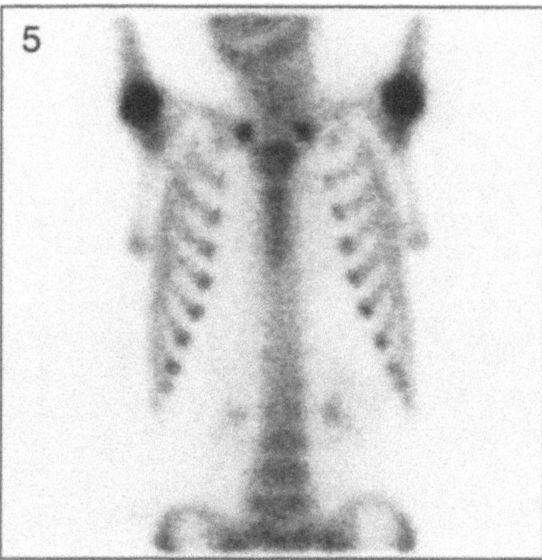

Fig. 2. Anterior view of thorax and spine

Fig. 5. Anterior view of thorax and spine

Technical Comments

- Note the differences in the sternum on these four images
- The lumbar spine is clearly seen in Figs. 1, 2 and 5
- The scapulae appear differently in Figs. 2 and 5 due to the different positions of the upper limbs

Fig. 4. Posterior view of thorax and spine

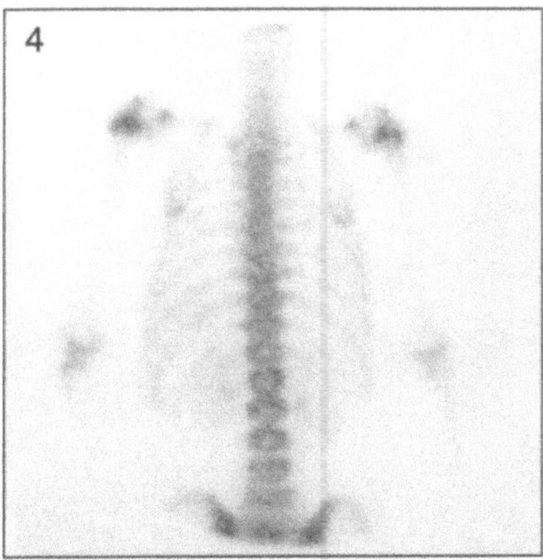

Fig. 2. Left anterior oblique view of thorax

Fig. 5. Posterior view of thorax, spine and pelvis

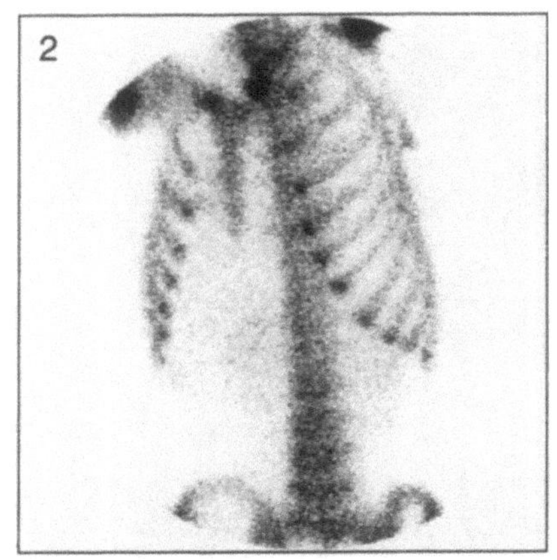

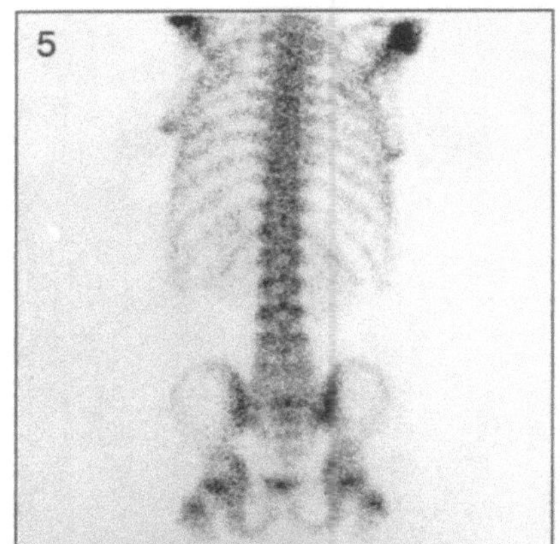

Fig. 6. Posterior view of thorax, spine and pelvis

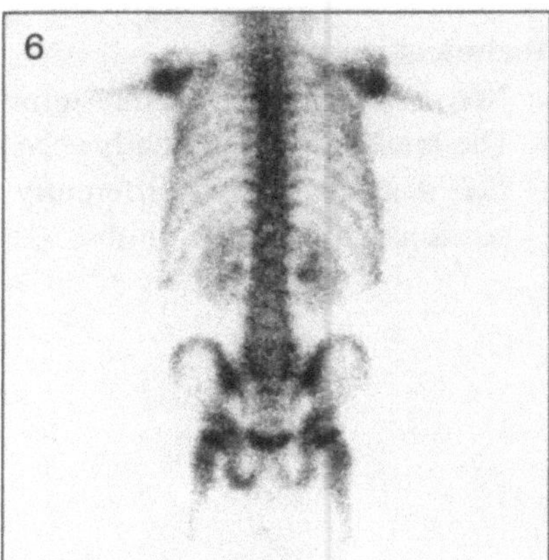

▶ **Potential Pitfall**
- Note increased activity in the left posterior pubic ramus and ischium in Fig. 6. This is an unusual variant of the normal synchondrosis

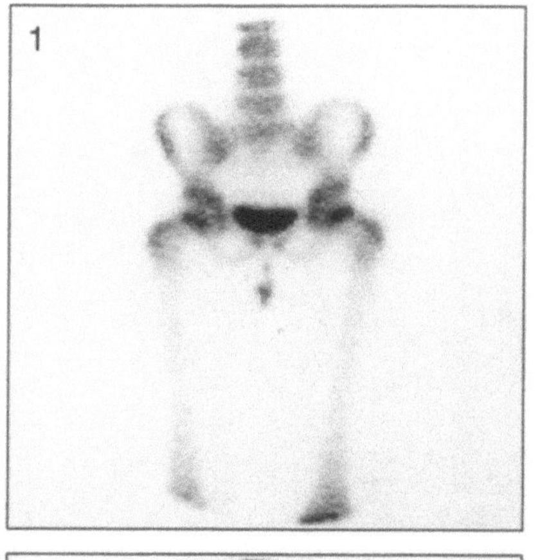

Fig. 1. Anterior view of spine, pelvis and femora

Fig. 4. Anterior view of spine and pelvis

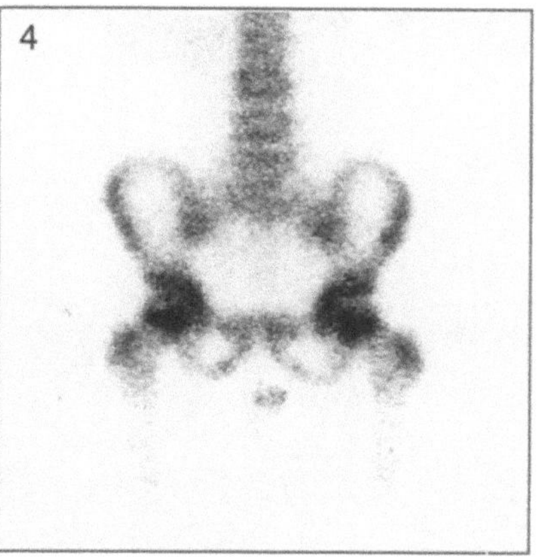

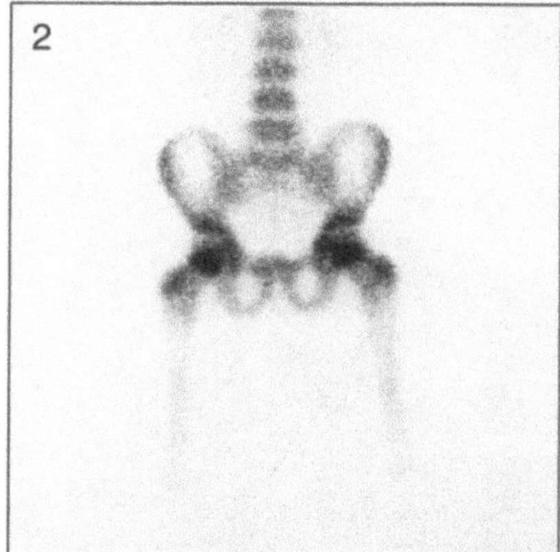

Fig. 2. Anterior view of spine, pelvis and femora

Fig. 5. Anterior view of pelvis and femora

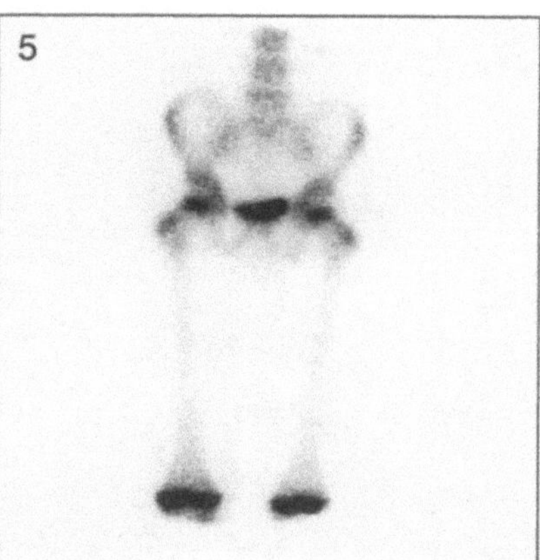

Technical Comments

- Note the clarity of lumbar spine on all views
- Urine contamination below the pelvis is seen in Figs. 1 and 4

Fig. 1. Posterior view of spine and pelvis

Fig. 4. Posterior view of spine, pelvis and femora

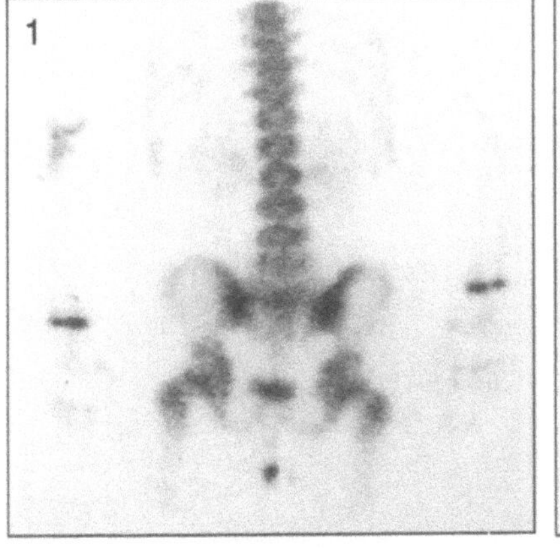

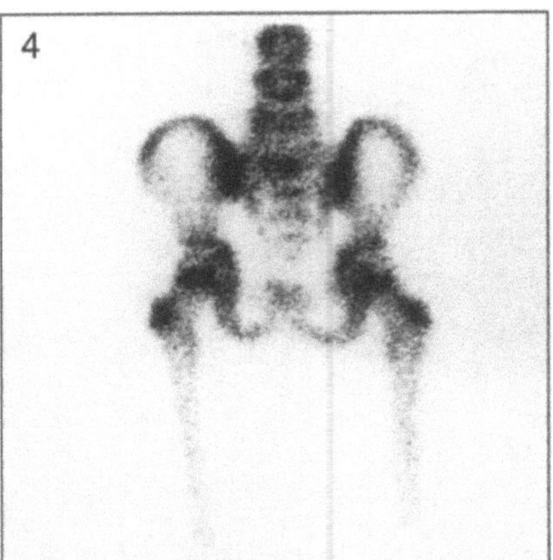

Fig. 2. Posterior view of spine and pelvis

Fig. 5. Posterior view of spine and pelvis

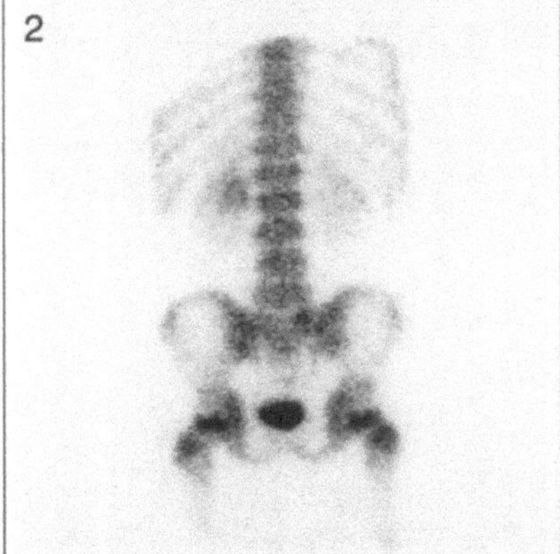

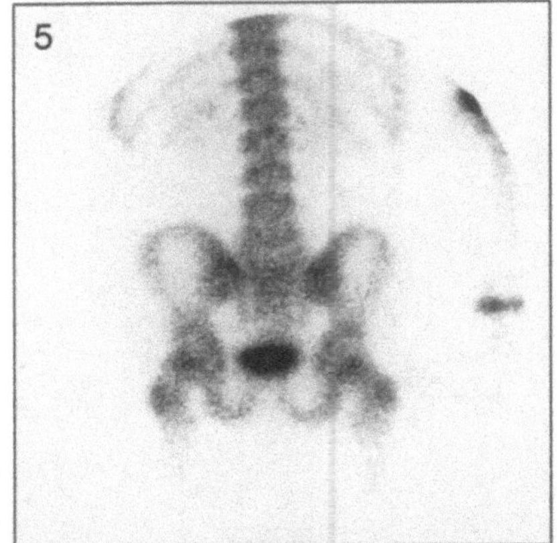

Fig. 6. Posterior view of spine and pelvis with lower limbs positioned in abduction

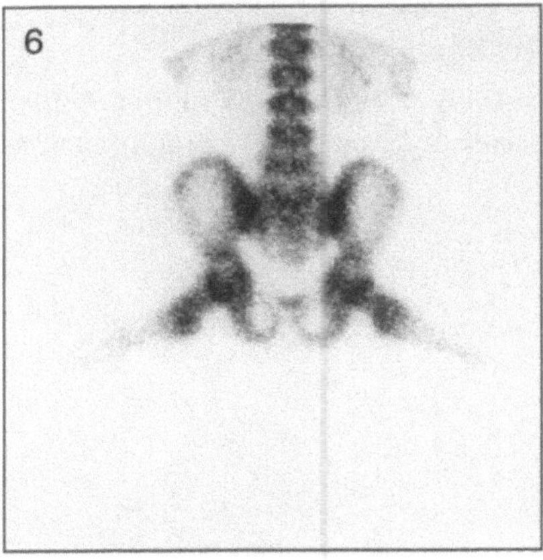

Technical Comments
- Urine contamination below the pelvis is seen in Fig. 1
- Fig. 6 shows the hips in abduction, an usual view used mainly to look for slipped capital femoral epiphysis

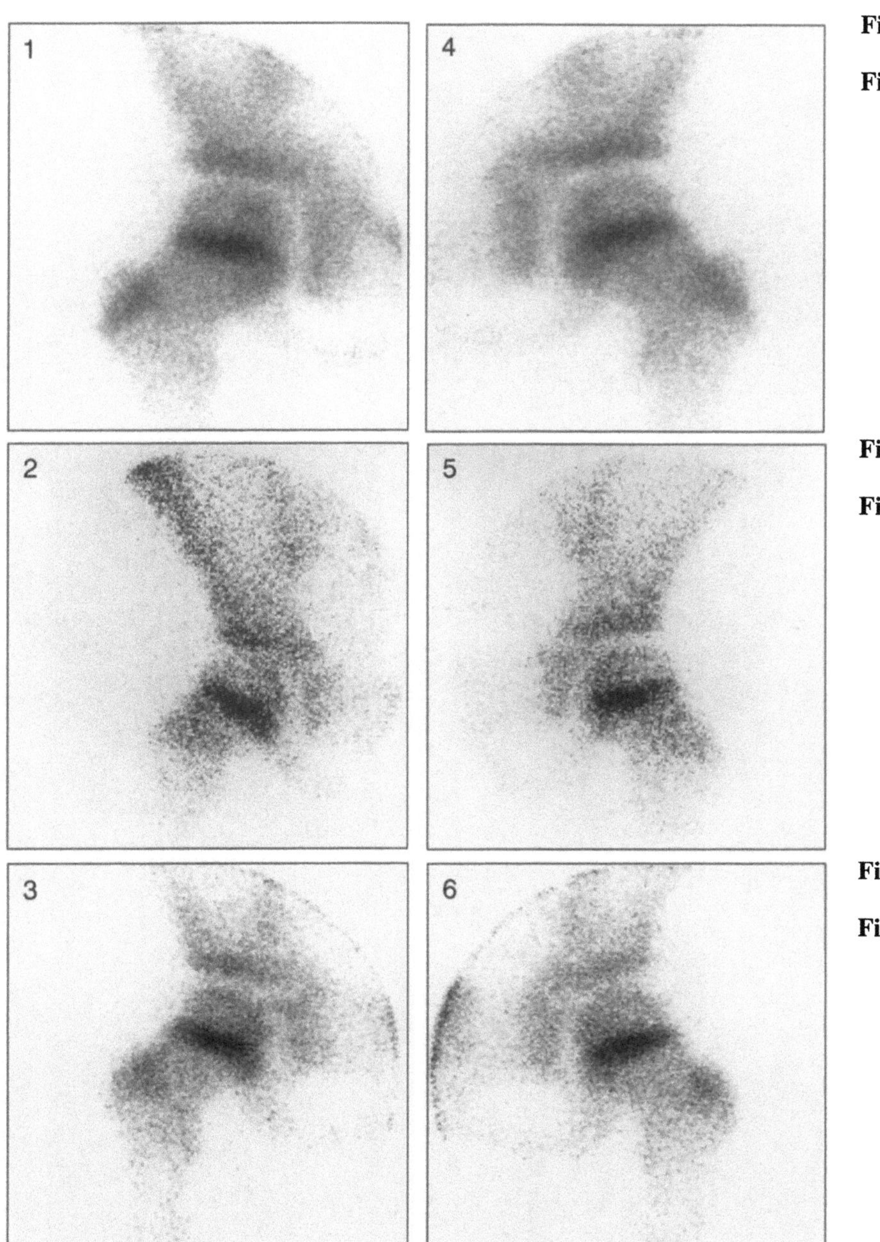

Fig. 1. Pinhole view of right hip

Fig. 4. Pinhole view of left hip

Fig. 2. Pinhole view of right hip

Fig. 5. Pinhole view of left hip

Fig. 3. Pinhole view of right hip

Fig. 6. Pinhole view of left hip

Fig. 1. Posterior view of lower limbs

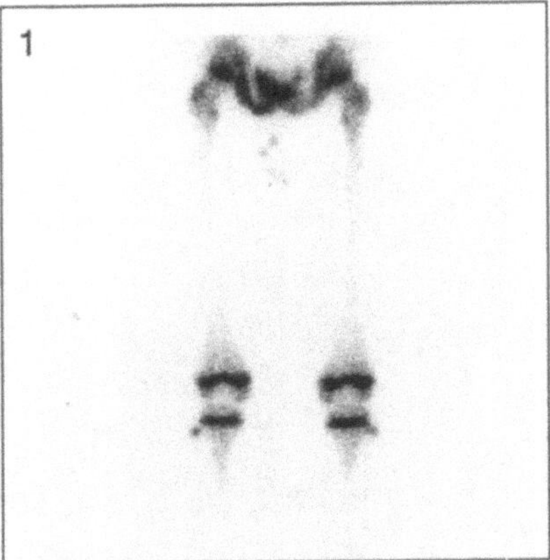

Fig. 2. Posterior view of knees, tibia, fibula and ankles

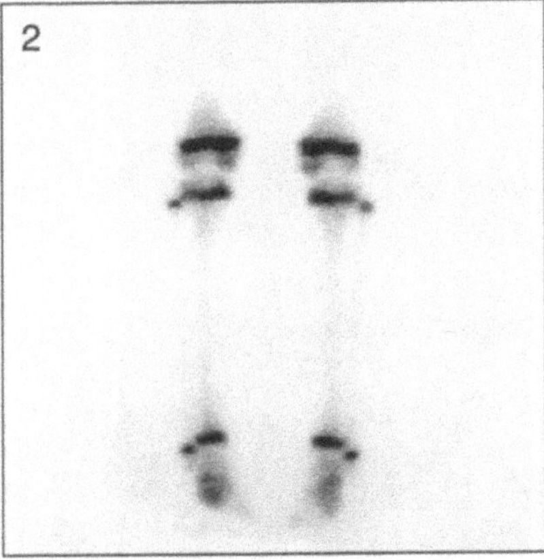

Fig. 3. Posterior view of knees, tibia, fibula and feet

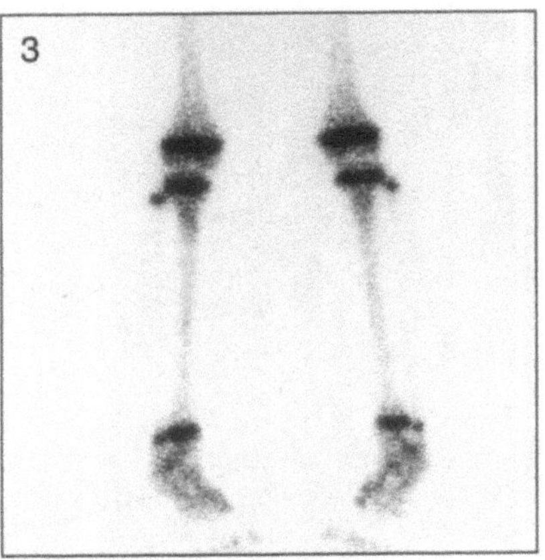

Technical Comment

– Note the "radiographic neutral positioning" of the feet in Figs. 1–3 resulting in clear visualization of the fibula

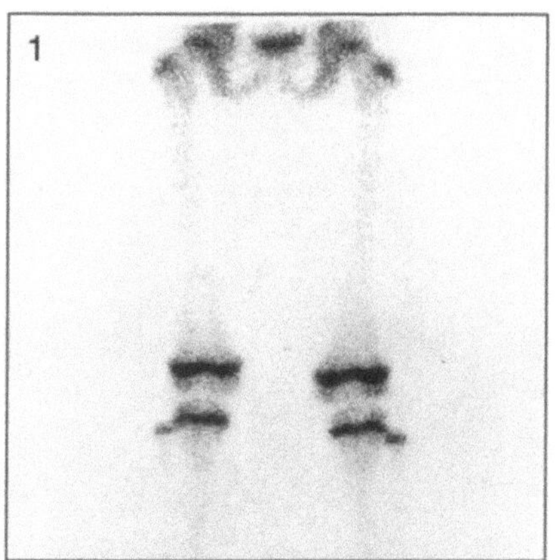

Fig. 1. Posterior view of knees

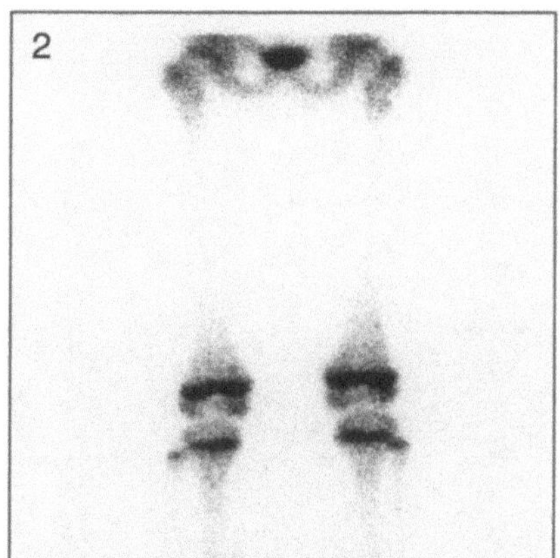

Fig. 2. Posterior view of knees

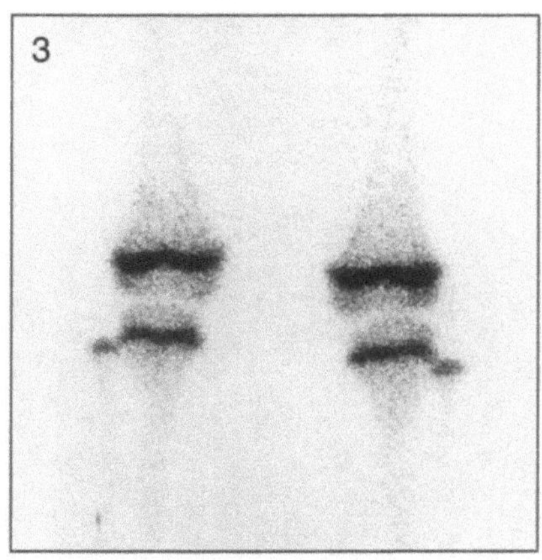

Fig. 3. Posterior view of knees

Technical Comments

− The magnified view (Fig. 3) shows the epiphyseal plates to best advantage
− The fibula is clearly seen in all figures because the toes were turned inward during image acquisition

Fig. 1. Lateral view of feet with straight knees

Fig. 4. Anterior view of feet

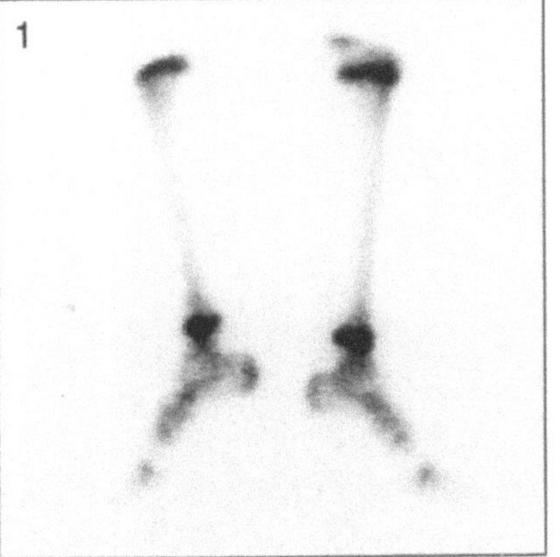

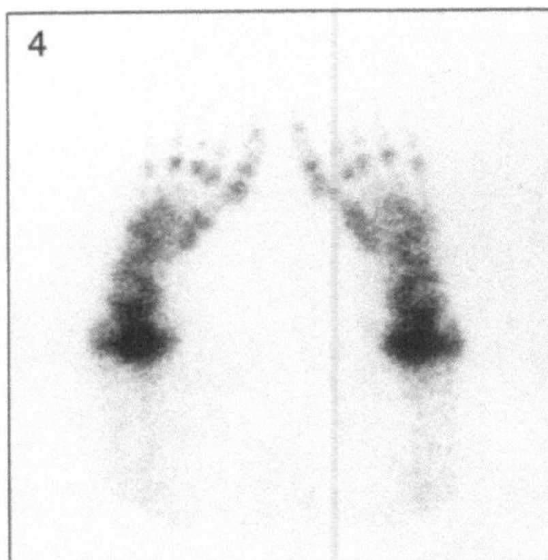

Fig. 2. Lateral view of feet with knees flexed and magnification

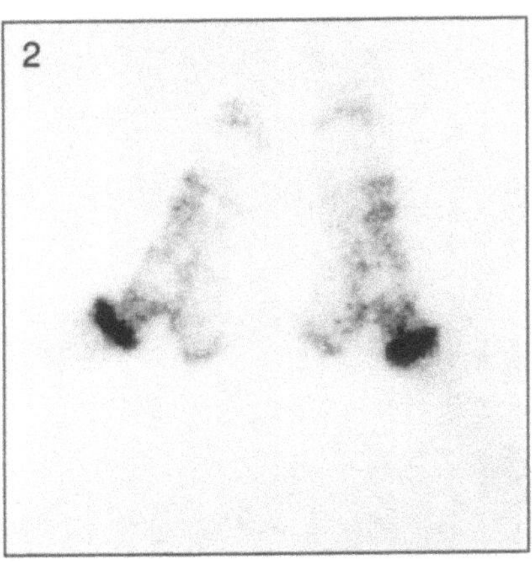

8: Age 6–7 Years

- A double headed whole body
  gamma camera was used
- Left image is the anterior view
- Right image is the posterior
  view

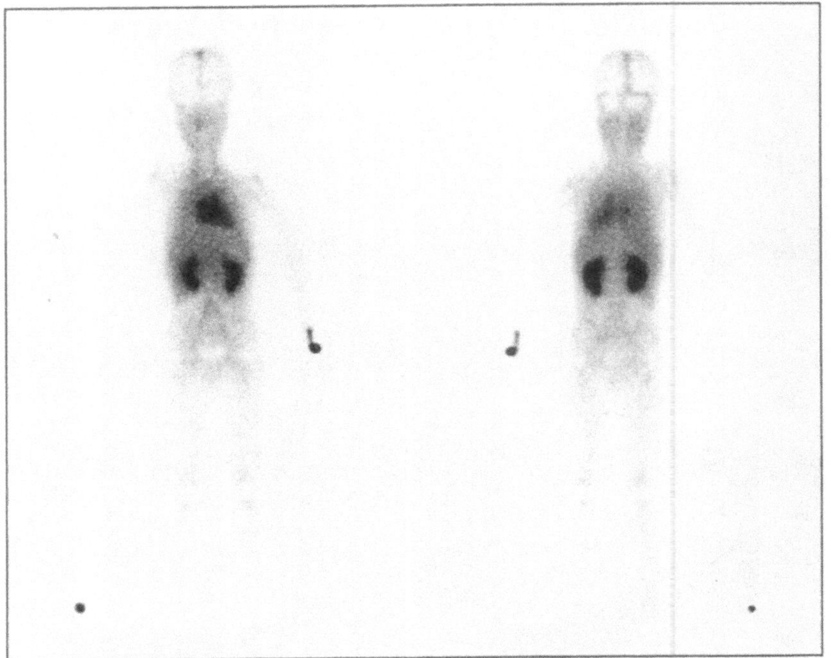

Technical Comments

- Note extravasation of isotope at the site of injection in the left hand
- Marker on child's right side

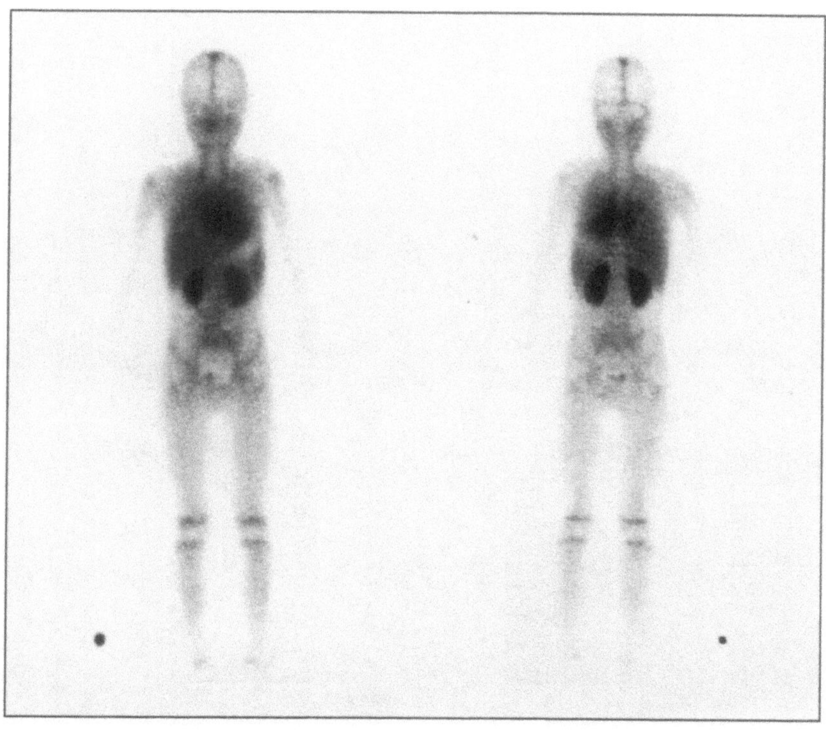

- A double headed whole body gamma camera was used
- Left image is the anterior view
- Right image is the posterior view

Technical Comment
- Marker on child's right side

- A double headed whole body
  gamma camera was used
- Left image is the anterior view
- Right image is the posterior
  view

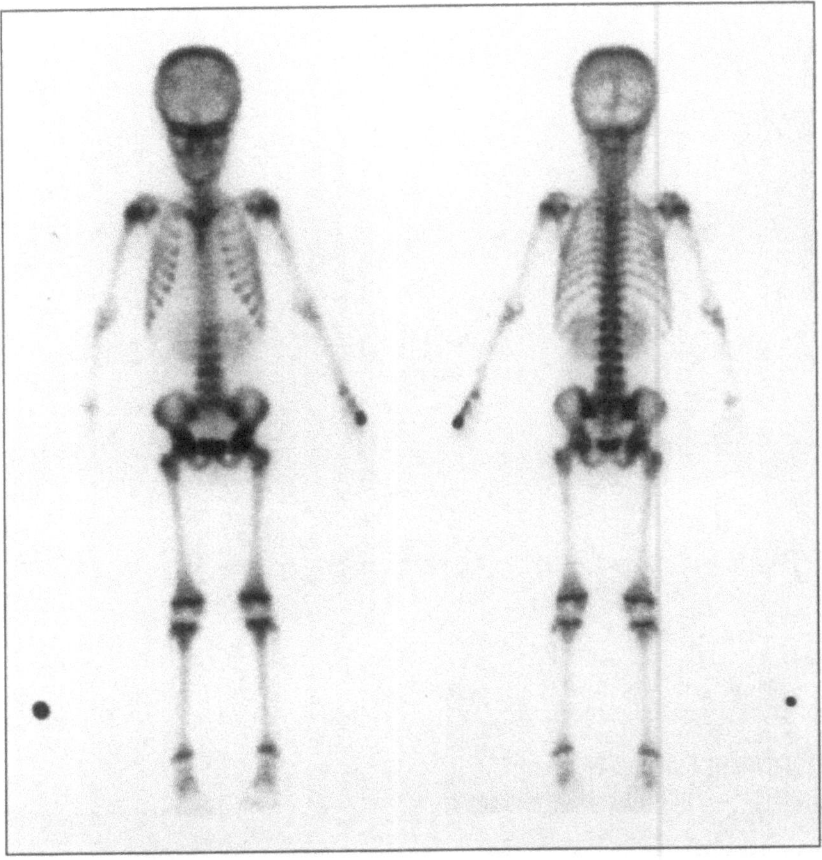

Technical Comments
- Note extravasation of isotope at the site of injection in the left hand
- Marker on child's right side

▶ **Potential Pitfall**
- Note band of apparent increased uptake of isotope in the region of the
  orbits on the anterior view. This is due to the positioning of the child

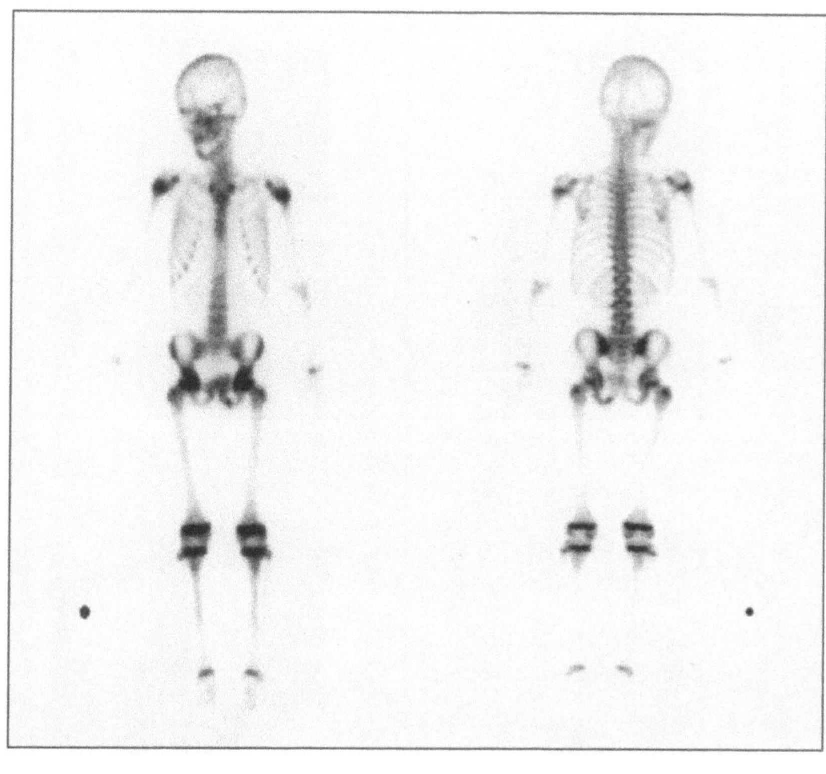

- A double headed whole body gamma camera was used
- Left image is the anterior view
- Right image is the posterior view

Technical Comments
- The child's head is rotated
- Marker on child's right side

▶ **Potential Pitfall**
- There is increased activity in the left inferior pubic ramus, best seen on the anterior view and is due to the normal synchondrosis

Fig. 1. Anterior view of skull and thorax

Fig. 4. Posterior view of skull and thorax

Fig. 2. Right lateral view of skull and right upper limb

Fig. 5. Left lateral view of skull and left upper limb

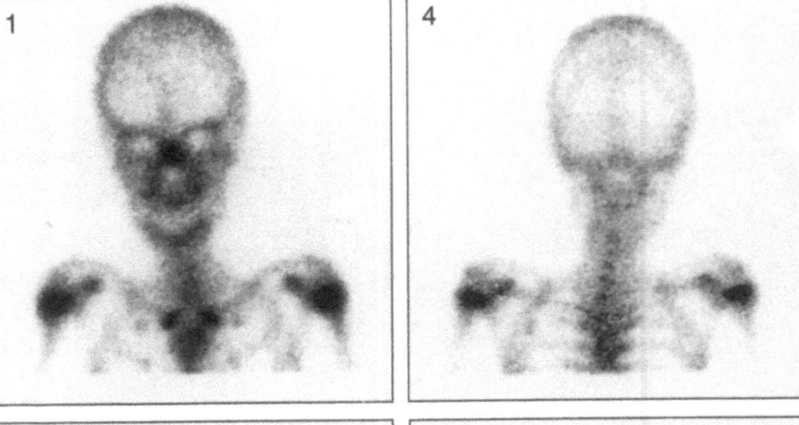

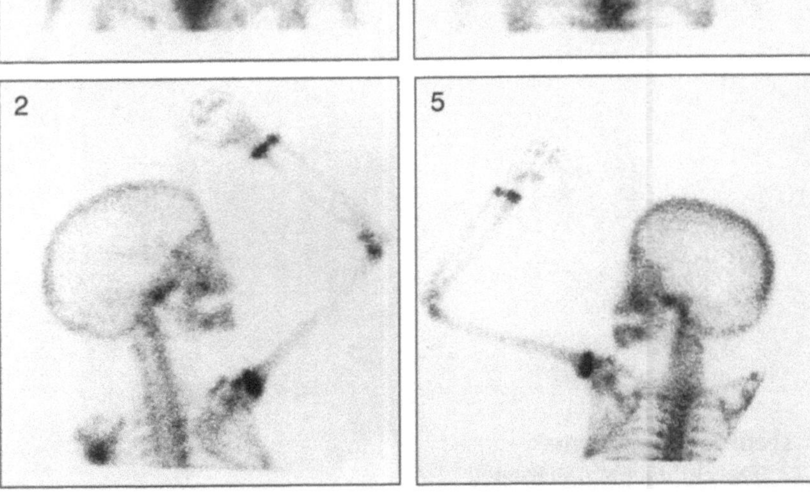

Technical Comment
– The lateral views of the skull (Figs. 2 and 5) were taken posteriorly

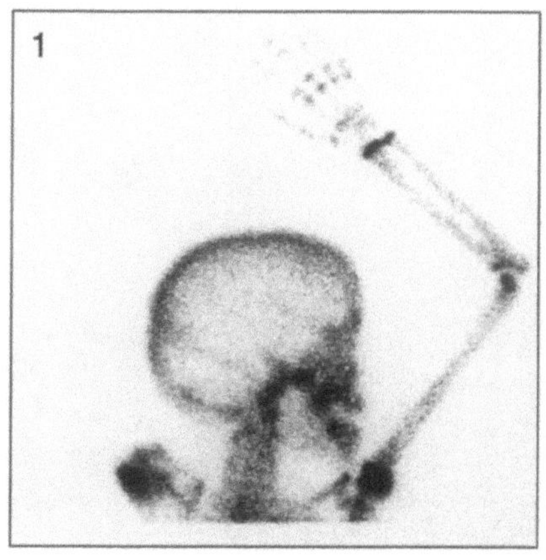

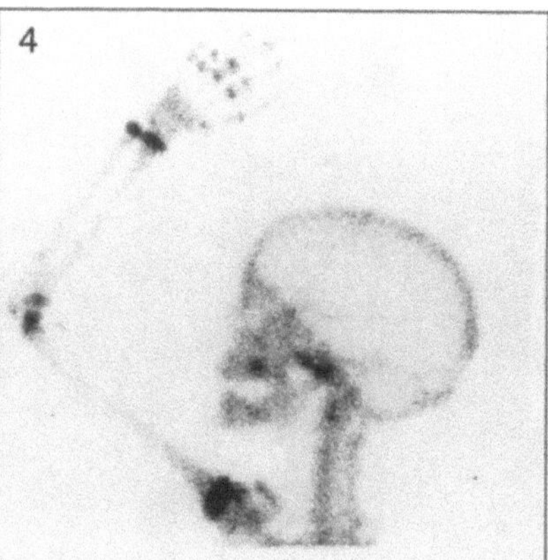

Fig. 1. Right lateral view of skull and right upper limb

Fig. 4. Left lateral view of skull and left upper limb

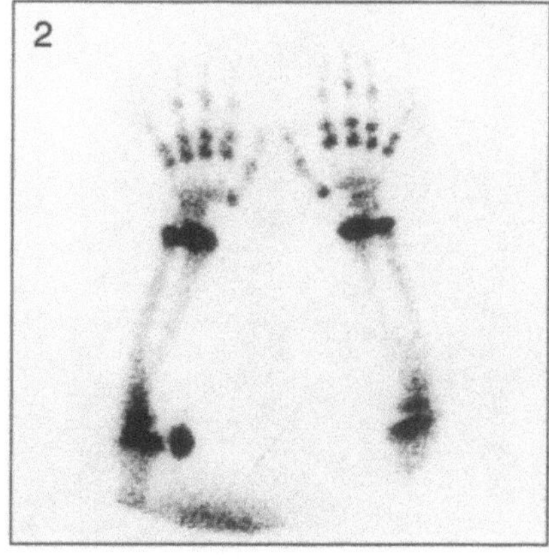

Fig. 2. Anterior view of both upper limbs

Technical Comments
– Note extravasation of isotope at the site of injection in the elbow in Fig. 2
– The lateral views of the skull (Figs. 1 and 4) were taken posteriorly

Fig. 1. Anterior view of thorax, spine and pelvis

Fig. 4. Anterior view of thorax and spine

Fig. 2. Anterior view of thorax, spine and part pelvis

Fig. 5. Anterior view of thorax and spine

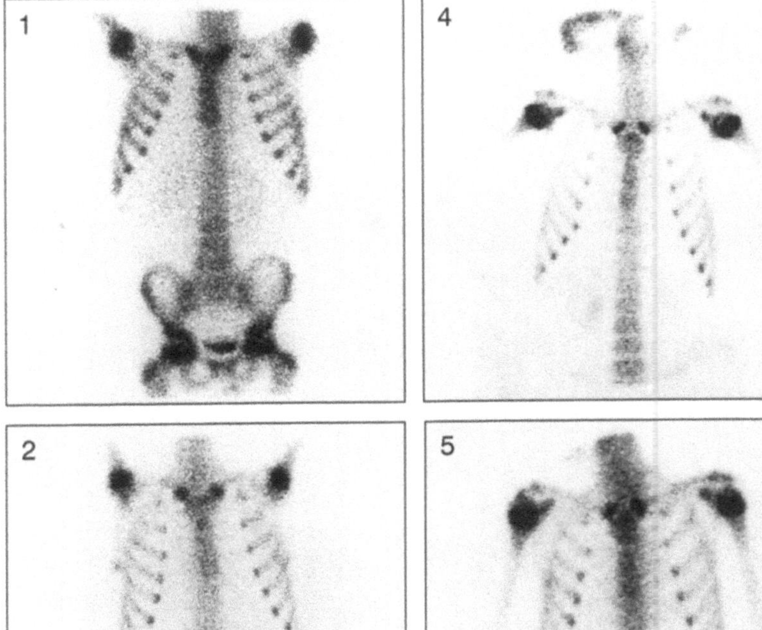

Technical Comments
- The lower lumbar spine is well seen on these anterior views
- The difference between the two kidneys in Fig. 5 is within normal limits

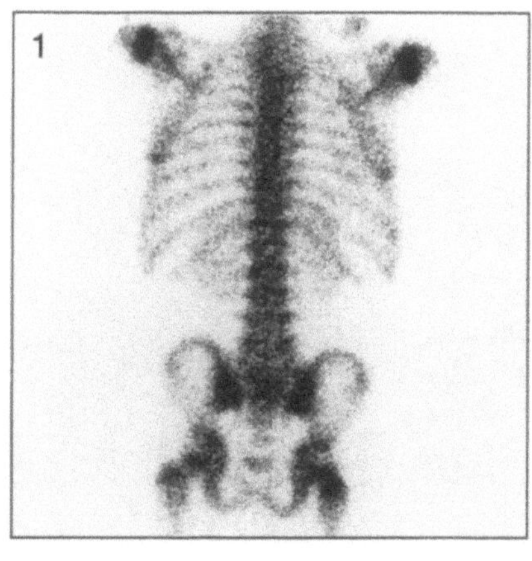

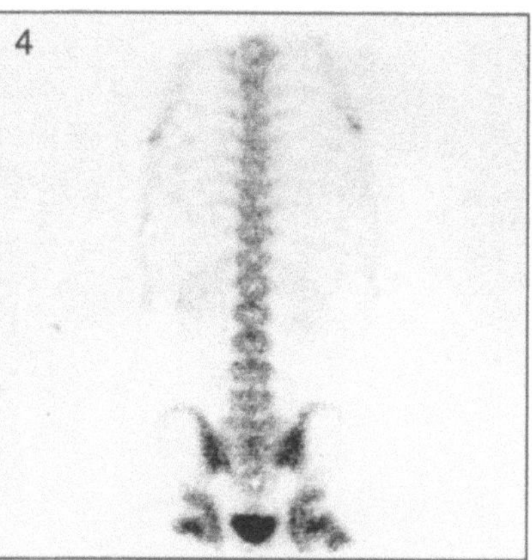

Fig. 1. Posterior view of thorax, spine and pelvis

Fig. 4. Posterior view of thorax, spine and part pelvis

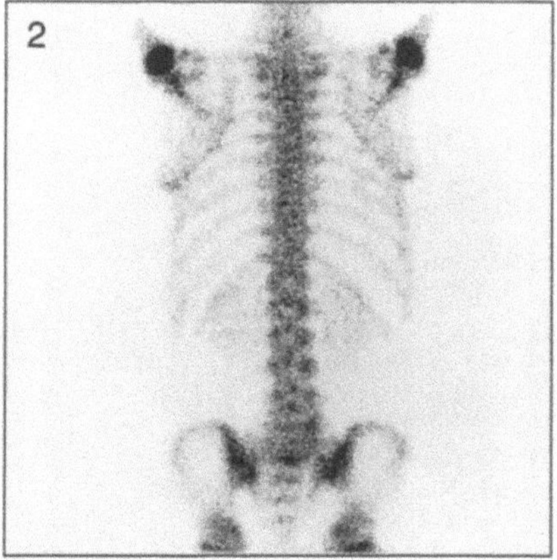

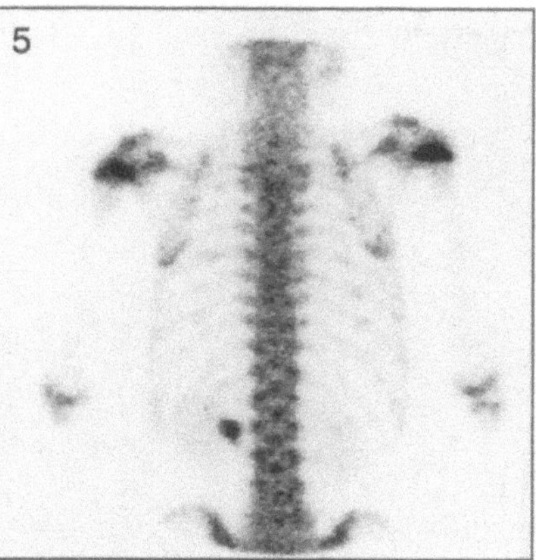

Fig. 2. Posterior view of thorax, spine and part pelvis

Fig. 5. Posterior view of thorax and spine

These images are from the same four children as seen in Figs. 1, 2, 4 and 5 on p. 110

Technical Comment
- There is accumulation of isotope in the renal pelvis of the left kidney in Fig. 5. This is still within normal limits (same child as in Fig. 5, p. 110)

▶ **Potential Pitfall**
- Increased activity over the line of the ribs in the posterior axillary portion, best seen in Fig. 5 on the right are not due to fractures but reflect "shine-through" from the anterior costo-chondral junctions

Fig. 1. Anterior view of spine, pelvis and femora

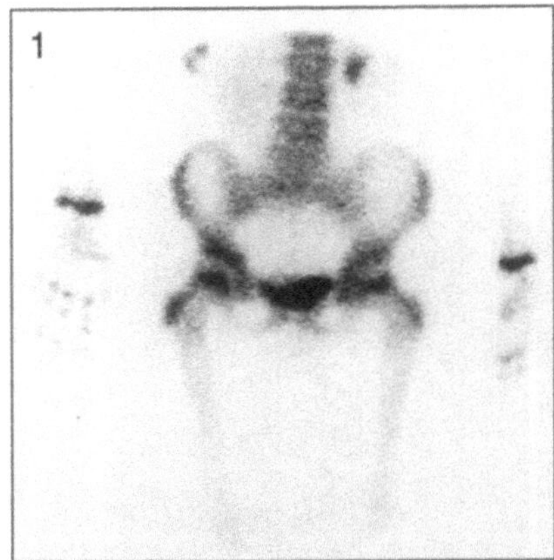

Fig. 2. Anterior view of spine, pelvis and femora

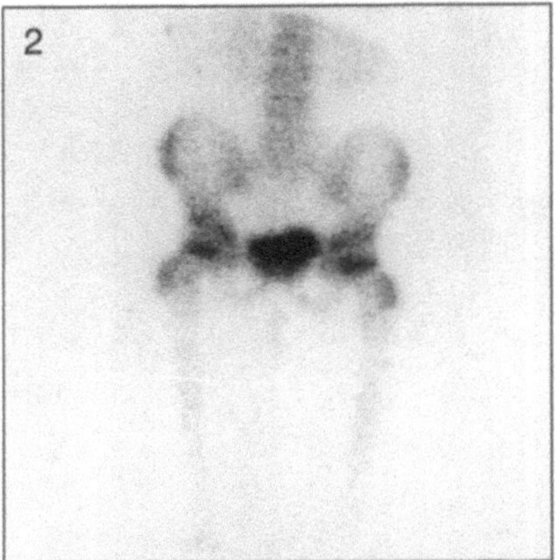

Fig. 3. Anterior view of spine, pelvis, femora and hands

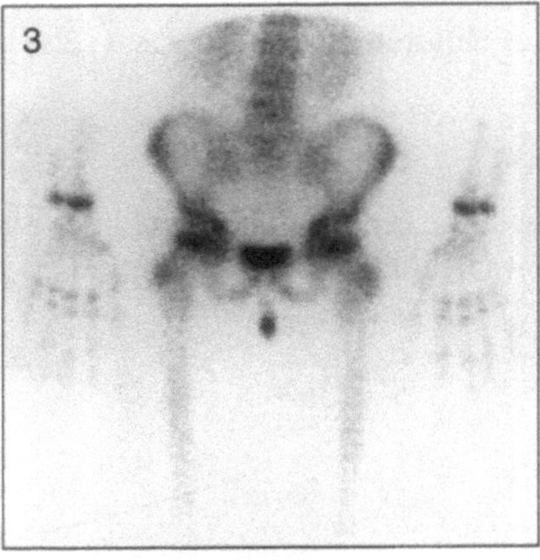

Technical Comment
– Urine contamination below the pelvis is seen in Fig. 3

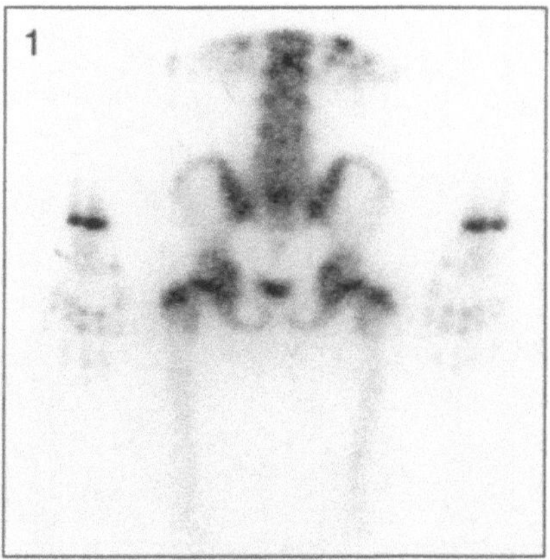

Fig. 1. Posterior view of spine, pelvis, femora and hands

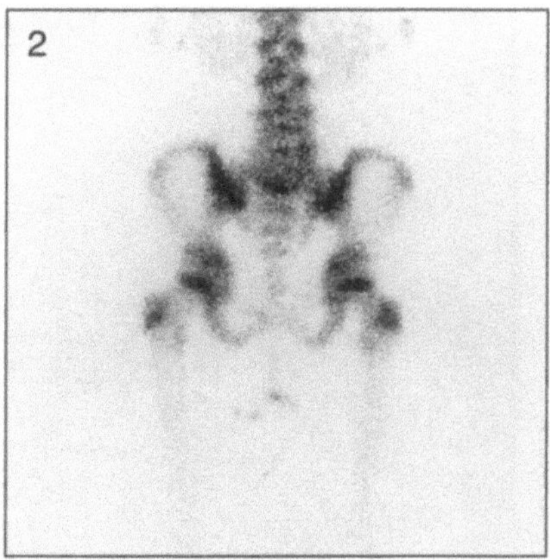

Fig. 2. Posterior view of spine, pelvis and femora

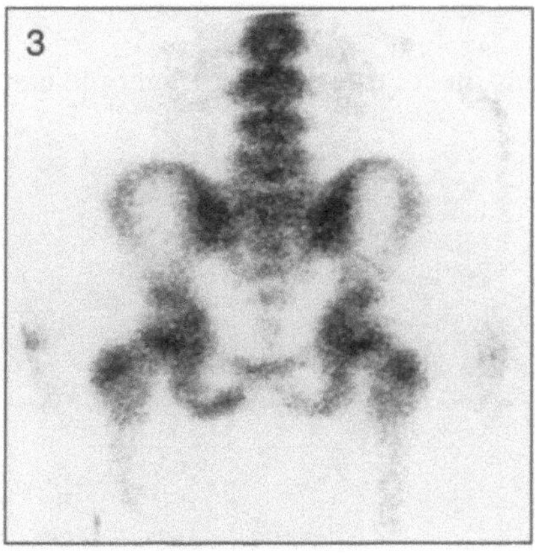

Fig. 3. Posterior view of spine and pelvis

Technical Comment
- Urine contamination below the pelvis is seen in Fig. 2

▶ Potential Pitfall
- In Fig. 3 increased activity is noted at the junction of the left ischial tuberosity and the posterior pubic ramus due to the normal synchondrosis at this site

Fig. 1. Pinhole view of right hip

Fig. 4. Pinhole view of left hip

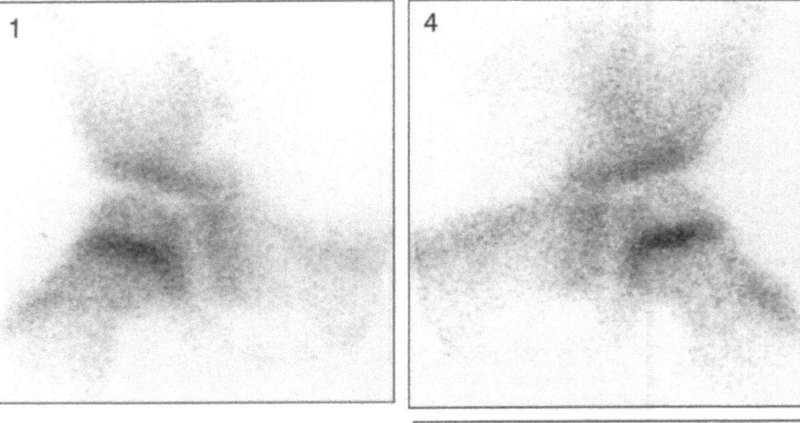

Fig. 5. Pinhole view of left hip

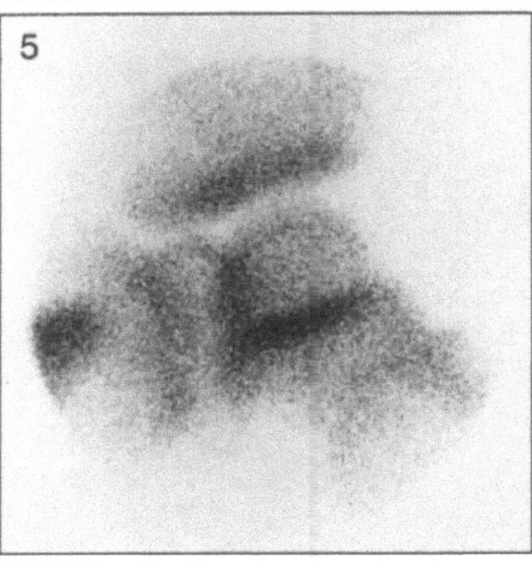

Technical Comment
- The larger size of the hip in Fig. 5 is due to the size of the pinhole insert used compared to Figs. 1 and 4

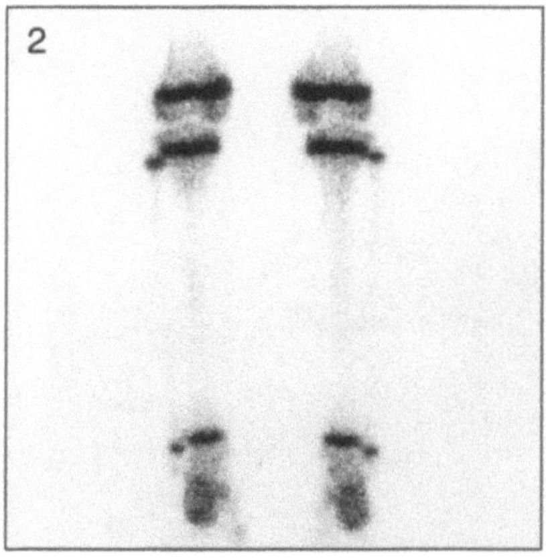

Fig. 1. Posterior view of femora and knees

Fig. 2. Posterior view of knees, tibia, fibula and ankles

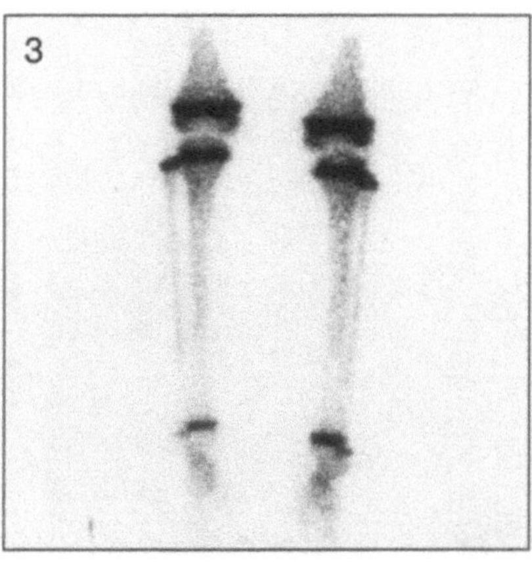

Fig. 3. Posterior view of knees, tibia, fibula and ankles

Technical Comment
- Note the good positioning of the feet resulting in clear separation of the fibula from the tibia in Fig. 2
- Note the different visualization of the ankles in Fig. 3. This is due to different positioning of the feet

▶ **Potential Pitfall**
- Note the slight but definite increased activity in the mid portion of the tibial shafts best seen in Fig. 3. As the child grows this will become more obvious and potentially confusing

Fig. 1. Posterior magnified view
of knees

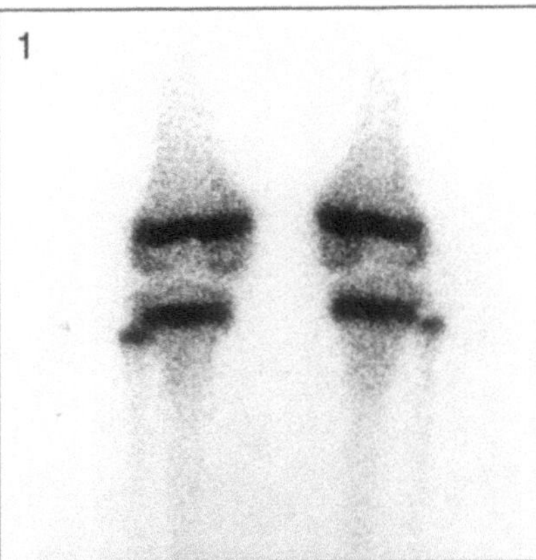

Fig. 2. Posterior view of knees

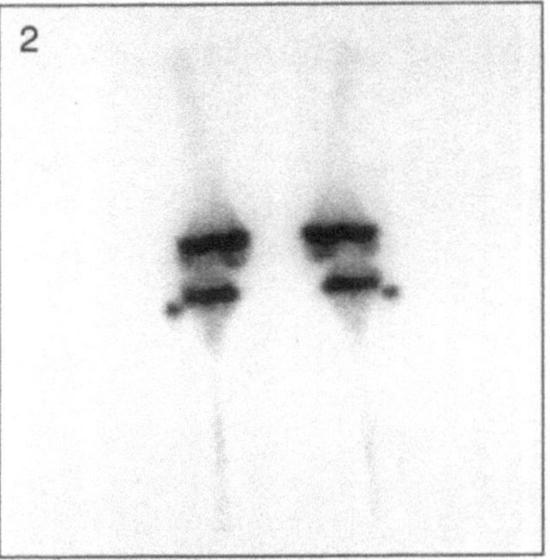

Technical Comment
– Note the clear definition of the growth plate from the adjacent meta-
  physes

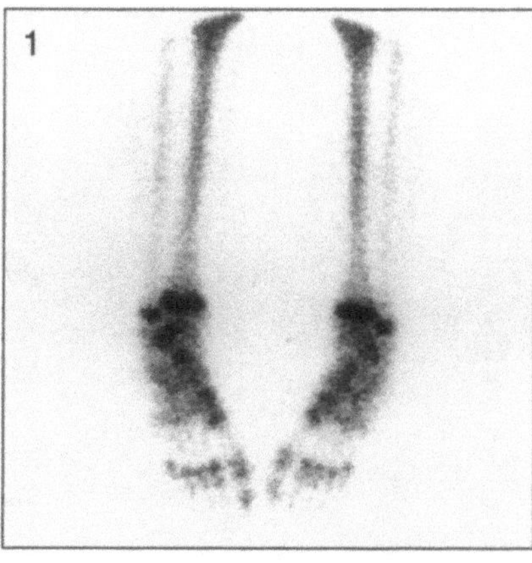

Fig. 1. Anterior view of ankles and feet

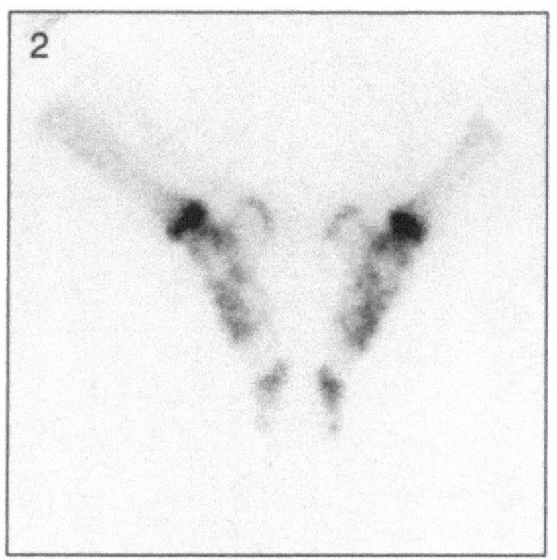

Fig. 2. Lateral view of feet

9: Age 7–8 Years

Fig. 1. Posterior view of thorax
and spine

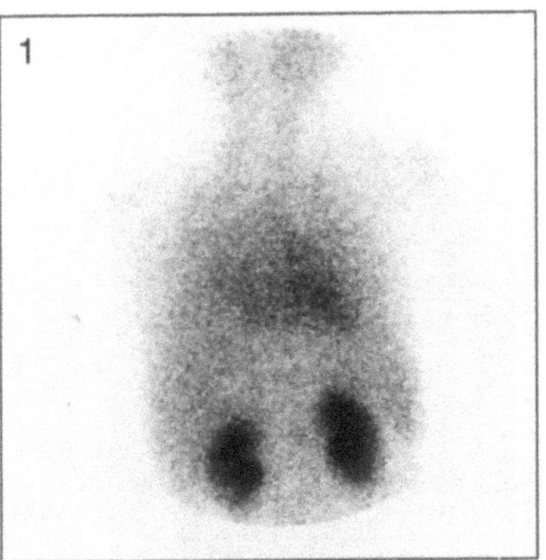

Fig. 2. Posterior view of spine
and pelvis

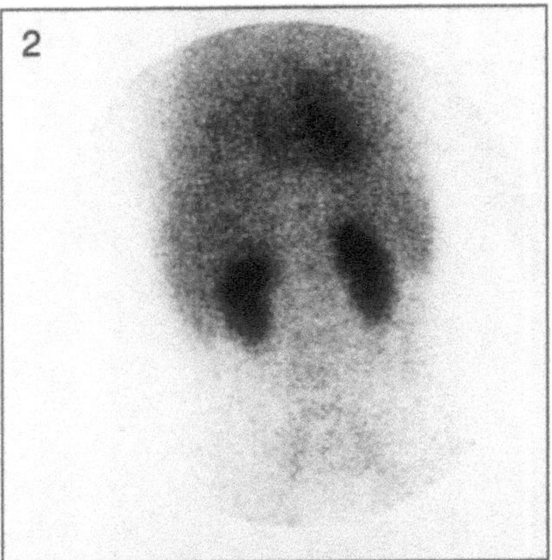

Fig. 3. Anterior view of pelvis
and femora

Fig. 6. Anterior view of pelvis
and femora

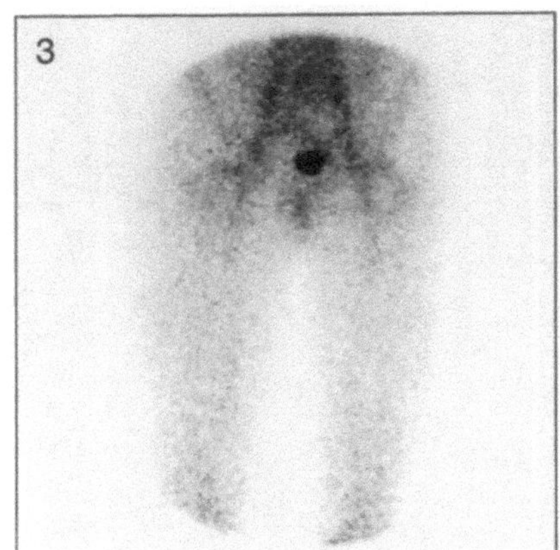
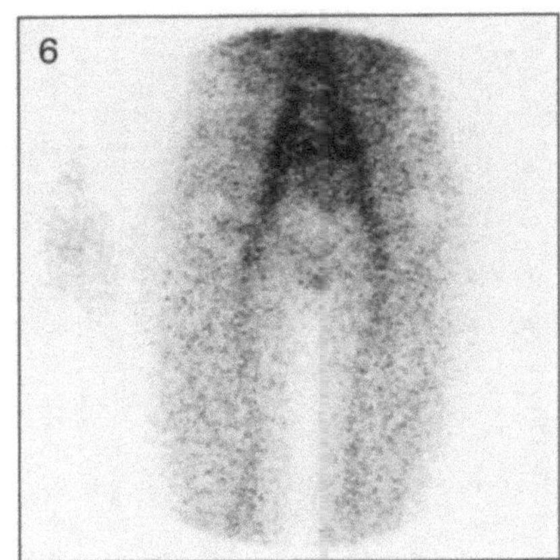

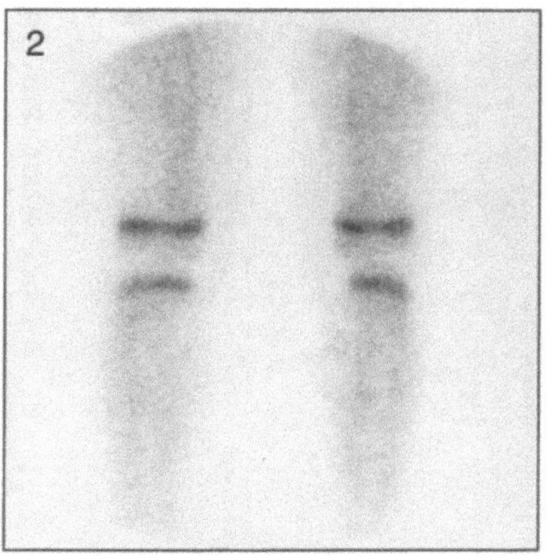

Fig. 1. Anterior view of both hands

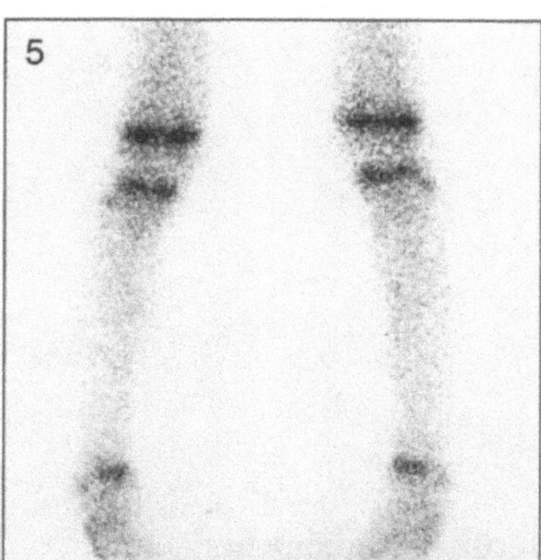

Fig. 2. Anterior view of knees

Fig. 5. Posterior view of tibia and fibula

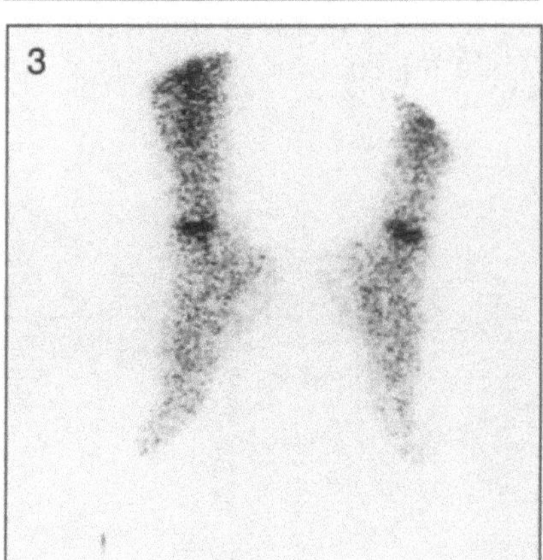

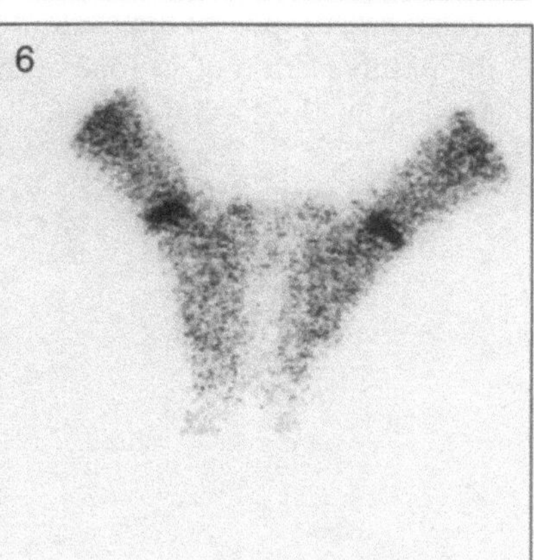

Fig. 3. Lateral view of feet

Fig. 6. Lateral view of feet

– A double headed whole body
  gamma camera was used
– Left image is the anterior view
– Right image is the posterior
  view

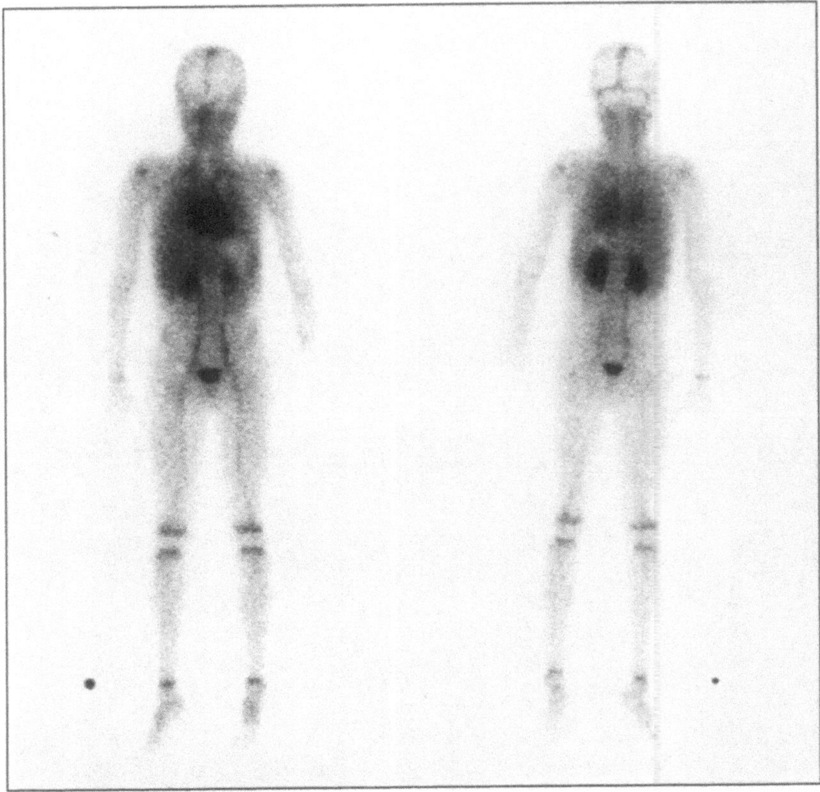

Technical Comments
– Distal end of the left upper limb is not seen due to a lead shield follow-
  ing extravasation of isotope at the site of injection
– Marker on child's right side

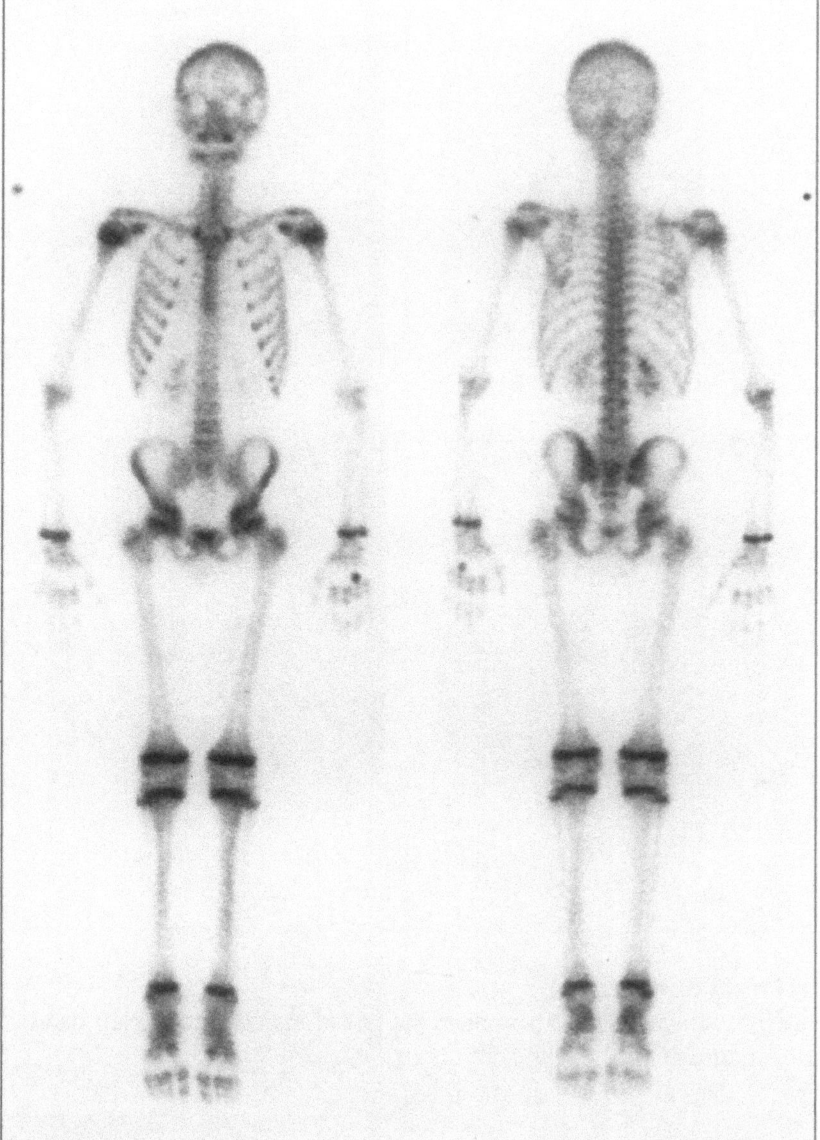

- A double headed whole body gamma camera was used
- Left image is the anterior view
- Right image is the posterior view

Technical Comments
- Note extravasation of isotope at the site of injection in the left hand
- Marker on child's right side

- A double headed whole body gamma camera was used
- Left image is the anterior view
- Right image is the posterior view

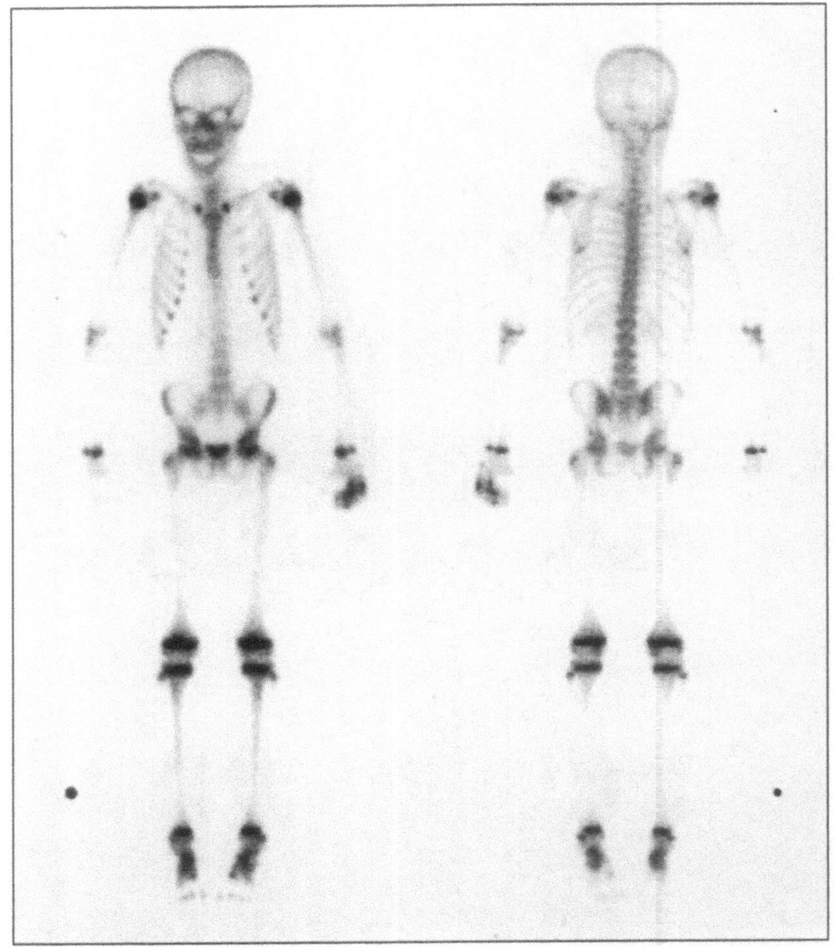

Technical Comments
- Note extravasation of isotope at the site of injection in the left hand
- Marker on child's right side

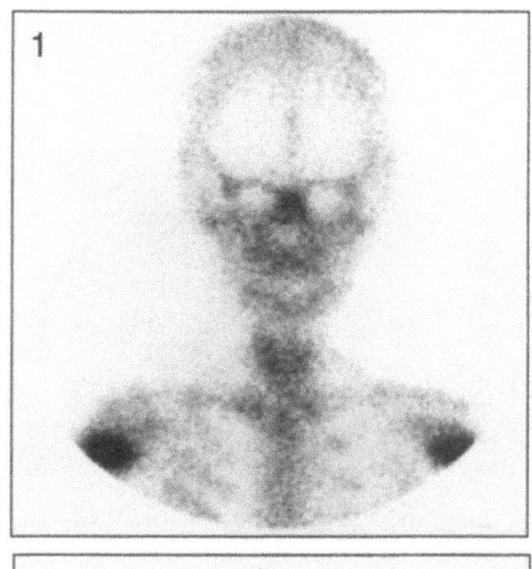

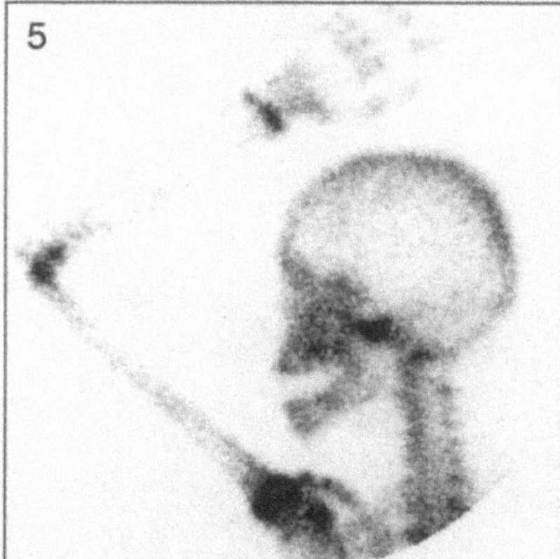

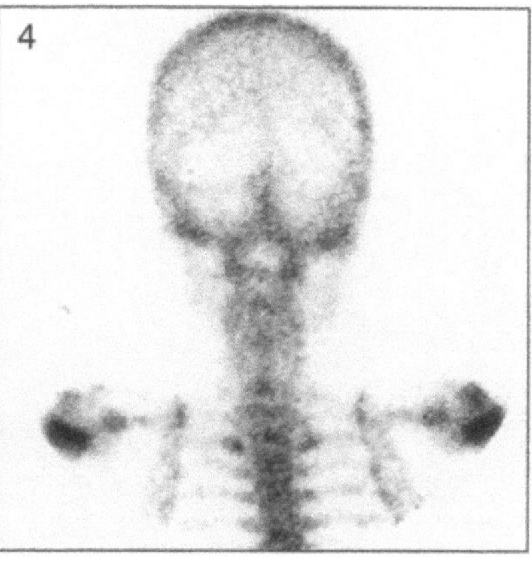

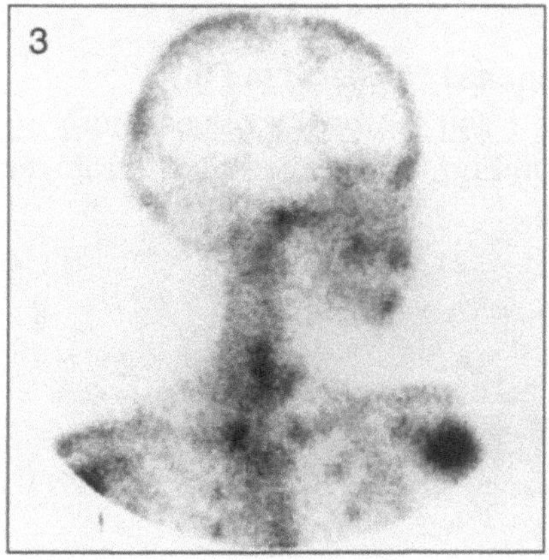

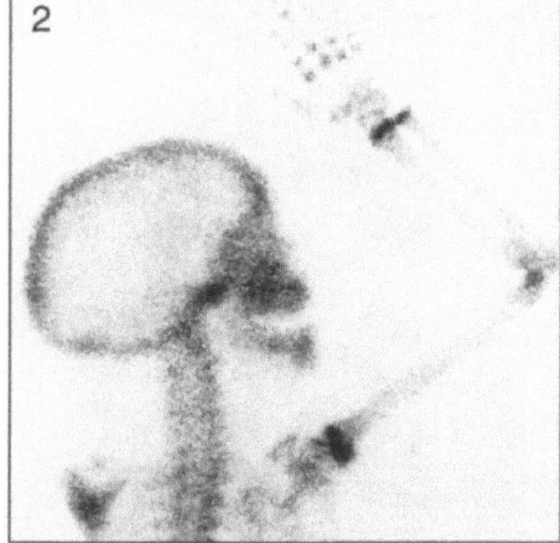

Fig. 1. Anterior view of skull and thorax

Fig. 4. Posterior view of skull and thorax

Fig. 2. Right lateral view of skull and right upper limb

Fig. 5. Left lateral view of skull and left upper limb

Fig. 3. Right lateral view of skull and anterior view of thorax

Technical Comments

– The coronal suture is clearly seen in Fig. 3. This is not the case in Figs. 2 and 5
– The thyroid gland is seen in Figs. 1 and 3 due to free 99m-Tc-pertechnetate
– The lateral views of the skull (Figs. 2 and 5) were taken posteriorly

Fig. 1. Right lateral view of skull and right upper limb

Fig. 4. Left lateral view of skull and left upper limb

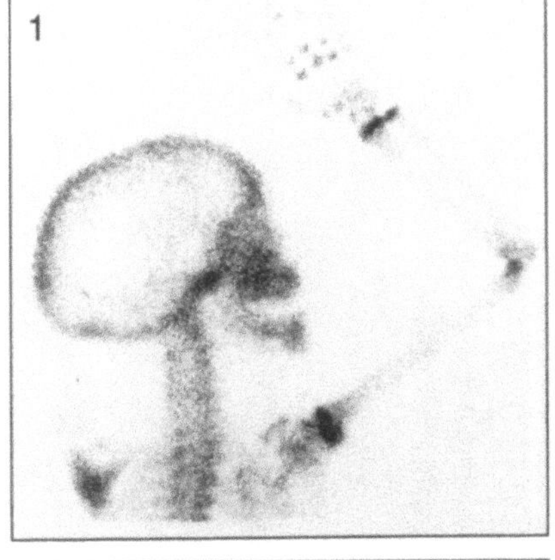

Fig. 2. Anterior view of upper limbs and hands

Fig. 5. Anterior view of hands

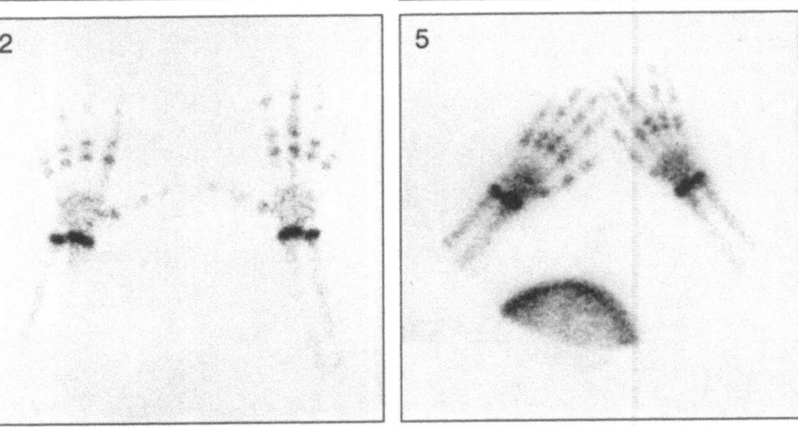

Technical Comments
- The hands in Fig. 5 are poorly positioned compared to Fig. 2
- The lateral views of the skull (Figs. 1 and 4) were taken posteriorly
- Note that part of the occipital bone in Fig. 4 has been excluded from the field of view

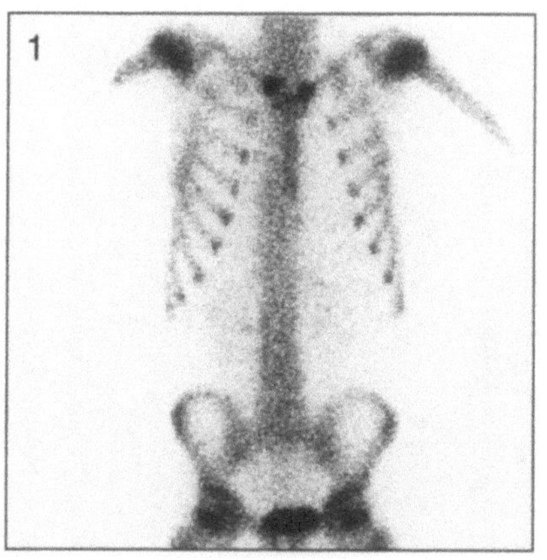

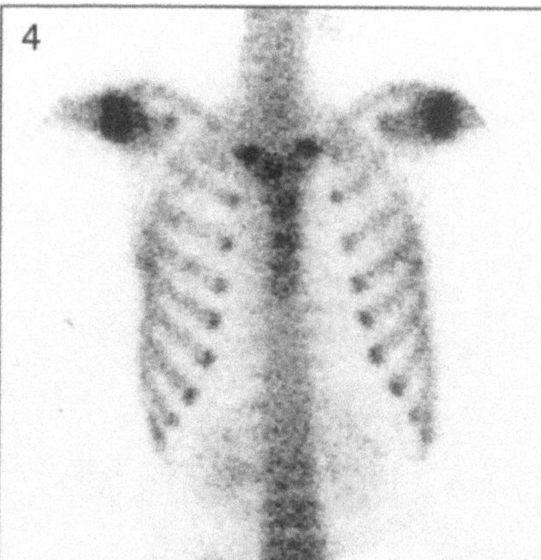

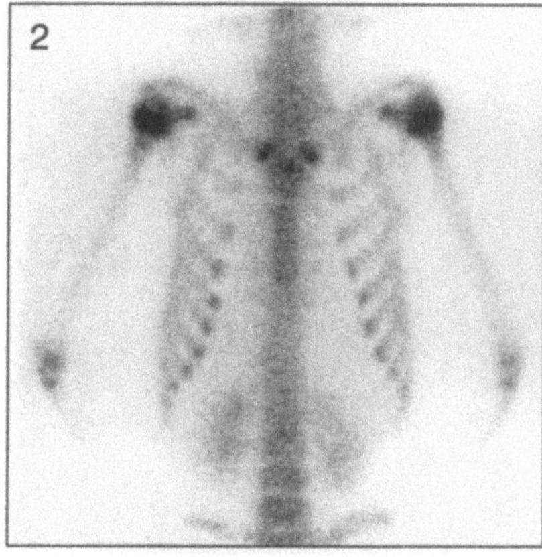

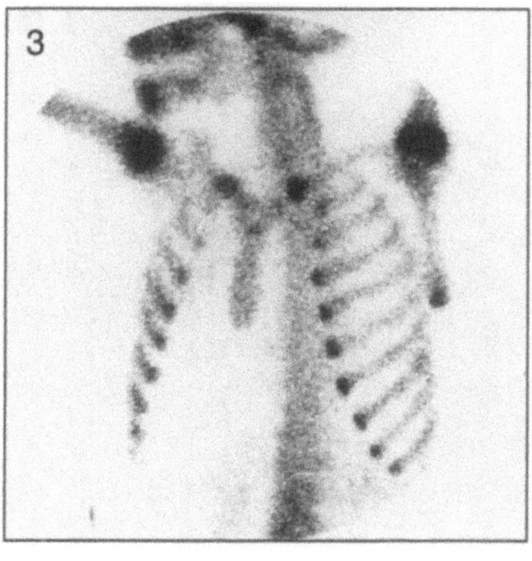

Fig. 1. Anterior view of thorax, spine and pelvis
Fig. 4. Anterior view of thorax and spine

Fig. 2. Anterior view of thorax and spine

Fig. 3. Left anterior oblique view of thorax

Technical Comments
– Note the clarity of the lumbar spine, especially in Fig. 2
– The sternum is best seen with slight obliquity of the thoracic cage as in Figs. 1 and 3

Fig. 1. Posterior view of thorax and spine

Fig. 4. Posterior view of thorax, spine and part pelvis

Fig. 2. Posterior view of thorax, spine and pelvis

Fig. 5. Posterior view of thorax, spine and pelvis

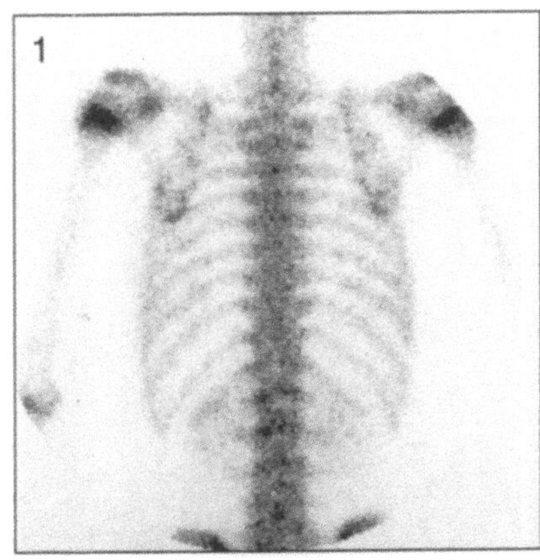

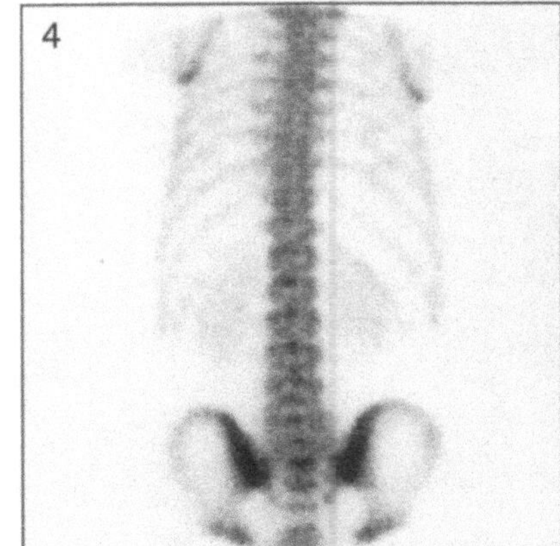

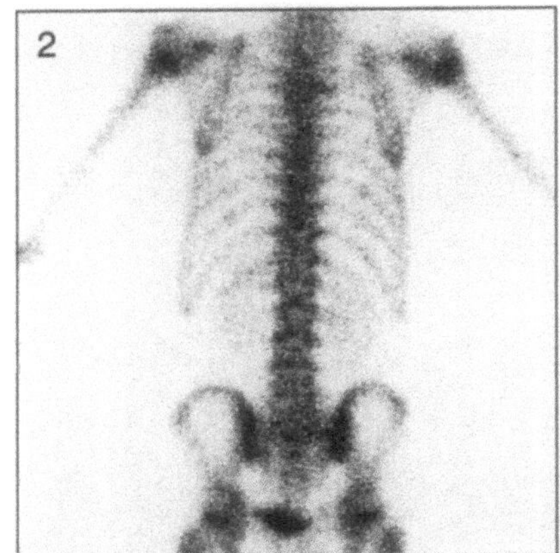

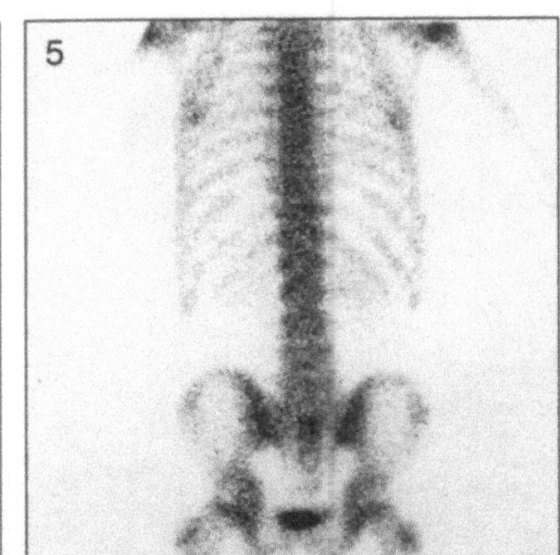

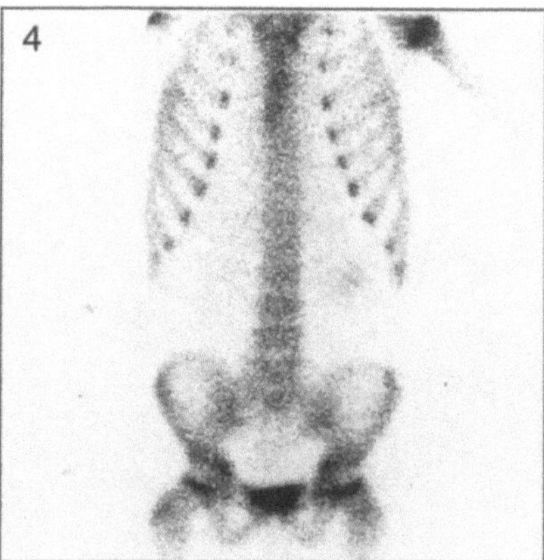

Fig. 1. Anterior view of pelvis and femora
Fig. 4. Anterior view of thorax, spine and pelvis

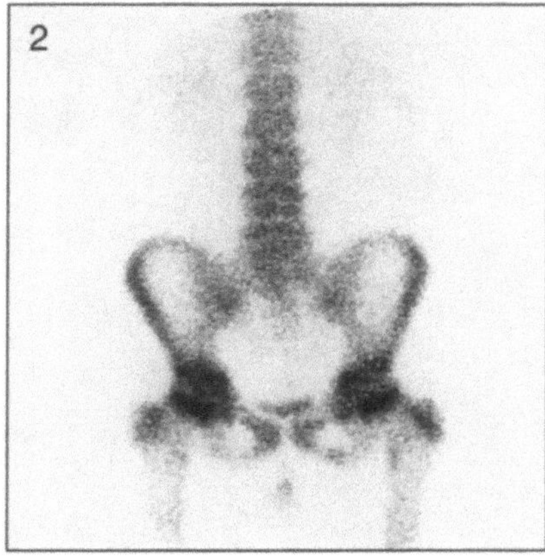

Fig. 2. Anterior view of spine and pelvis

Technical Comments
- Note that the bladder is virtually empty on all the images
- Urine contamination below the pelvis is seen in Fig. 2

Fig. 1. Posterior view of spine, pelvis and femora

Fig. 4. Posterior view of spine and pelvis

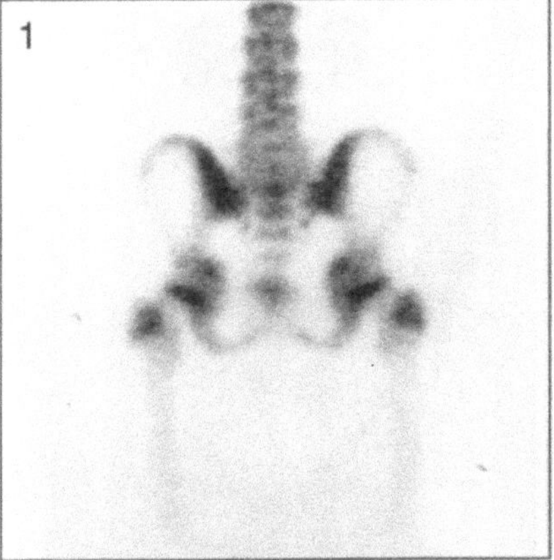

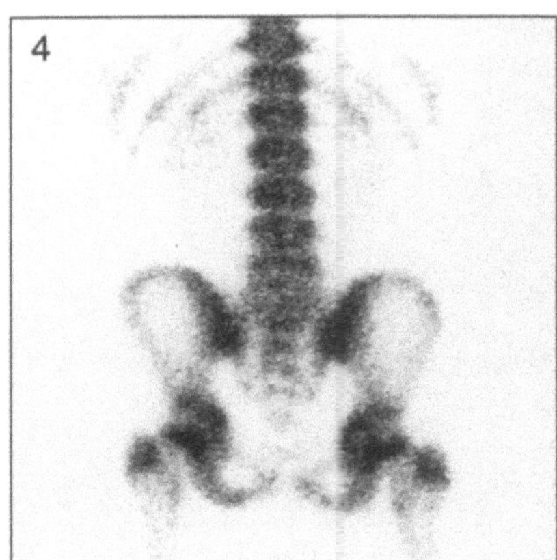

Fig. 2. Posterior view of pelvis and femora

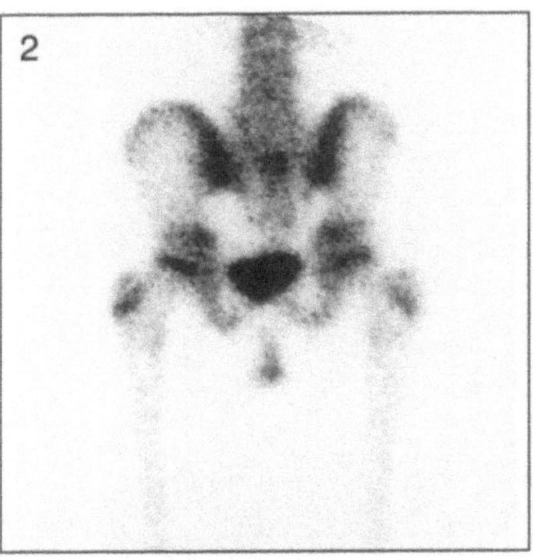

Technical Comment

– Urine contamination below the pelvis is seen in Fig. 2

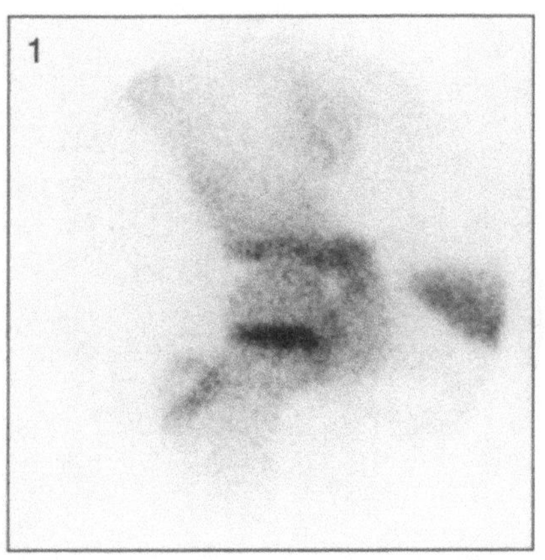
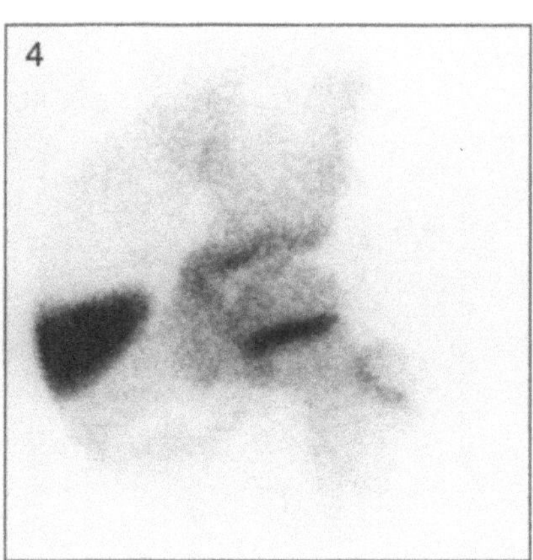

Fig. 1. Pinhole view of right hip

Fig. 4. Pinhole view of left hip

Fig. 1. Posterior view of femora and knees

Fig. 4. Posterior view of hips, femora and knees

Fig. 2. Posterior view of tibia, fibula and ankles

Fig. 5. Posterior view of knees, tibia, fibula and ankles

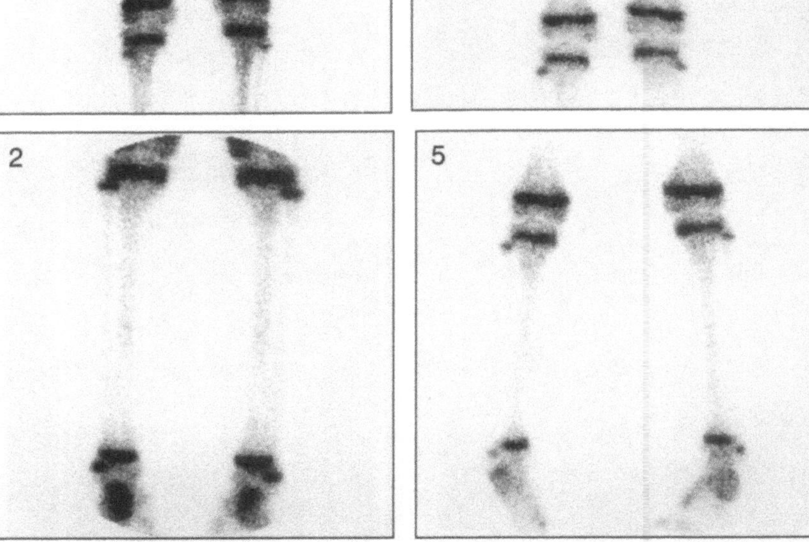

Technical Comments
- Note the clarity and the separation of the epiphysis of the fibula from the tibia due to the good positioning of the feet. This is not true for the left knee in Fig. 1, especially when compared to the right knee
- Urine contamination below the pelvis is seen in Fig. 1

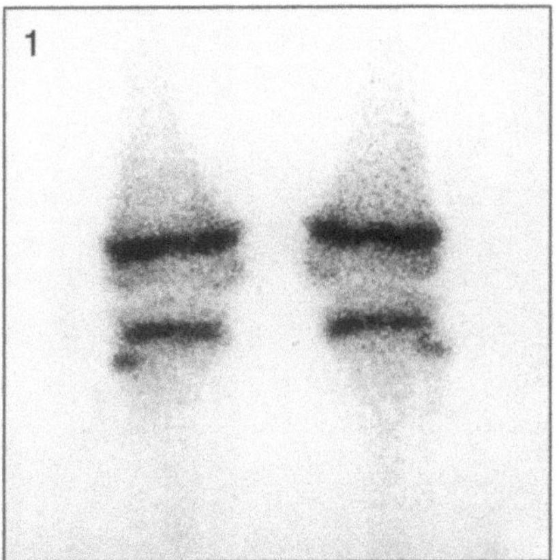

Fig. 1. Posterior magnified view of knees

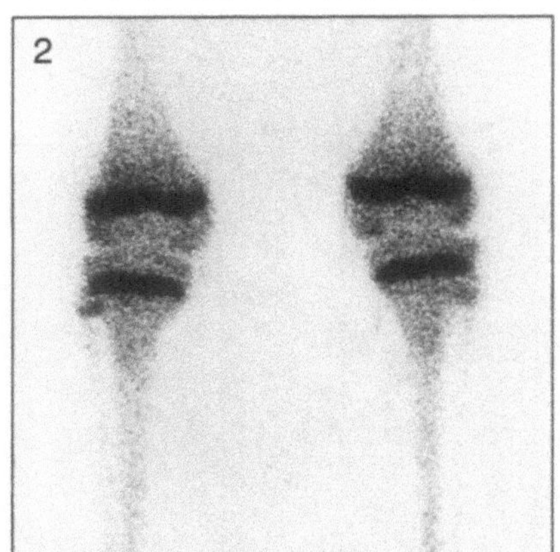

Fig. 2. Posterior magnified view of knees

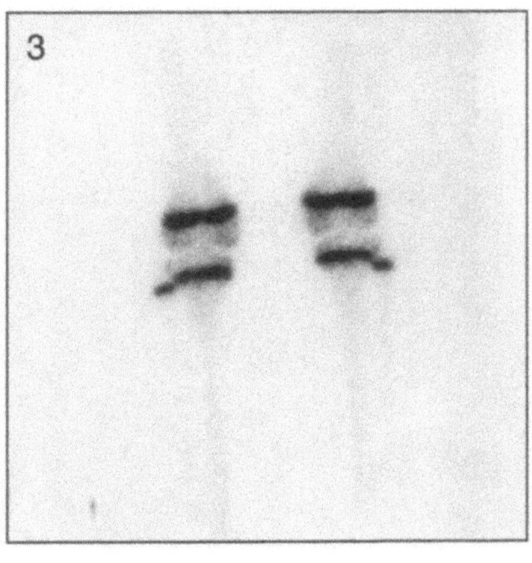

Fig. 3. Posterior view of knees

Technical Comment

– There is clear separation between the upper tibia and the fibula due to good positioning of the feet in all images, apart from the right knee in Fig. 2 where the fibula cannot be clearly seen

Fig. 1. Posterior view of feet

Fig. 4. Lateral view of feet

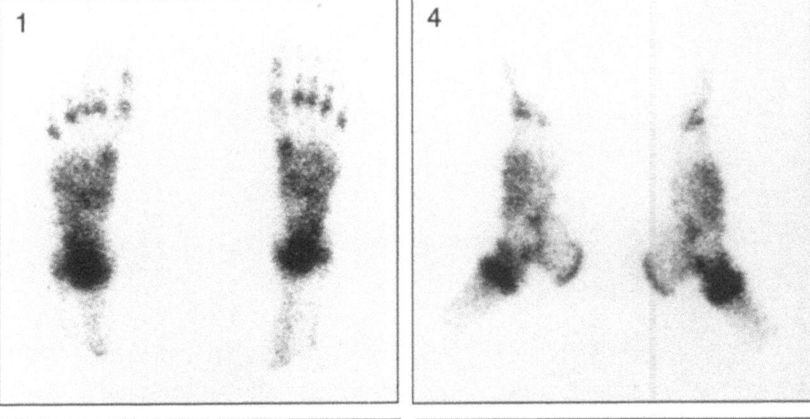

Fig. 2. Posterior view of feet

Fig. 5. Lateral view of feet

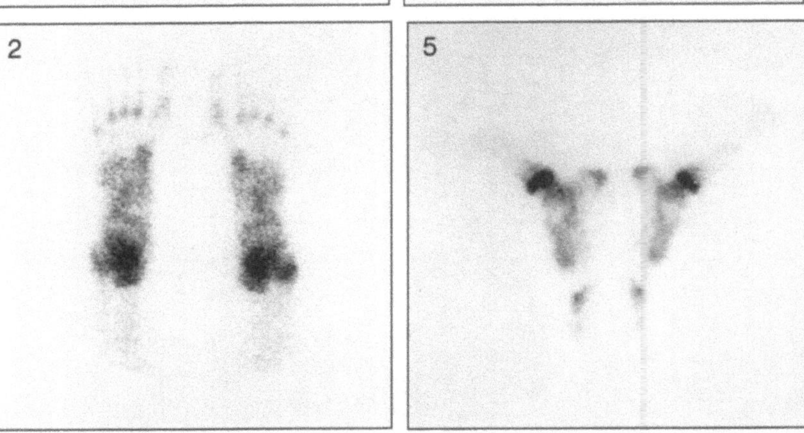

10: Age 8–9 Years

Fig. 1. Anterior view of pelvis

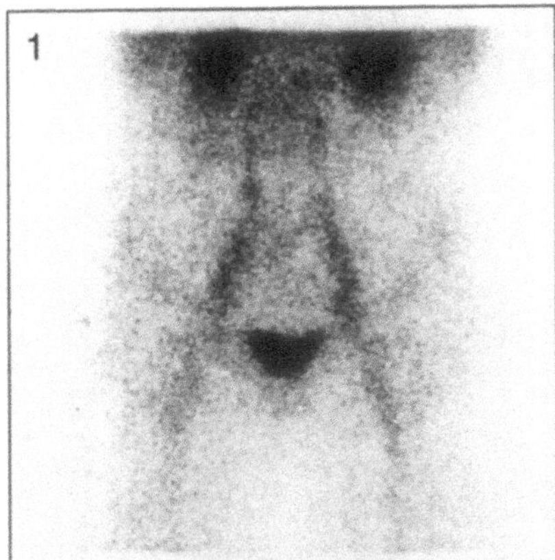

Fig. 2. Posterior view of knees

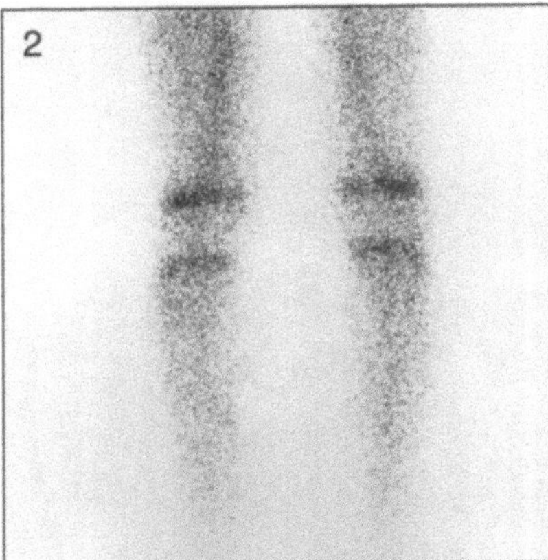

Fig. 3. Anterior view of tibia, fibula and feet

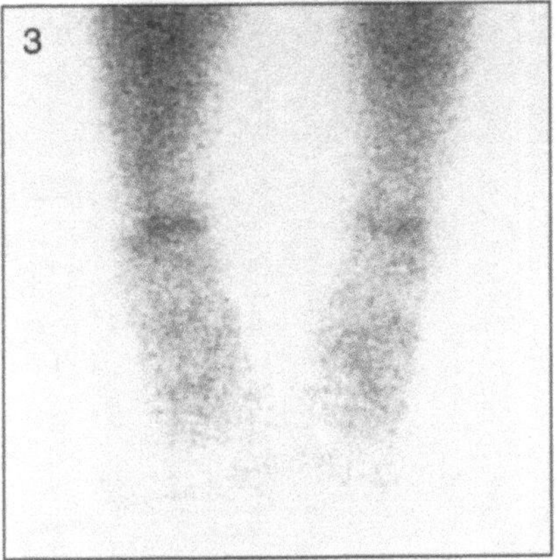

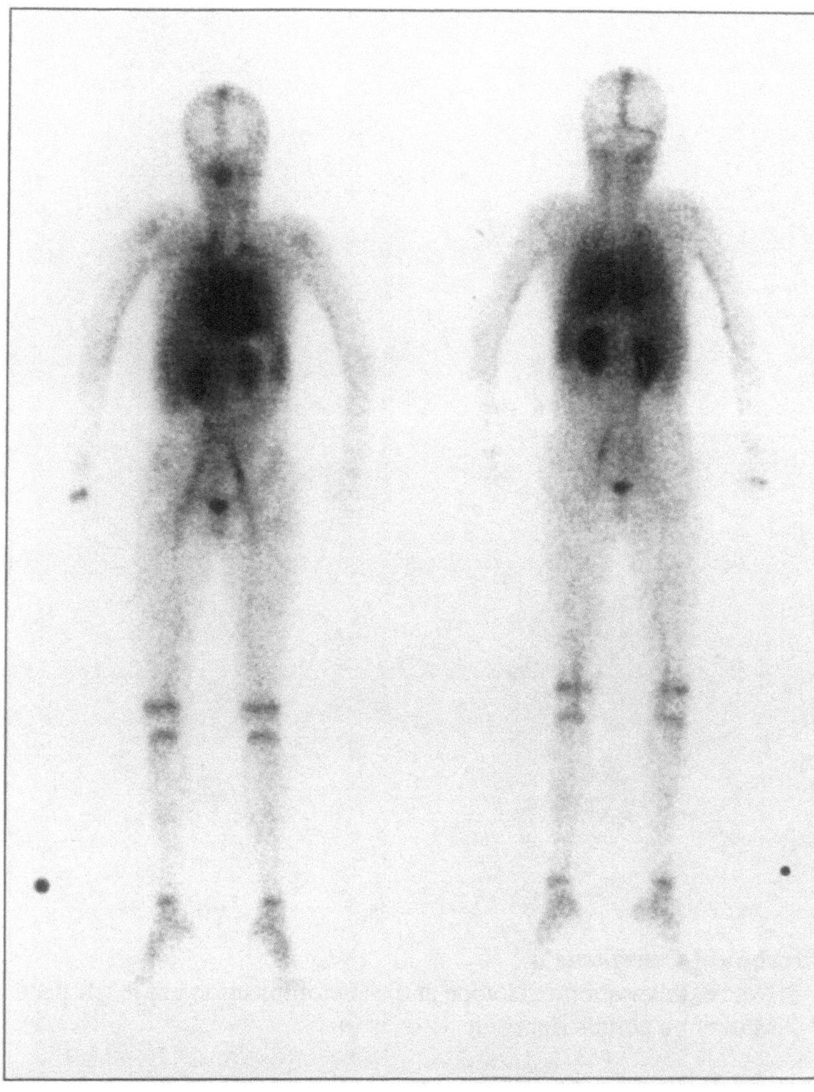

- A double headed whole body gamma camera was used
- Left image is the anterior view
- Right image is the posterior view

Technical Comments

- Note on the posterior view the difference between the two straight sinuses with more isotope entering the right than the left. This is a variation of normality
- Note extravasation of isotope at the site of injection in the right hand
- Marker on child's right side

► **Potential Pitfall**

- Focal increased uptake of isotope is seen on the anterior view in the region of the maxillary sinus. This is probably due to sinusitis

– A double headed whole body
  gamma camera was used
– Left image is the anterior view
– Right image is the posterior
  view

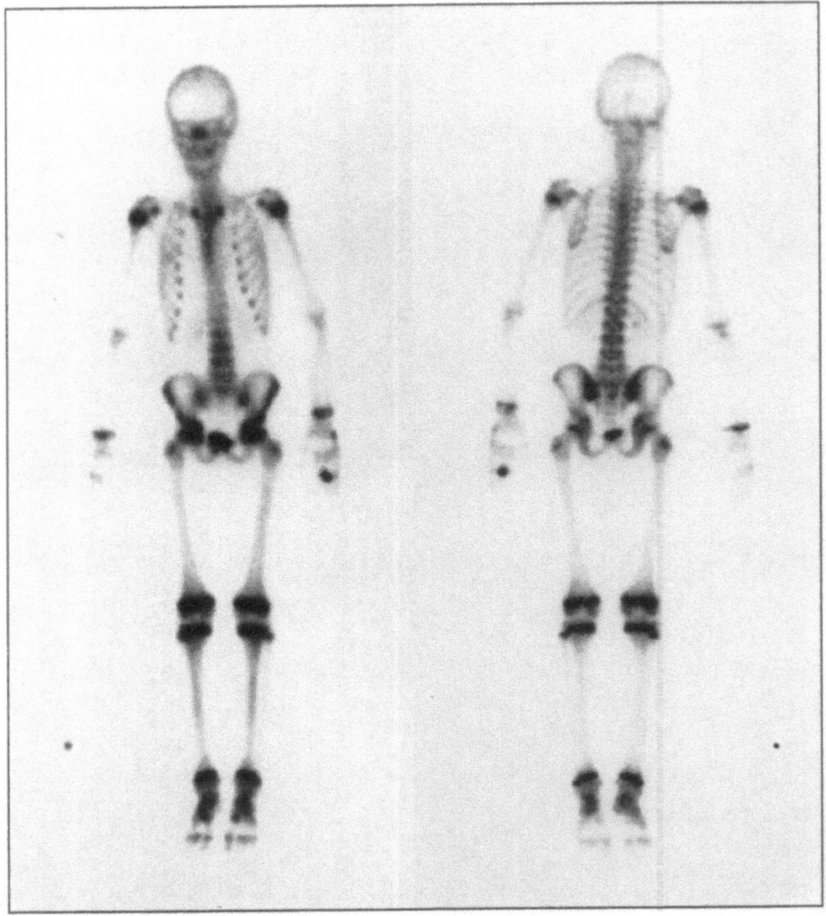

Technical Comments
– Note extravasation of isotope at the site of injection in the left hand
– Marker on child's right side

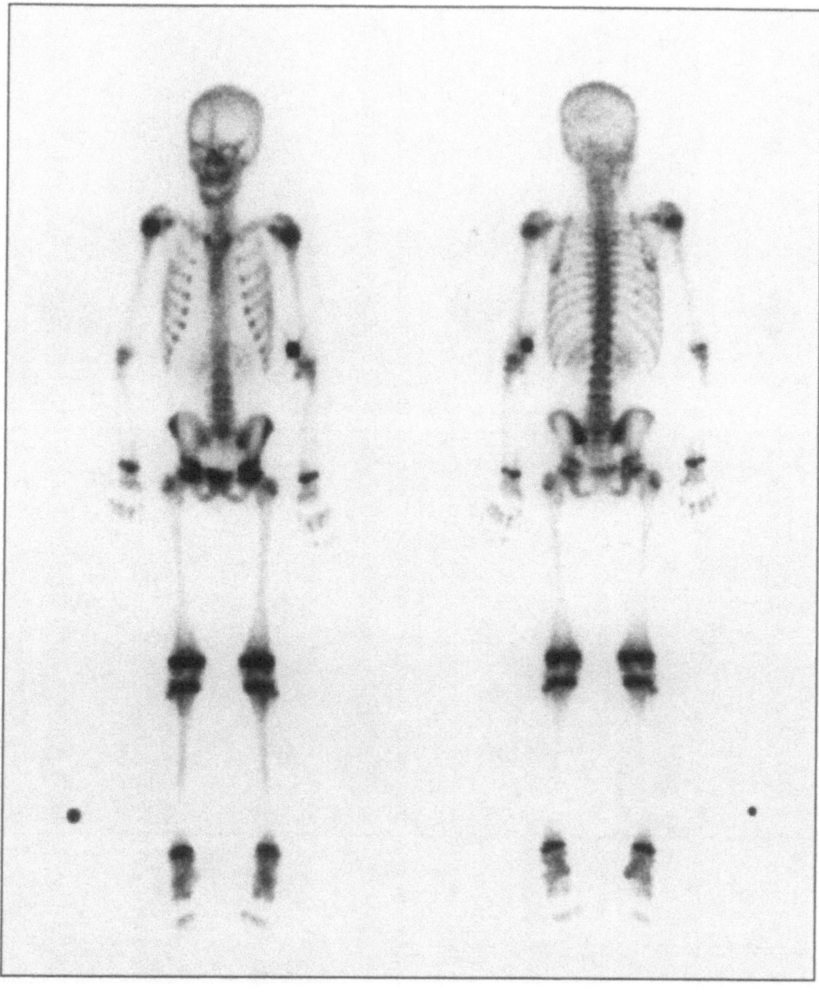

- A double headed whole body gamma camera was used
- Left image is the anterior view
- Right image is the posterior view

Technical Comments

- The child's head is rotated causing asymmetry especially to the facial bones
- Note extravasation of isotope at the site of injection in the left elbow
- Marker on child's right side

Fig. 1. Anterior view of skull and thorax

Fig. 4. Posterior view of skull and thorax

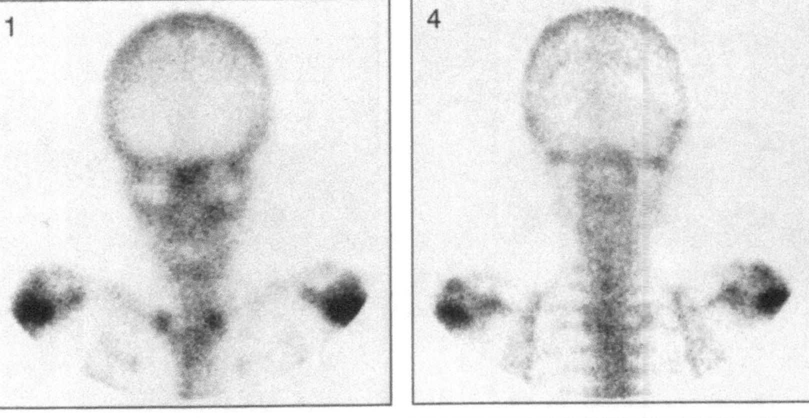

Fig. 2. Right lateral view of skull

Fig. 5. Left lateral view of skull

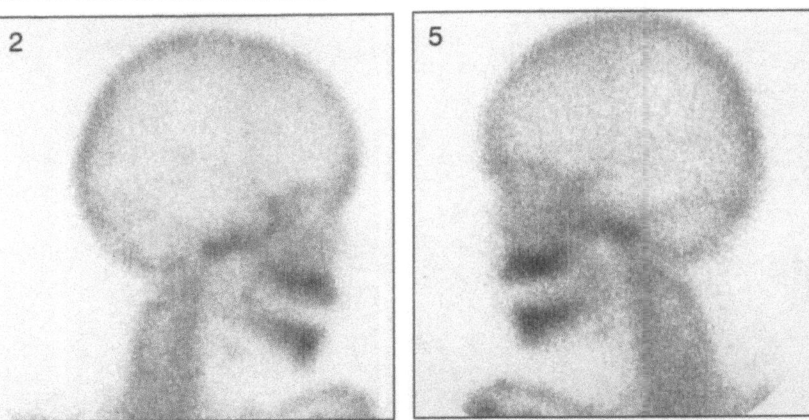

Fig. 6. Left lateral view of skull and anterior view of thorax

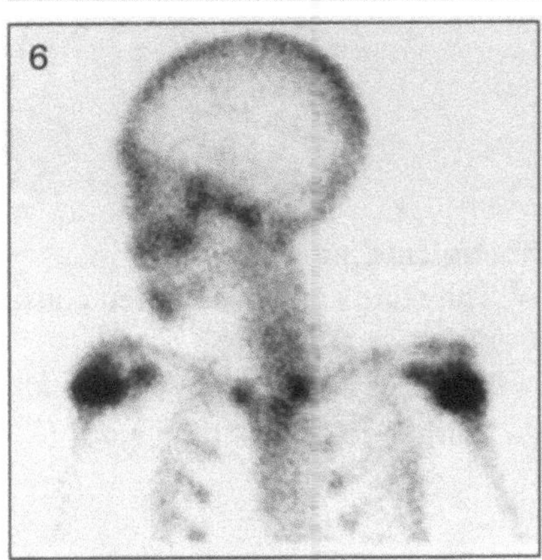

Technical Comments
- Note the sterno-clavicular joints in Figs. 1 and 6, the asymmetry of the joints is due to rotation of the patient
- The lateral views of the skull (Figs. 2 and 5) were taken anteriorly

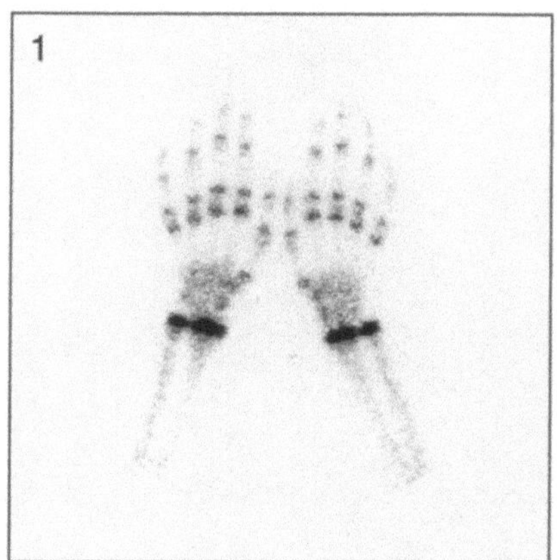 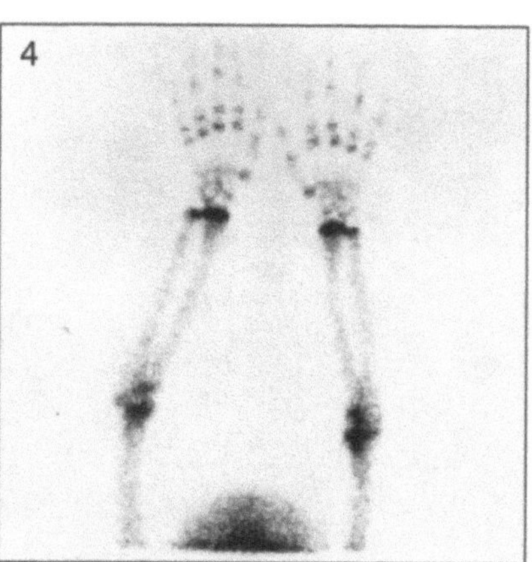

Fig. 1. Anterior view of hands

Fig. 4. Anterior view of hands and upper limbs

Fig. 1. Anterior view of thorax and spine

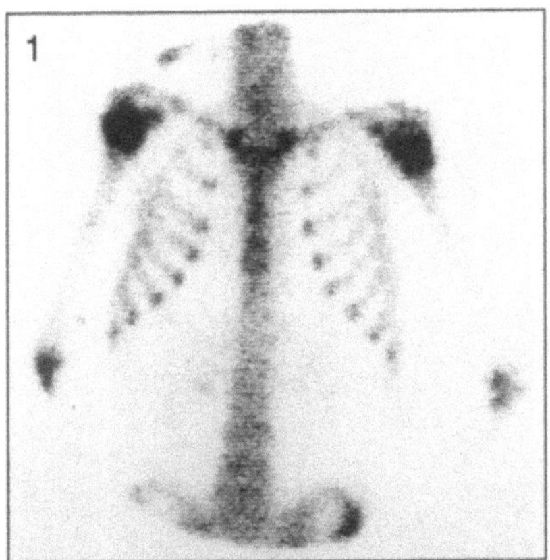

Fig. 2. Anterior view of thorax and spine

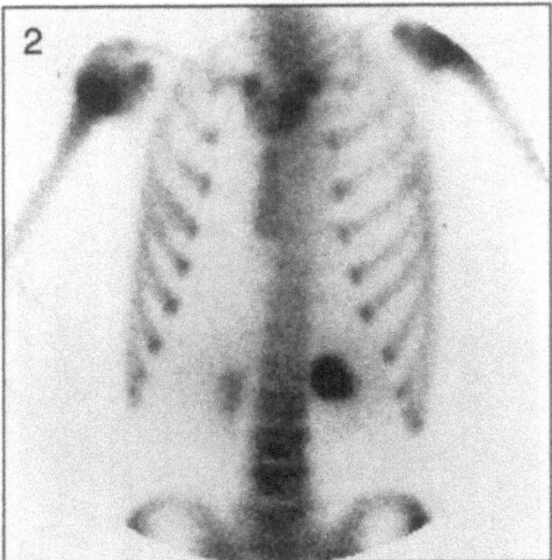

Fig. 3. Right anterior oblique view of thorax

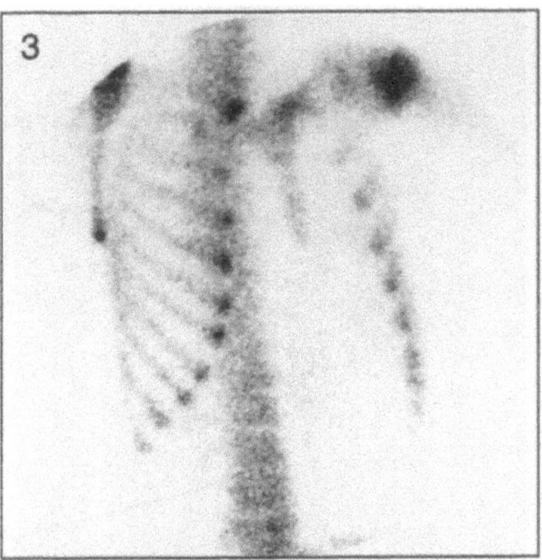

Technical Comments
- The lumbar spine in Fig. 2 is clearly seen
- The asymmetry of the sterno-clavicular joints in Fig. 2 is due to the slight rotation
- Abnormal accumulation of isotope in a dilated left renal pelvis is noted in Fig. 2

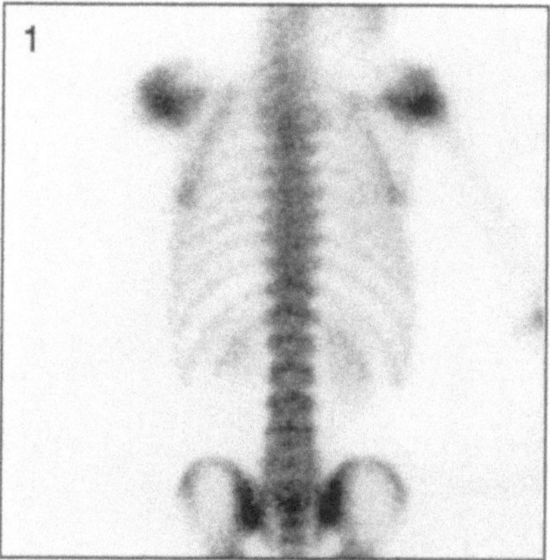

Fig. 1. Posterior view of thorax, spine and part pelvis

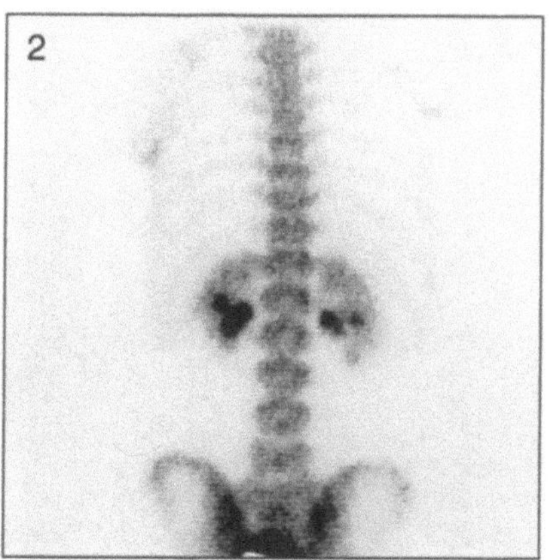

Fig. 2. Posterior view of thorax, spine and part pelvis

► **Potential Pitfall**
  – The tracer is seen in abnormal calyces and renal pelvis of both kidneys
    in Fig. 2

Fig. 1. Anterior view of spine, pelvis and femora

Fig. 4. Anterior view of spine, pelvis and femora

Fig. 2. Anterior view of spine and pelvis

Fig. 5. Anterior view of spine, pelvis and femora

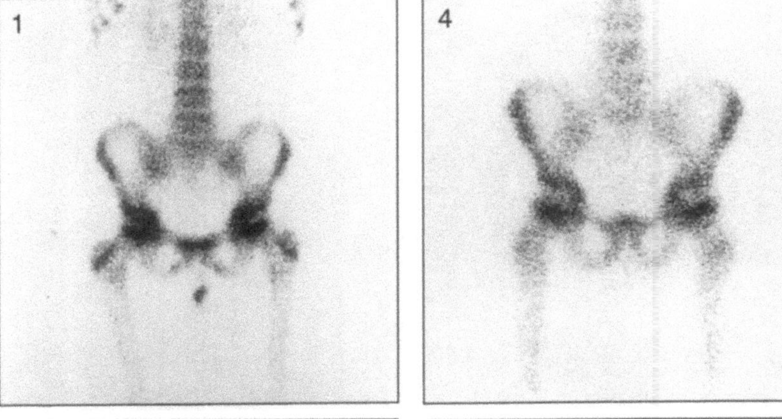

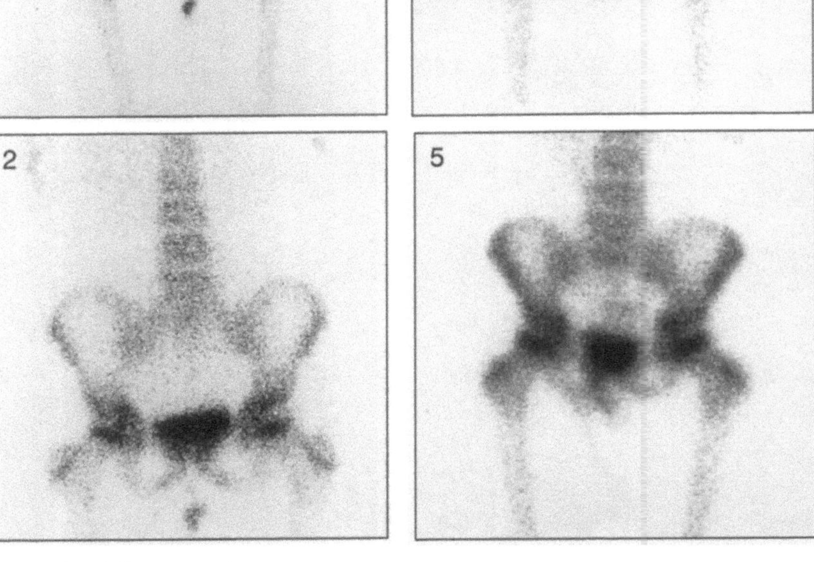

Technical Comment
– Urine contamination below the pelvis is seen in Figs. 1, 2 and 5

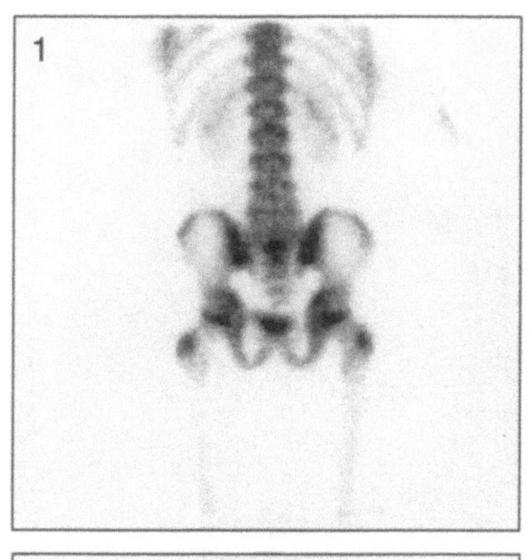

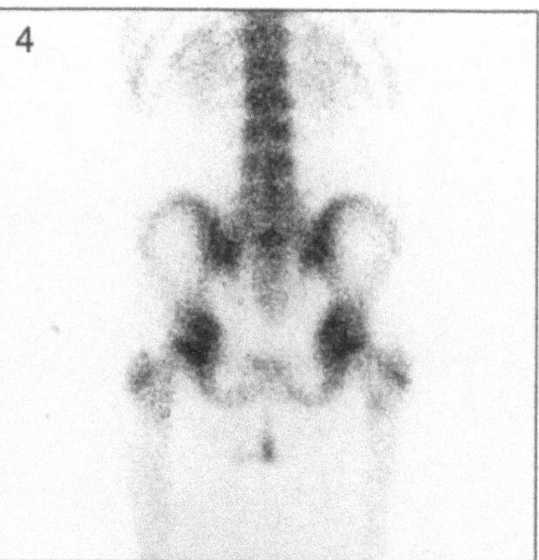

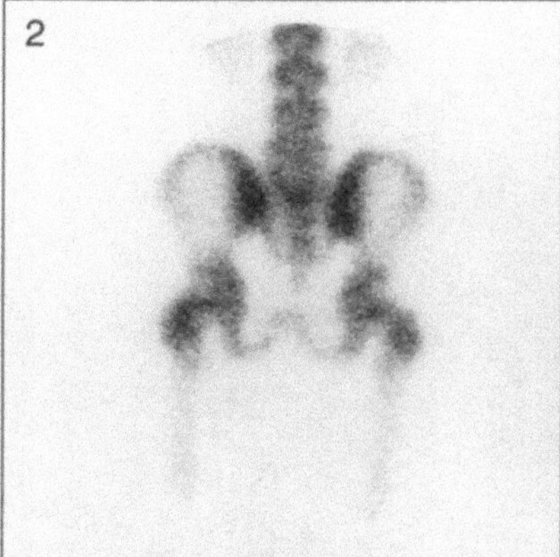

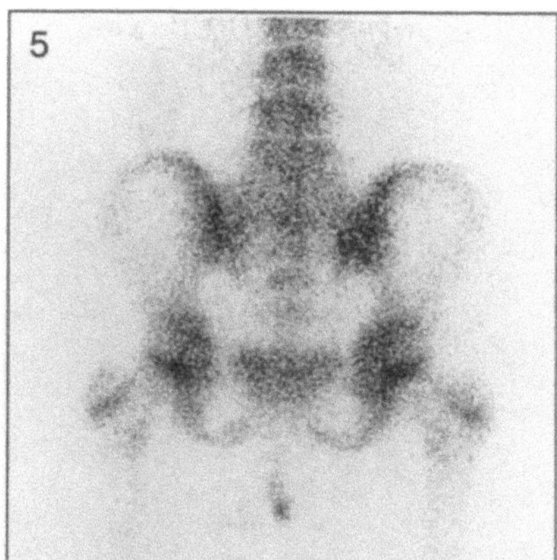

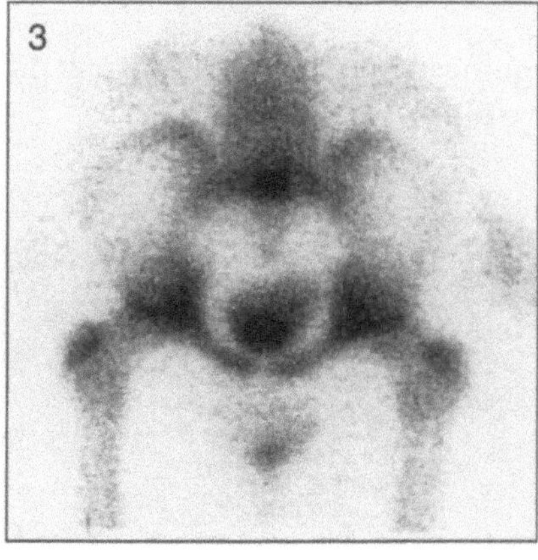

Fig. 1. Posterior view of spine, pelvis and femora

Fig. 4. Posterior view of spine, pelvis and femora

Fig. 2. Posterior view of spine, pelvis and femora

Fig. 5. Posterior view of spine, pelvis and femora

Fig. 3. Pelvic inlet view. This view is infrequently required in paediatrics

Technical Comment
– Urine contamination below the pelvis is seen in Figs. 3–5

Fig. 1. Pinhole view of right hip

Fig. 4. Pinhole view of left hip

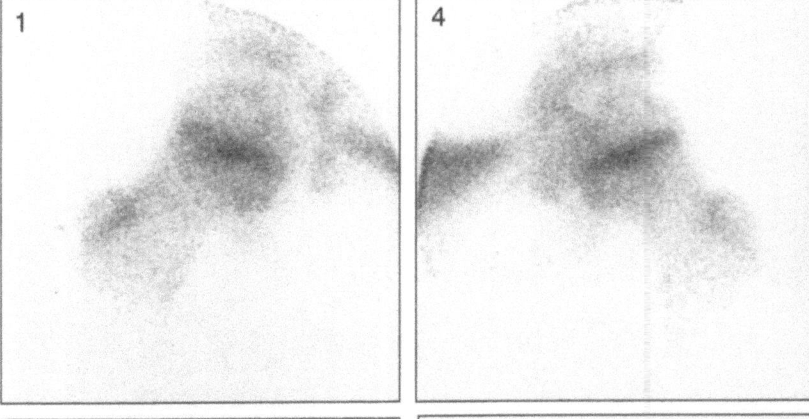

Fig. 2. Pinhole view of right hip

Fig. 5. Pinhole view of left hip

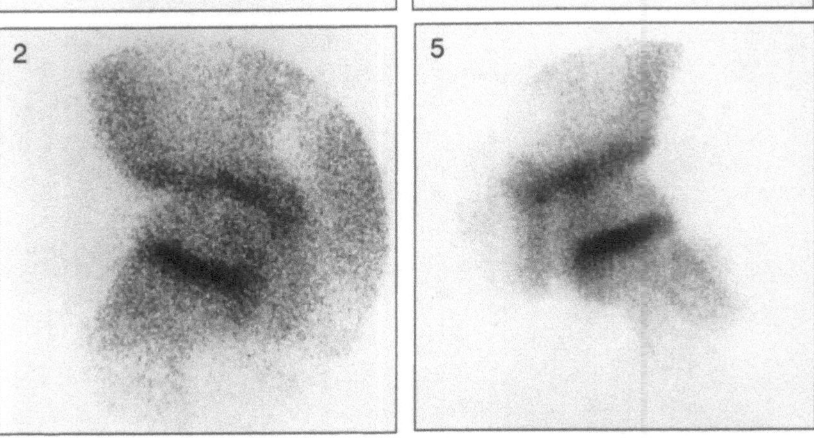

Technical Comment
- Figs. 2 and 5 show a change in magnification creating an apparent difference between the sizes of the two hips

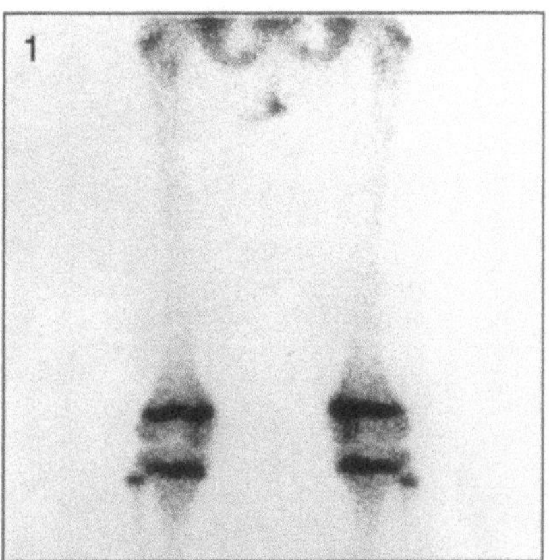

Fig. 1. Posterior view of femora and knees

Fig. 2. Posterior view of knees, tibia and fibula

Technical Comment
– Urine contamination below the pelvis is seen in Fig. 1

Fig. 1. Posterior view of knees

Fig. 4. Anterior view of knees

Fig. 2. Medial lateral view of left knee

Fig. 5. Medial lateral view of right knee

Fig. 3. Lateral lateral view of right knee

Fig. 6. Lateral lateral view of left knee

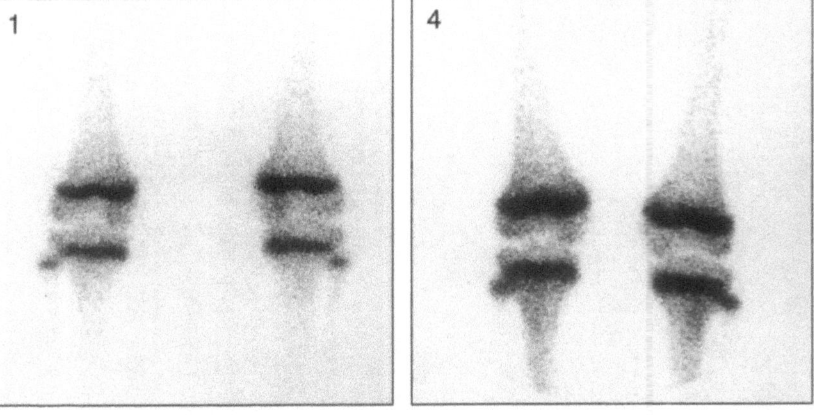

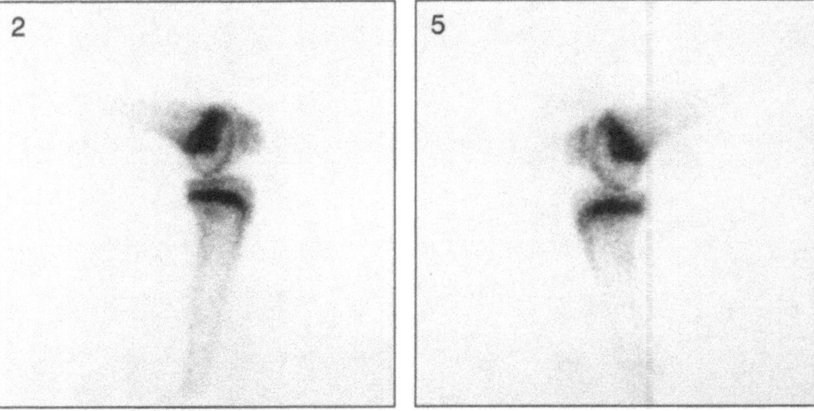

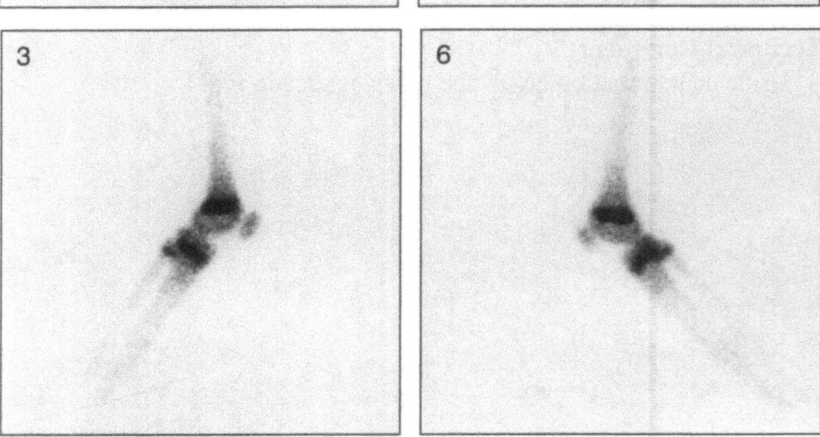

Technical Comment

– The pair of lateral knees in Figs. 2 and 5 represent medial lateral knees while Figs. 3 and 6 represent lateral knees in the lateral projection. This accounts for the clarity of the fibula in the lower series and the difficulty in seeing the fibula in the upper series

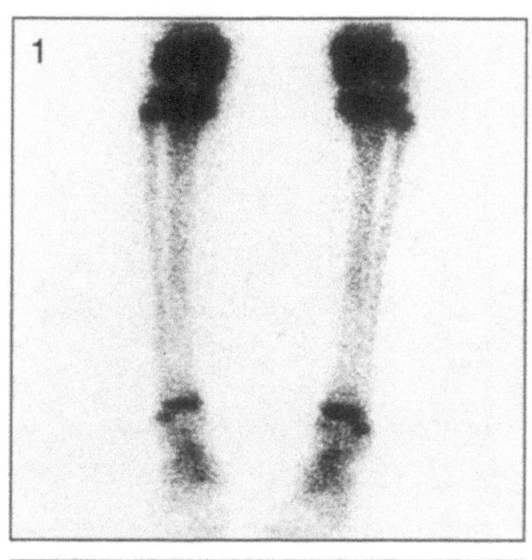

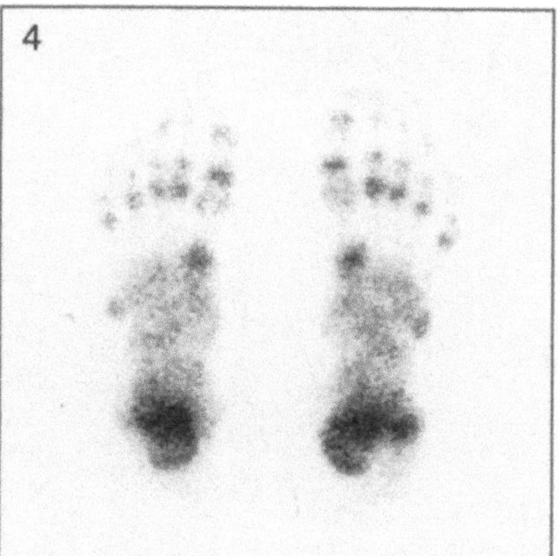

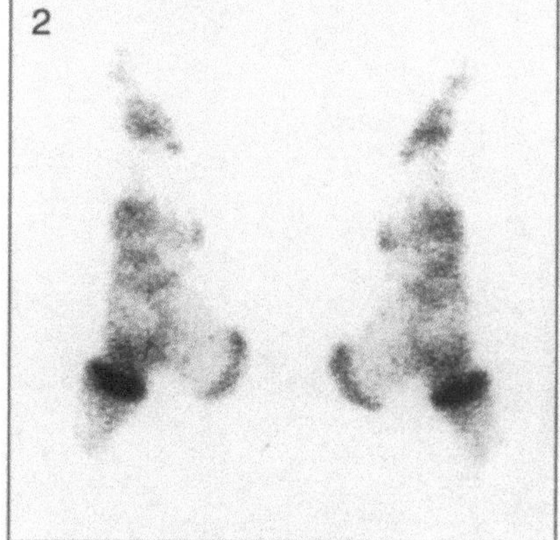

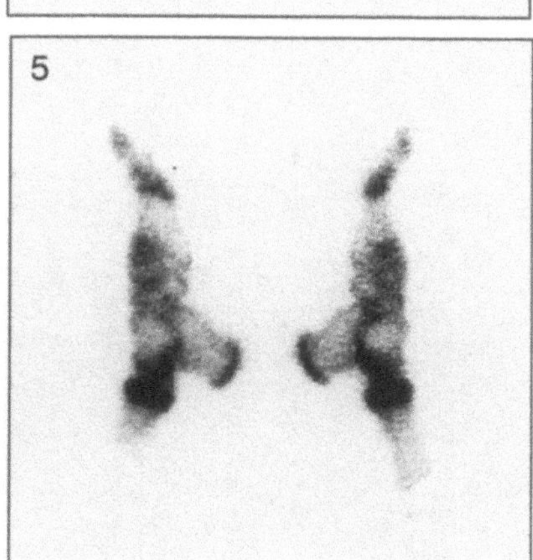

Fig. 1. Posterior view of tibia, fibula and ankles

Fig. 4. Posterior view of feet

Fig. 2. Lateral view of feet

Fig. 5. Lateral view of feet

► **Potential Pitfalls**
- Note the slightly increased activity in the mid-shaft of the tibia more marked on the right than on the left in Fig. 1. This is a variation of normality
- The lack of clarity of the epiphyseal plates around the knees in Fig. 1 is due to overexposure during the acquisition

11: Age 9–10 Years

Fig. 1. Anterior view of thorax, spine and pelvis

Fig. 4. Posterior view of thorax and spine

Fig. 2. Anterior view of pelvis and femora

Fig. 5. Posterior view of pelvis and femora

Fig. 3. Anterior view of hands

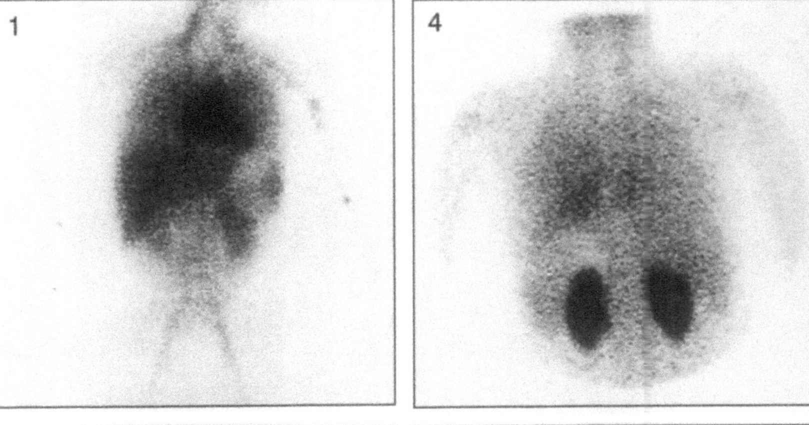

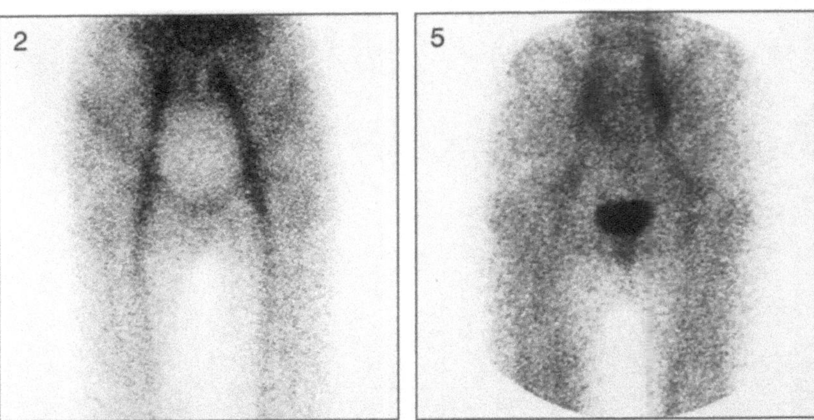

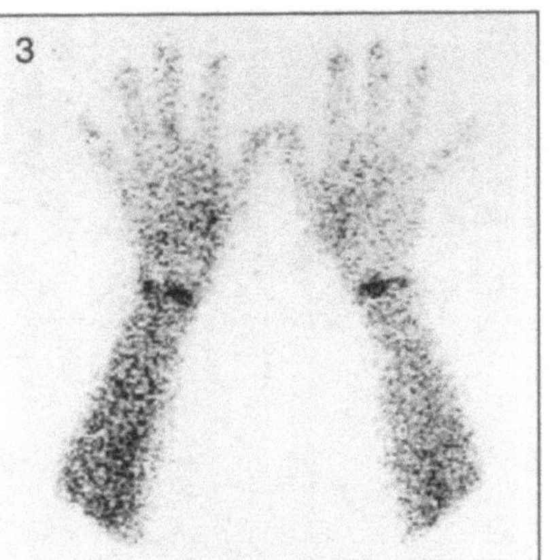

Technical Comments

- Fig. 2 shows a photon deficient area in the region of the bladder. This is due to a bladder full of urine at the time of injection of radiotracer
- The photon deficient area above the left kidney in Figs. 1 and 4 is due to a full stomach

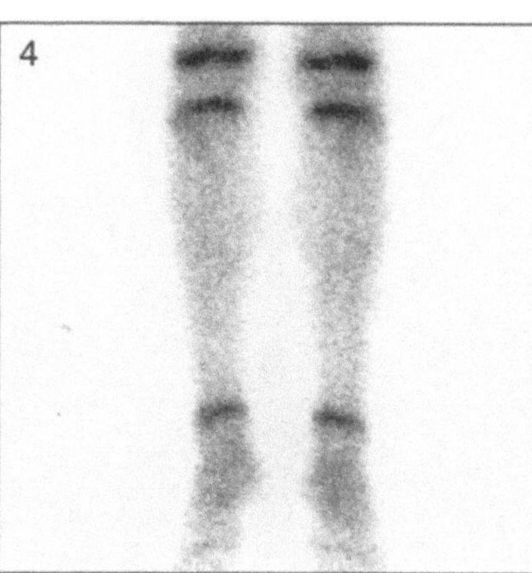

Fig. 1. Posterior view of knees

Fig. 4. Anterior view of knees, tibia, fibula and ankles

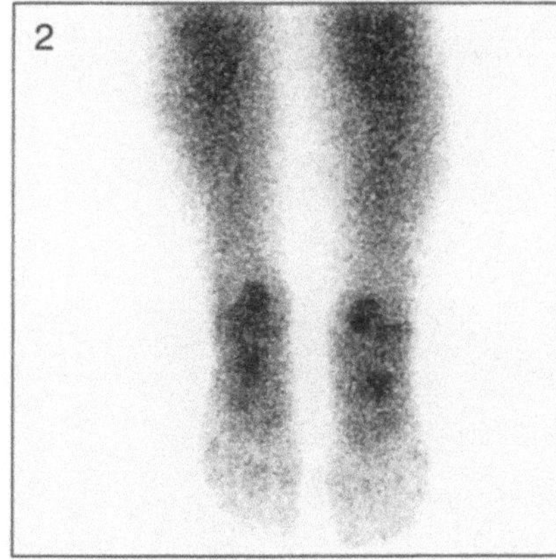

Fig. 2. Posterior view of feet

– A double headed whole body
  gamma camera was used
– Left image is the anterior view
– Right image is the posterior
  view

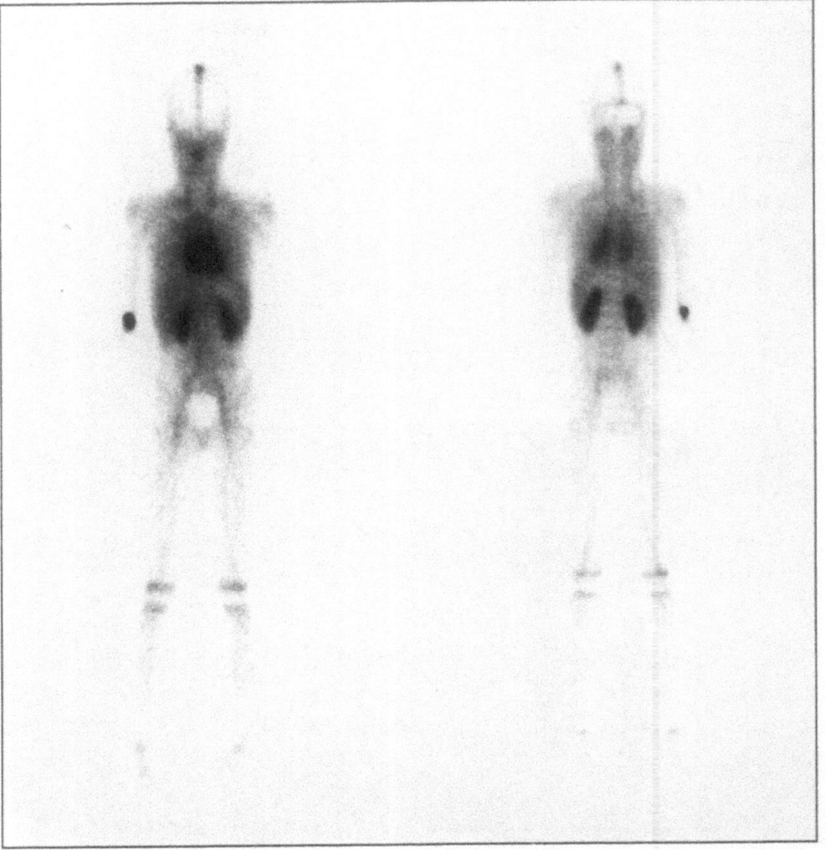

Technical Comments

– Note extravasation of isotope at the site of injection in the right elbow
– The photon deficient area in the region of the bladder on the anterior
  view is due to a bladder full of urine at the time of injection of radiotracer
– Marker on child's right side

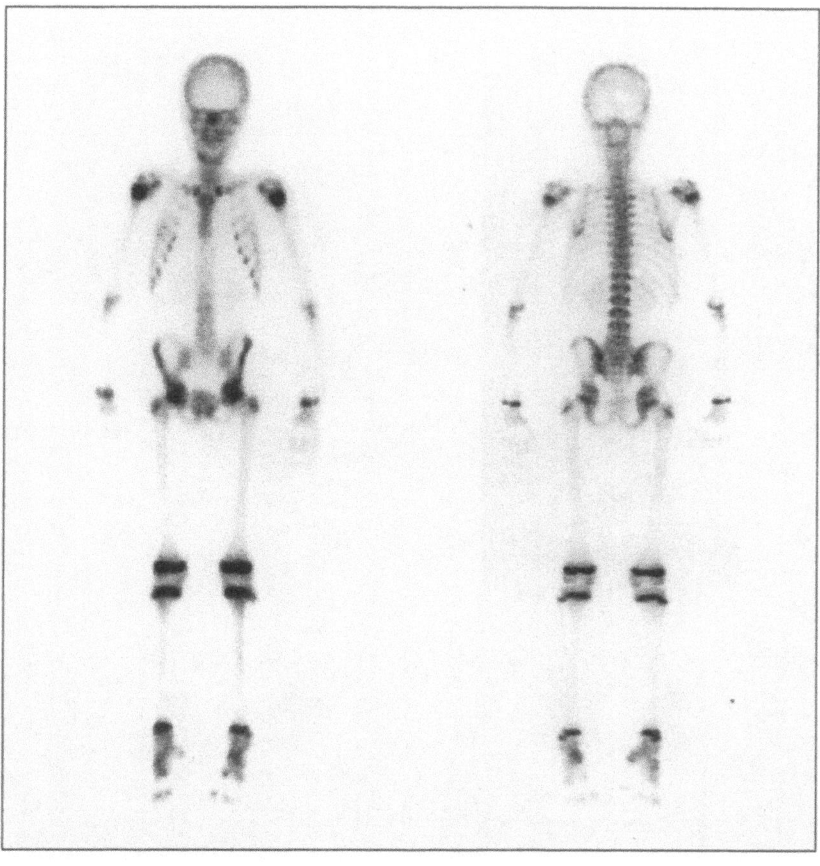

- A double headed whole body gamma camera was used
- Left image is the anterior view
- Right image is the posterior view

Technical Comments

- There is better positioning of the left foot, with the toes turned inward compared to the right foot. The consequent improved images of the left knee are noted.
- The child's pelvis is rotated causing asymmetry
- Marker on child's right side, posterior view only

- A double headed whole body
  gamma camera was used
- Left image is the anterior view
- Right image is the posterior
  view

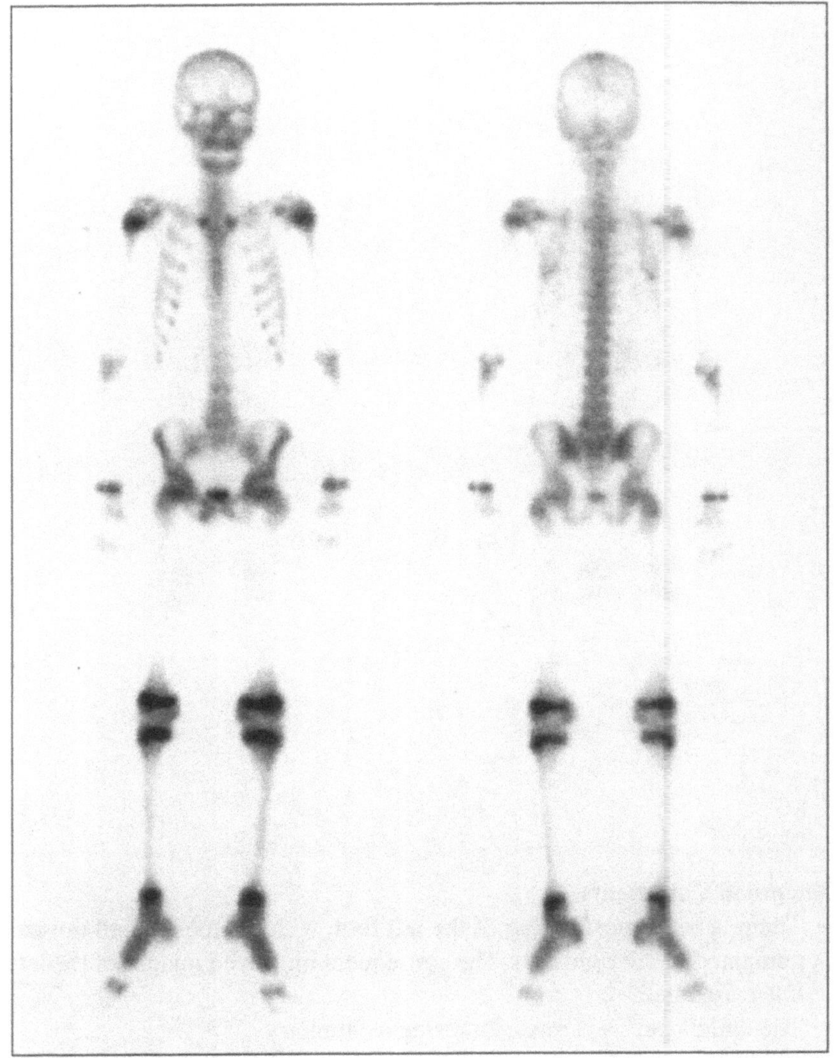

Technical Comment
- There is poor positioning of the feet, with the toes turned outward

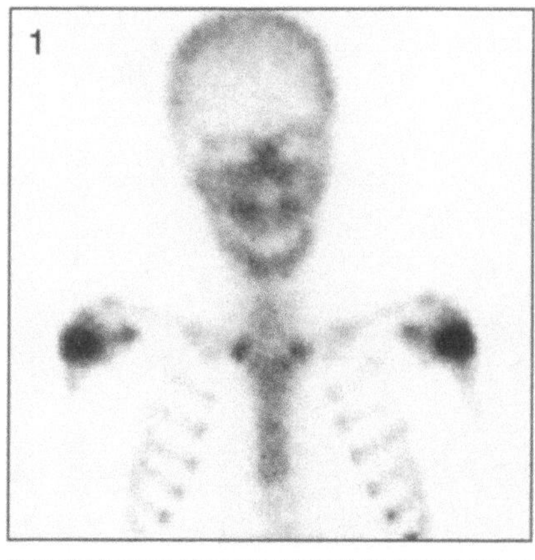

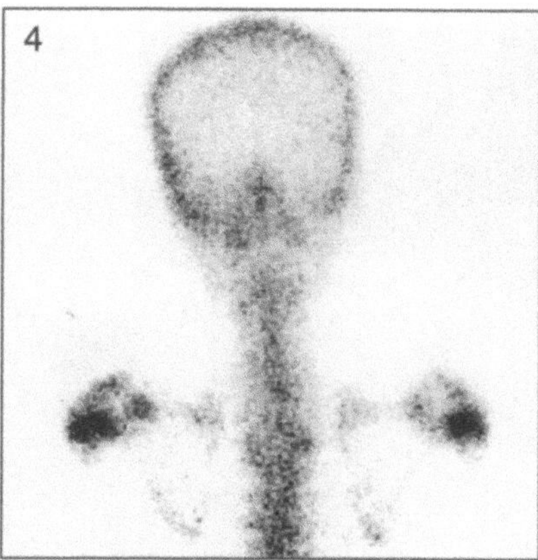

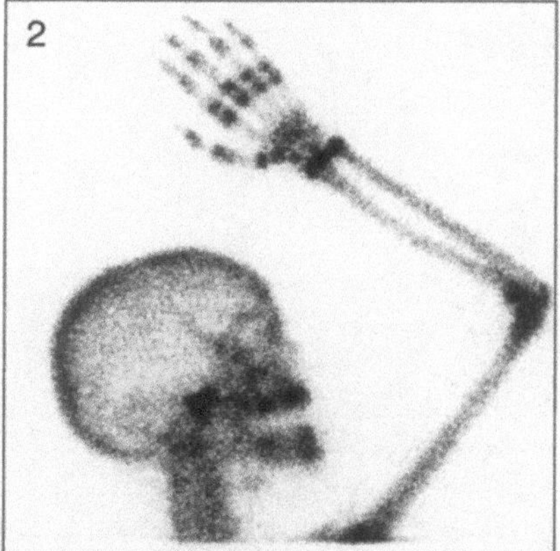

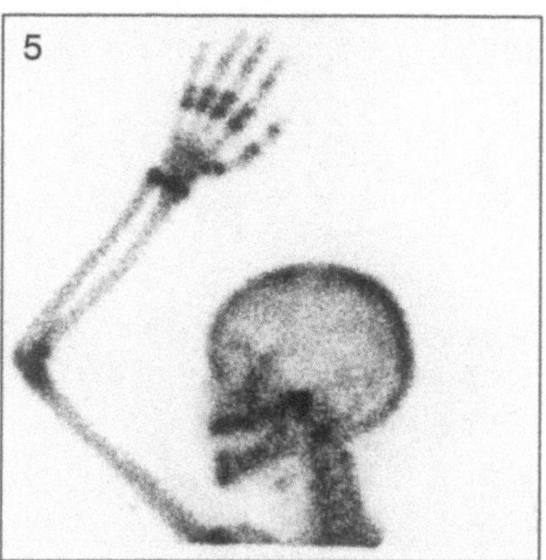

Fig. 1. Anterior view of skull and thorax

Fig. 4. Posterior view of skull and thorax

Fig. 2. Right lateral view of skull and right upper limb

Fig. 5. Left lateral view of skull and left upper limb

Technical Comments

- There is slightly increased activity in the right maxillary region in Fig. 1, this is due to rotation of the head
- The lateral views of the skull (Figs. 2 and 5) were taken posteriorly

Fig. 1. Anterior view of skull and thorax

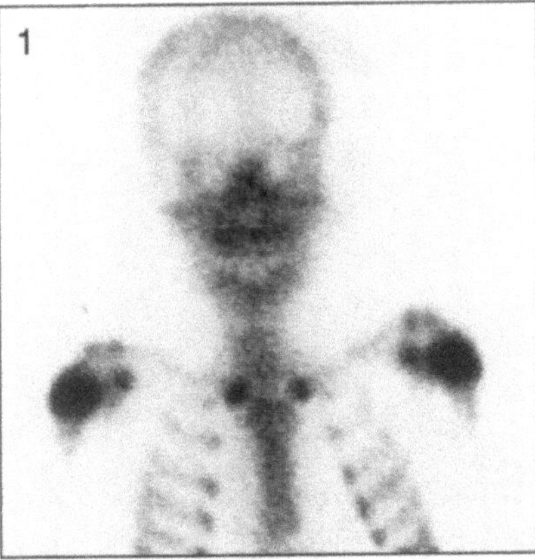

Fig. 2. Right lateral view of skull and right upper limb

Fig. 5. Left lateral view of skull and left upper limb

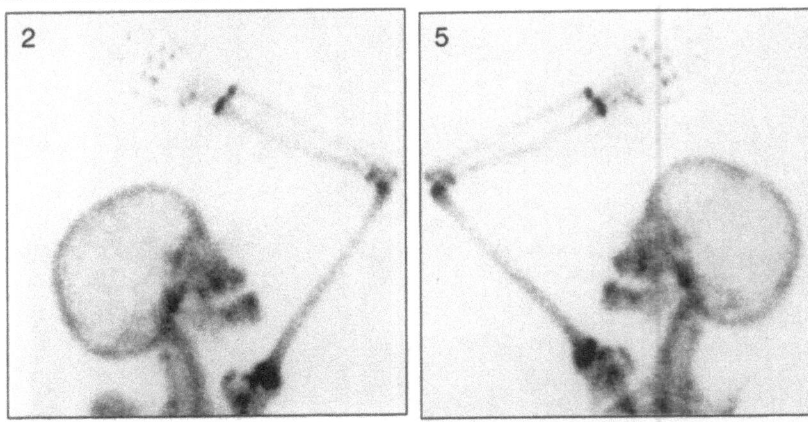

Technical Comment
– The lateral views of the skull (Figs. 2 and 5) were taken posteriorly

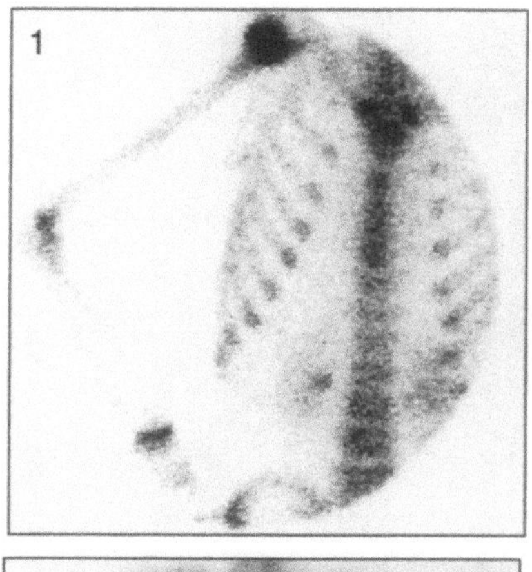

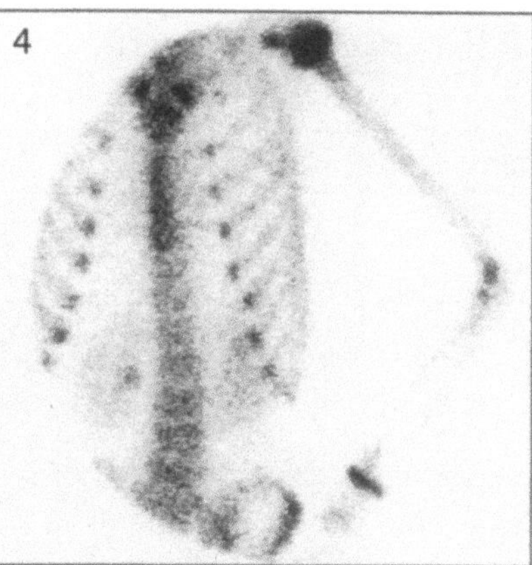

Fig. 1. Anterior view of right hemithorax and right upper limb

Fig. 4. Anterior view of left hemithorax and left upper limb

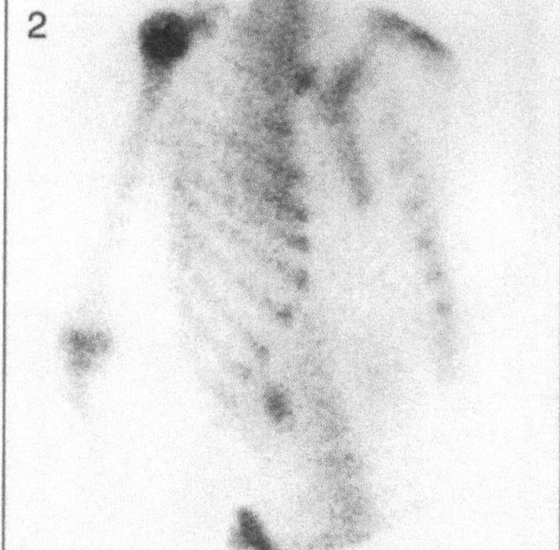

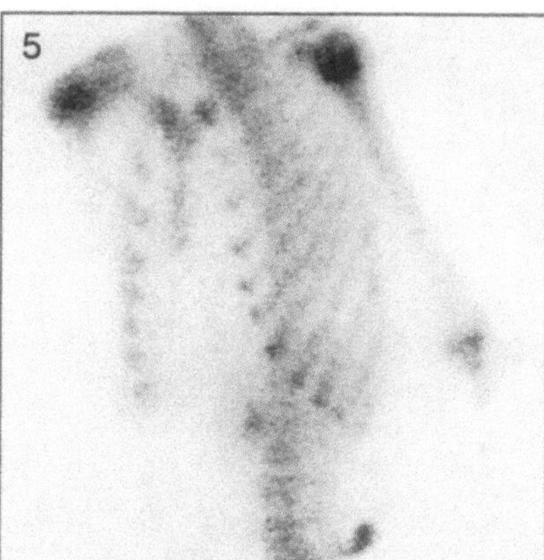

Fig. 2. Right anterior oblique view of thorax

Fig. 5. Left anterior oblique view of thorax

Technical Comments

– Figs. 1 and 4 show the difficulty in obtaining good images of the upper limbs with the anterior thoracic views. The upper limbs should be imaged at the time of lateral skull images
– Retention of tracer is noted in the dilated right renal pelvis in Fig. 2

Fig. 1. Anterior view of thorax, spine and part pelvis

Fig. 4. Anterior view of thorax, spine and pelvis

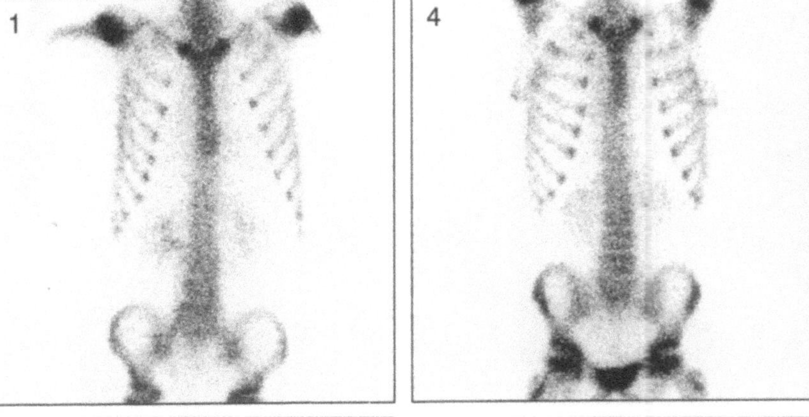

Fig. 2. Anterior view of thorax, spine and part pelvis

Fig. 5. Anterior view of thorax, spine and pelvis

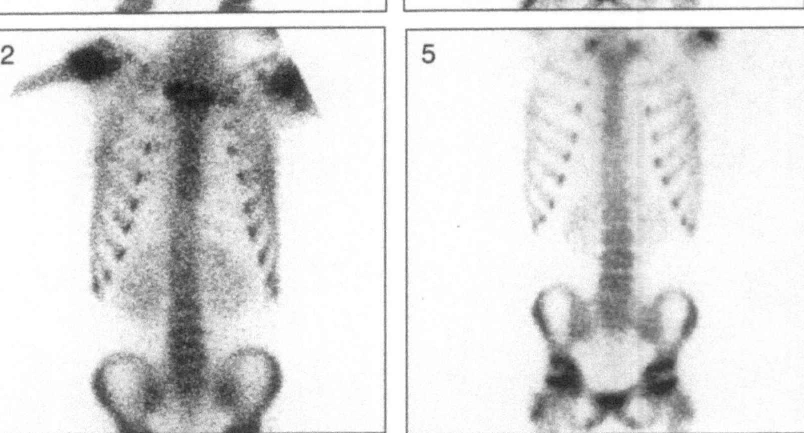

Technical Comment

– The lower lumbar spine is clearly seen on the anterior views in Figs. 2, 4 and 5

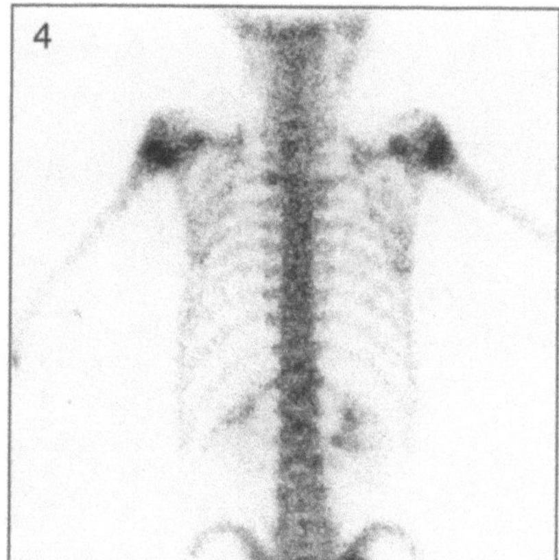

Fig. 1. Posterior view of thorax, spine and part pelvis

Fig. 4. Posterior view of thorax and spine

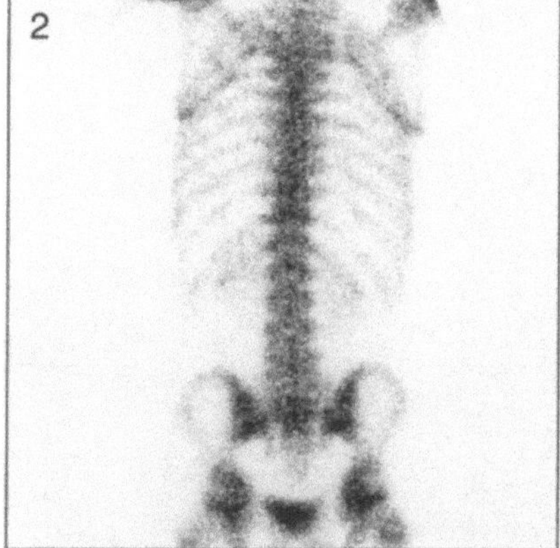

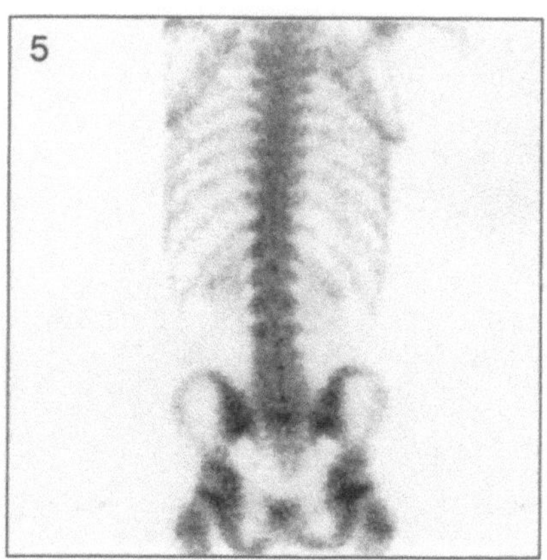

Fig. 2. Posterior view of thorax, spine and pelvis

Fig. 5. Posterior view of thorax, spine and pelvis

Technical Comment

- The scapula is clearly seen in all images, the differences are due to the varying positions of the upper limbs

▶ Potential Pitfall

- Focal increased activity is seen in the region of the posterior aspect of the left 3rd rib in Fig. 4. This is due to the activity from the overlying left sterno-clavicular joint "shining through"

Fig. 1. Anterior view of pelvis and femora

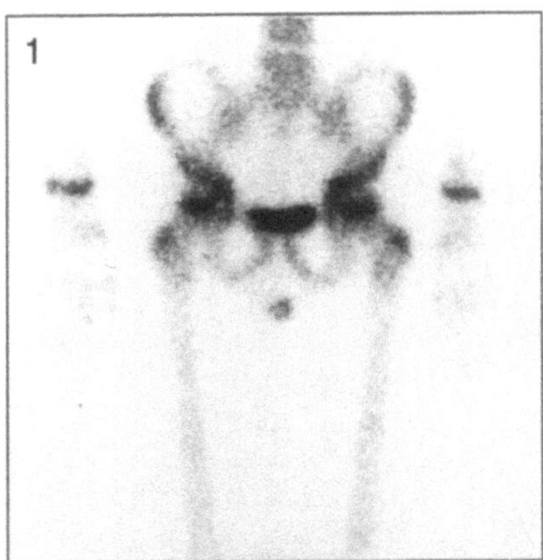

Fig. 2. Anterior view of spine, pelvis and femora

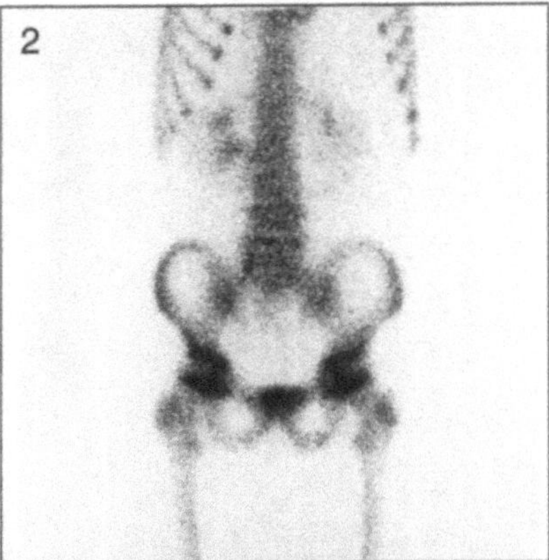

Fig. 3. Anterior magnified view of pelvis and femora

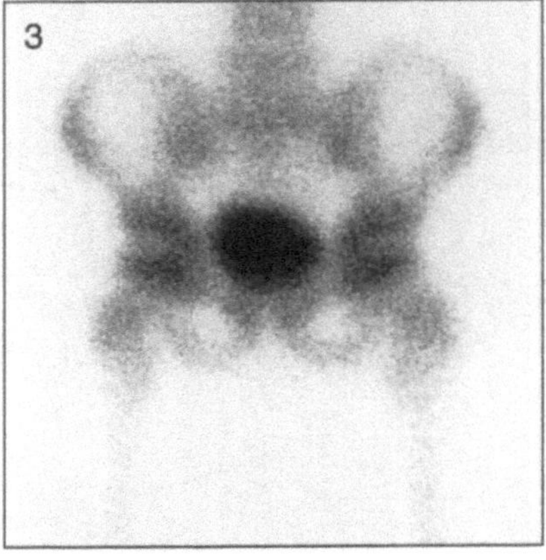

Technical Comments
- Urine contamination below the pelvis is seen in Fig. 1
- The bladder is full of isotope in Fig. 3

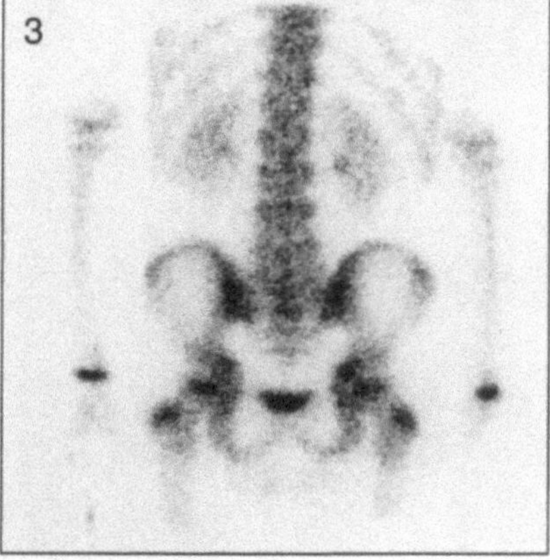

Fig. 1. Posterior view of spine and pelvis

Fig. 2. Posterior view of spine and pelvis

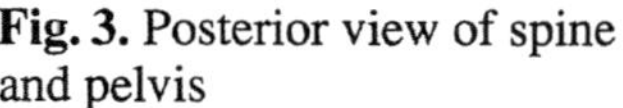

Fig. 3. Posterior view of spine and pelvis

Technical Comment

– In Fig. 2 notice the slight different rotation of the greater trochanter which is not separately seen on the left but is seen to overlie the femoral neck. This is due to poor positioning of the left knee and foot

► Potential Pitfall

– The increased activity seen over the lower ribs in Fig. 2 is due to retention of tracer in the collecting system of the horseshoe-kidney

Fig. 1. Pinhole view of right hip

Fig. 4. Pinhole view of left hip

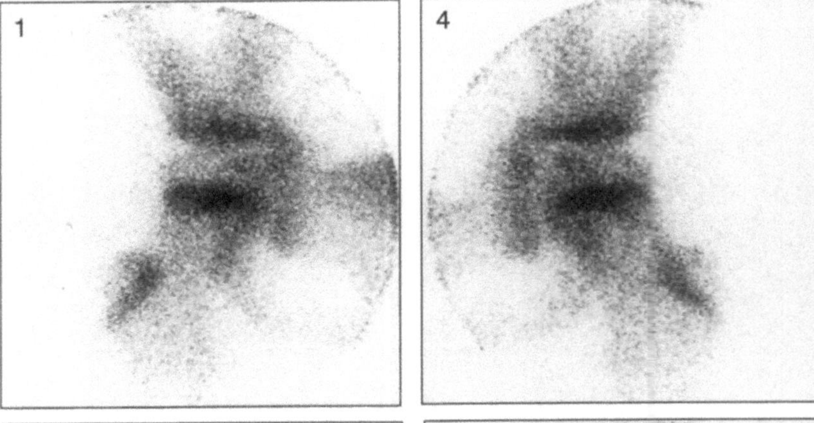

Fig. 2. Pinhole view of right hip

Fig. 5. Pinhole view of left hip

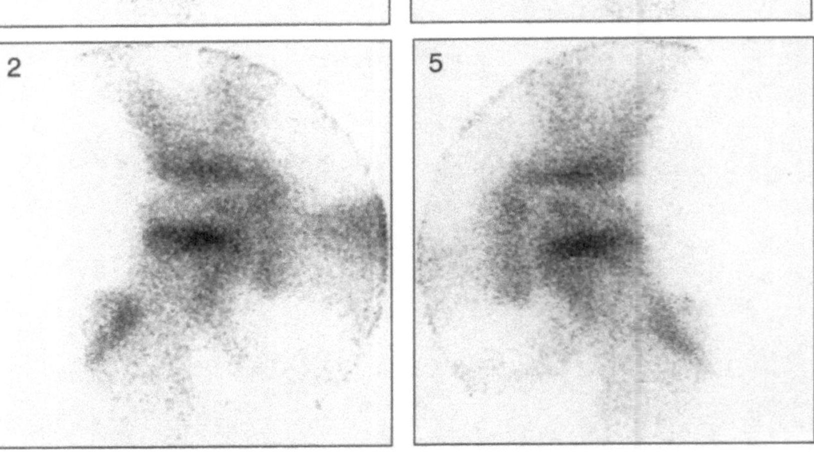

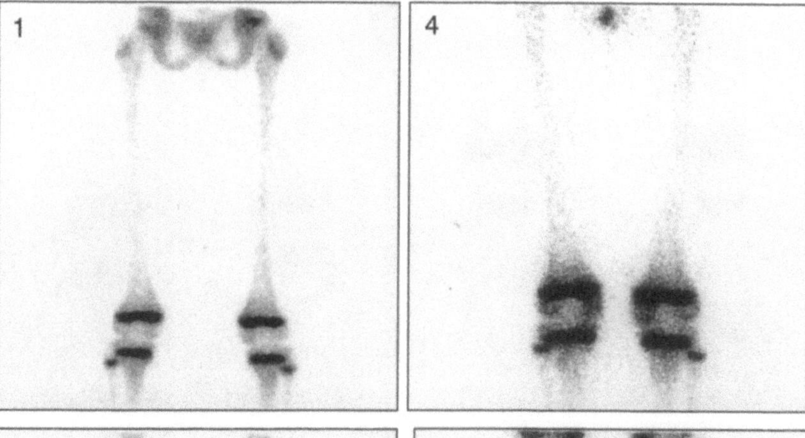

Fig. 1. Posterior view of femora and knees

Fig. 4. Posterior view of femora and knees

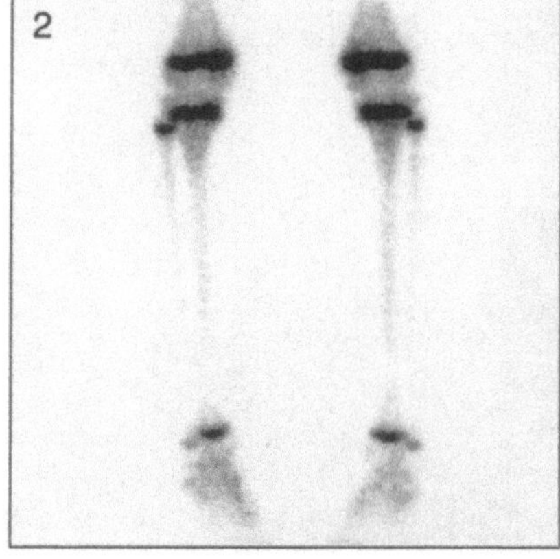
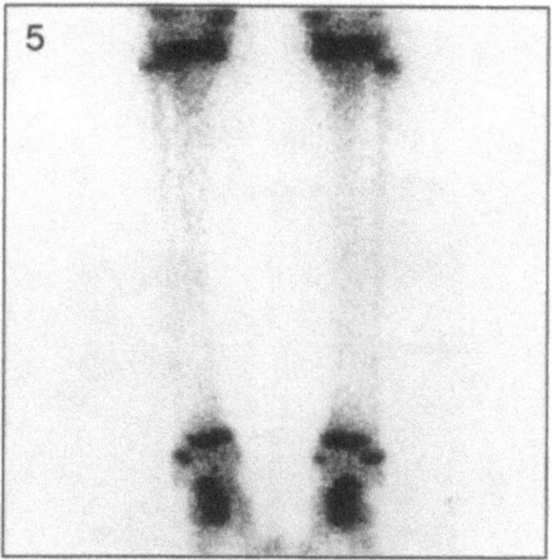

Fig. 2. Posterior view of knees, tibia, fibula and ankles

Fig. 5. Posterior view of tibia, fibula and ankles

Technical Comments

- Note the in-turning of the feet in Figs. 2 and 5 into the "radiographic neutral position"
- The difference between the greater trochanters in Fig. 1 is a variation of normality
- Urine contamination between the femora is seen in Fig. 4

Fig. 1. Posterior view of knees

Fig. 4. Posterior view of knees

Fig. 2. Posterior view of knees

Fig. 3. Anterior oblique view of knees

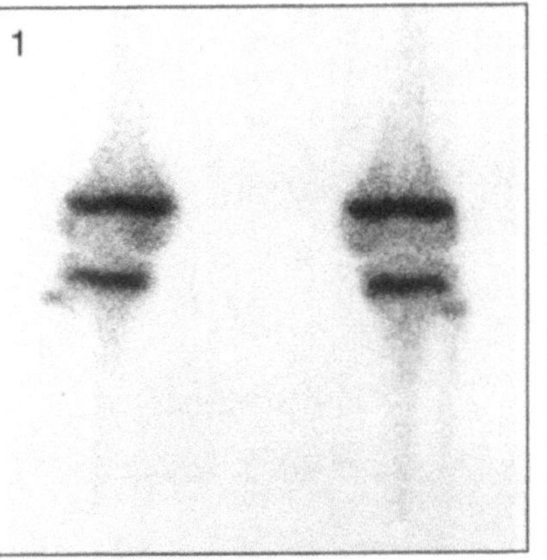

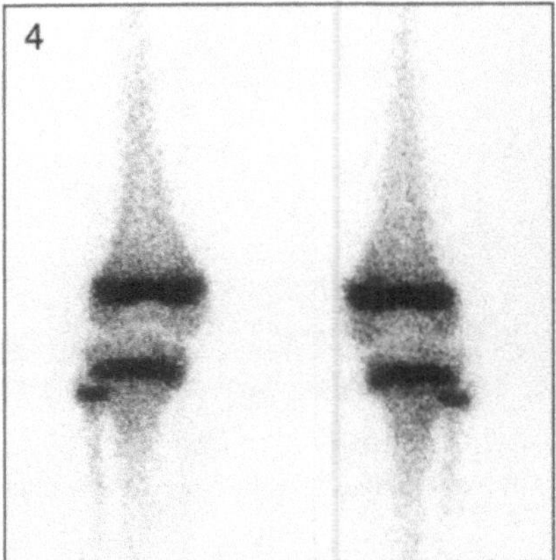

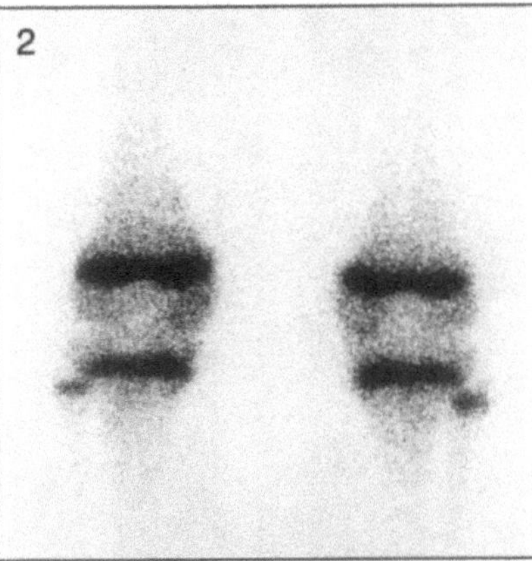

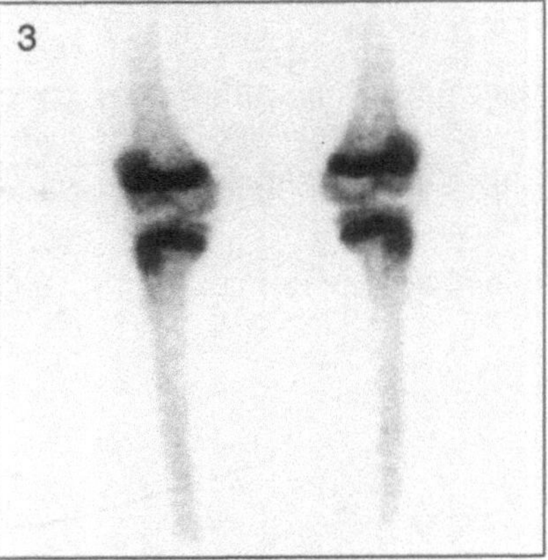

Technical Comments
- Fig. 3 is an oblique view of the knees obtained when the medial femoral condyles need to be imaged, not frequently used in paediatrics
- The patella is seen overlying the lateral aspect of the femoral epiphyseal plate in Fig. 3

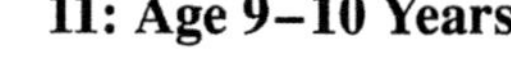

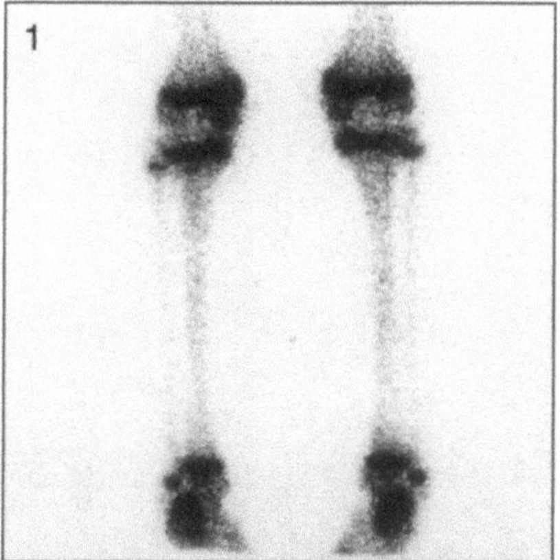

Fig. 1. Posterior view of knees, tibia, fibula and ankles

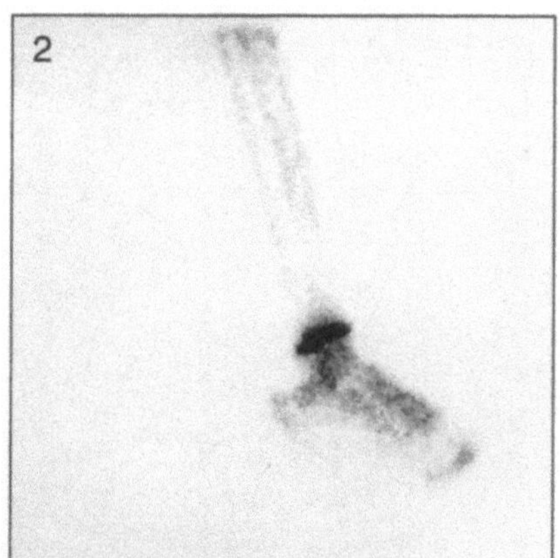

Fig. 2. Lateral view of foot

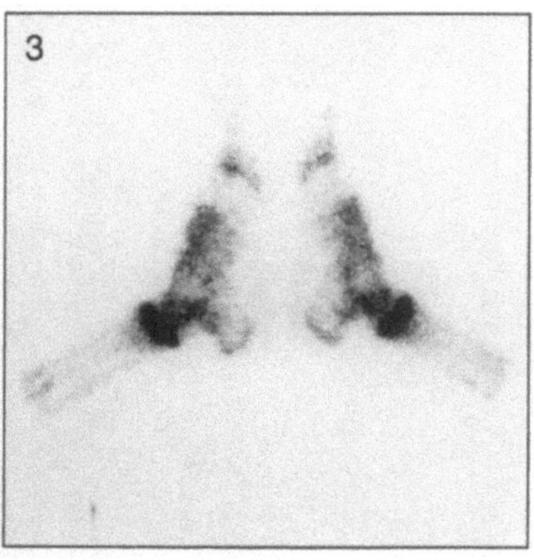

Fig. 3. Lateral view of feet

12: Age 10–11 Years

Fig. 1. Posterior view of skull and thorax

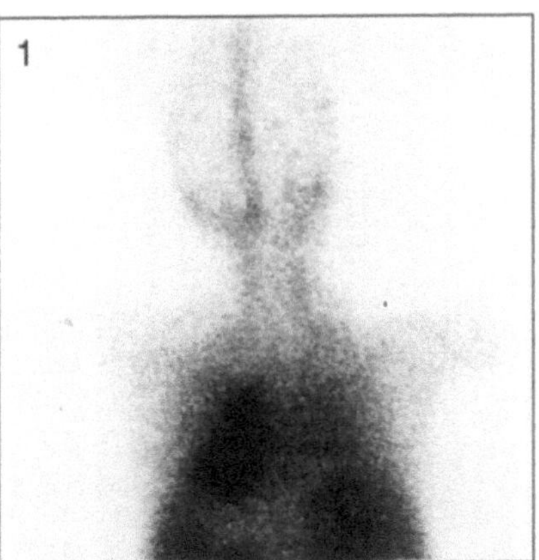

Fig. 2. Anterior view of pelvis

Fig. 5. Anterior view of lower limbs

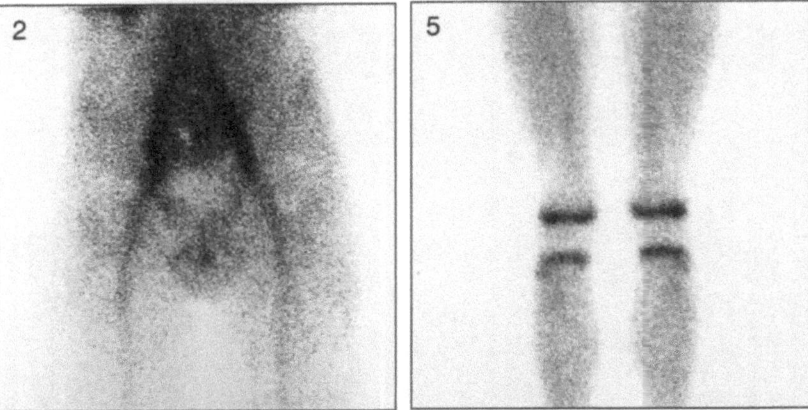

Fig. 3. Posterior view of spine and pelvis

Fig. 6. Anterior view of lower limbs

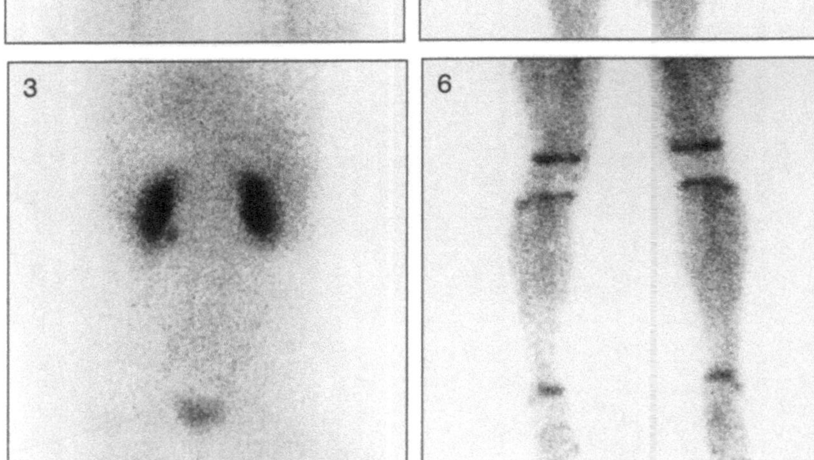

Technical Comment

– The skull is rotated in Fig. 1 causing the sagittal sinus to appear off center

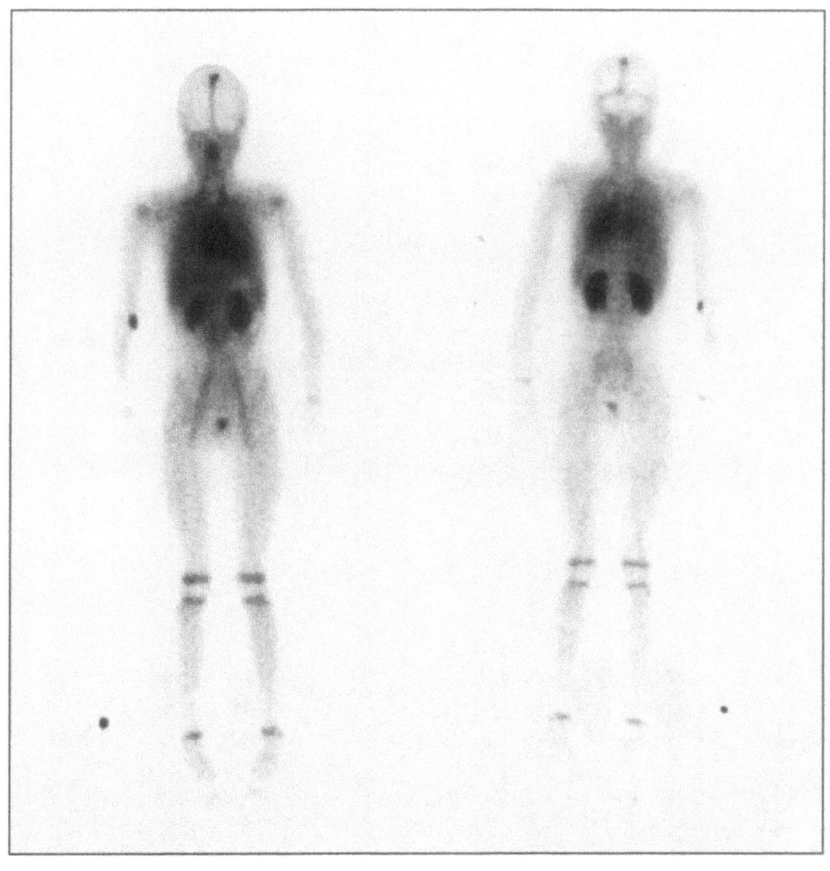

- A double headed whole body gamma camera was used
- Left image is the anterior view
- Right image is the posterior view

Technical Comments

- Note extravasation of isotope at the site of injection in the right elbow
- Marker on child's right side

– A double headed whole body
  gamma camera was used
– Left image is the anterior view
– Right image is the posterior
  view

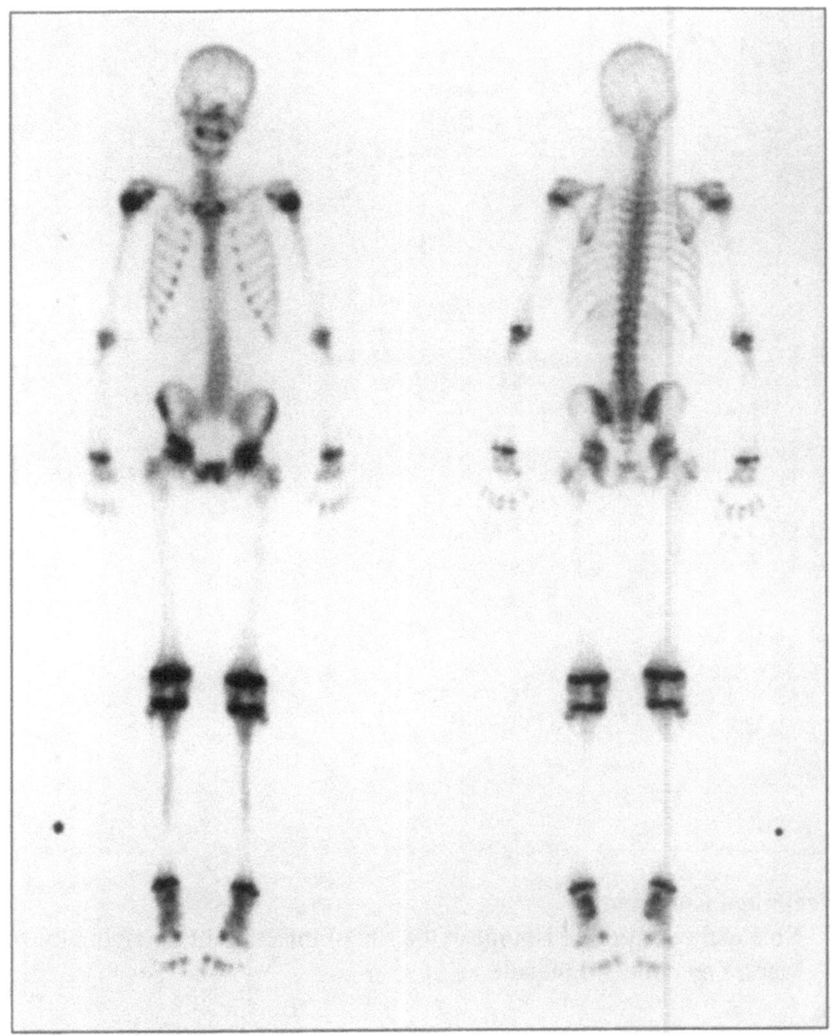

Technical Comment
– Marker on child's right side

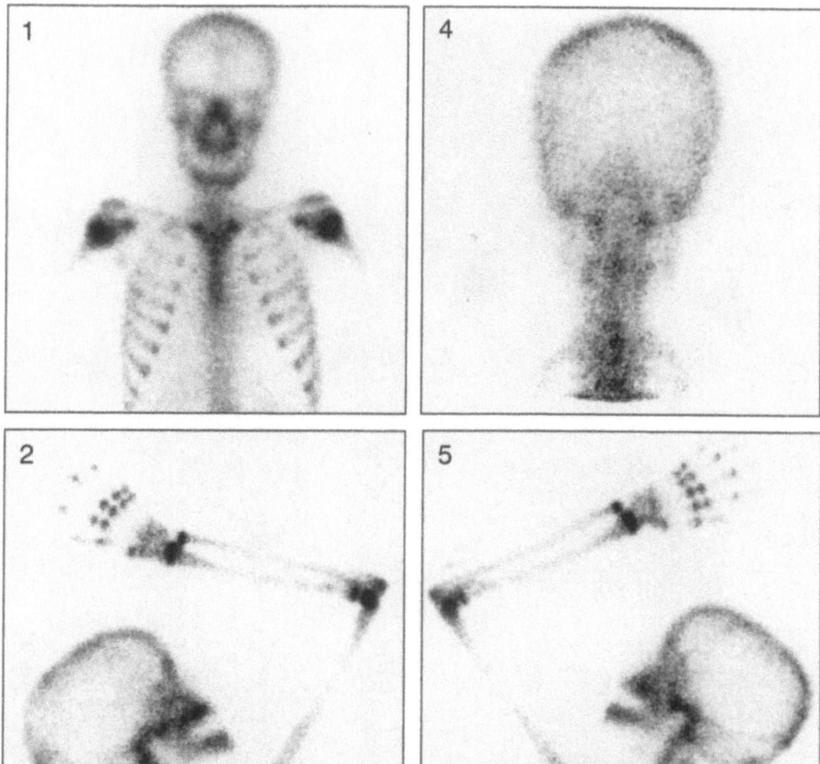

Fig. 1. Anterior view of skull and thorax

Fig. 4. Posterior view of skull

Fig. 2. Right lateral view of skull and right upper limb

Fig. 5. Left lateral view of skull and left upper limb

Technical Comment
– The lateral views of the skull (Figs. 2 and 5) were taken posteriorly

Fig. 1. Anterior view of skull and thorax

Fig. 4. Posterior view of skull

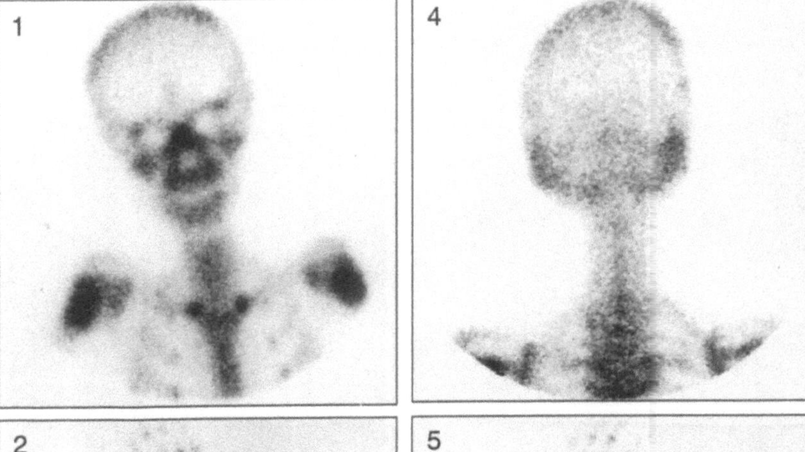

Fig. 2. Right lateral view of skull and right upper limb

Fig. 5. Left lateral view of skull and left upper limb

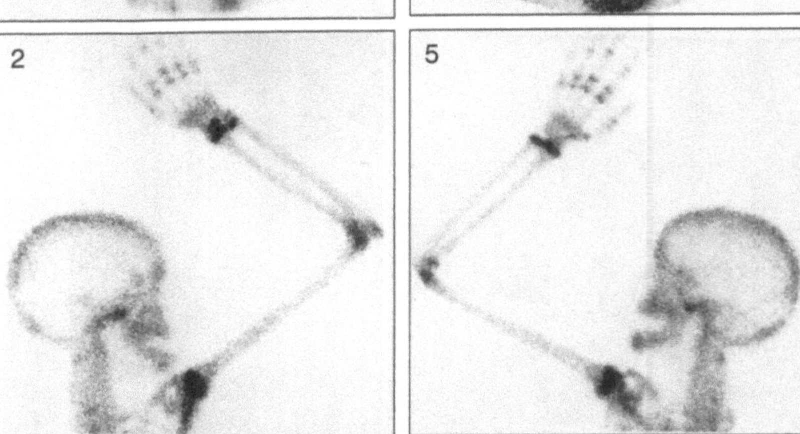

Fig. 3. Forearms

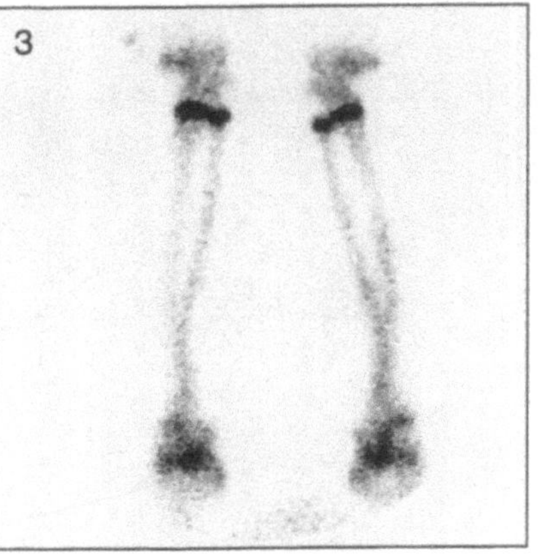

Technical Comments

- The image of the forearms in Fig. 3 reveals overlap of the upper radius and ulna; this should be compared to Figs. 2 and 5 where there is clear separation of radius from ulna
- The differences between the mastoid areas in Fig. 4 are due to rotation
- The lateral views of the skull (Figs. 2 and 5) were taken posteriorly

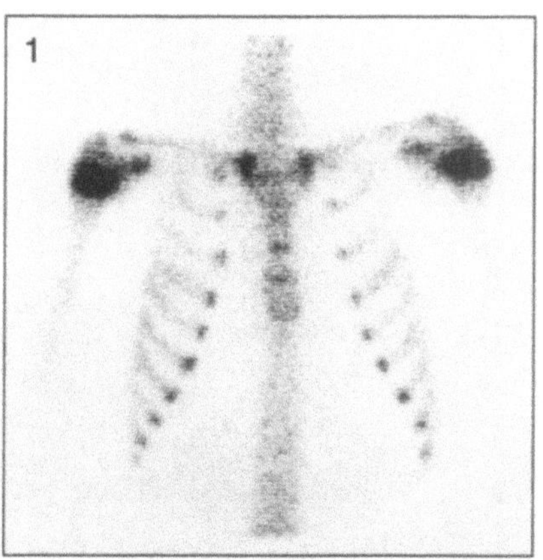

Fig. 1. Anterior view of thorax

Fig. 2. Anterior view of thorax,
spine and part pelvis

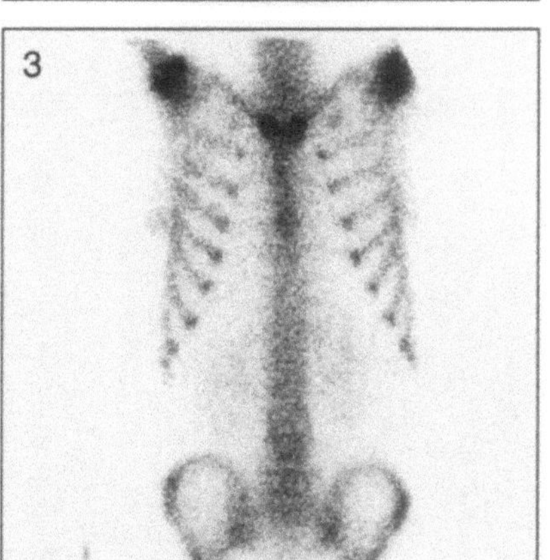

Fig. 3. Anterior view of thorax,
spine and part pelvis

▶ **Potential Pitfall**
   – Note the sternum in Fig. 1 where increased activity is noted in two
     areas. This represents normal growth centers of the sternum, and
     should not be confused with a fracture

Fig. 1. Right anterior oblique view of thorax

Fig. 4. Left anterior oblique view of thorax

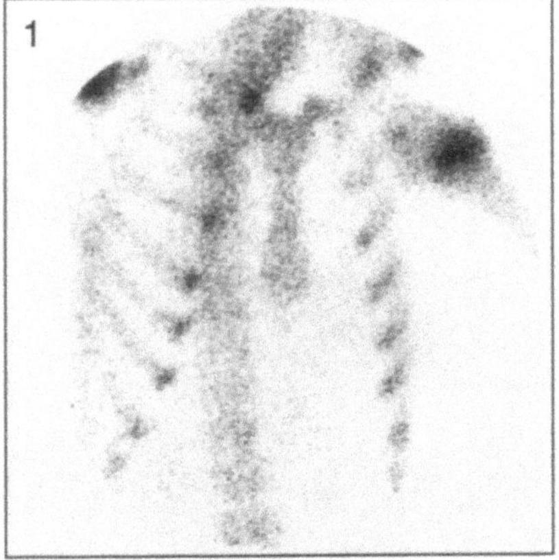

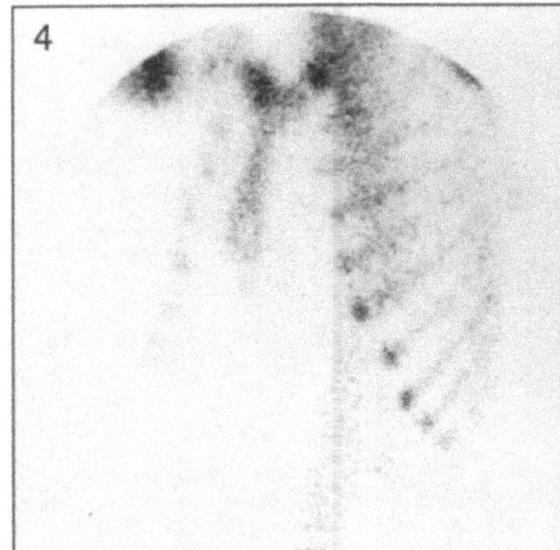

Fig. 2. Right anterior oblique view of thorax

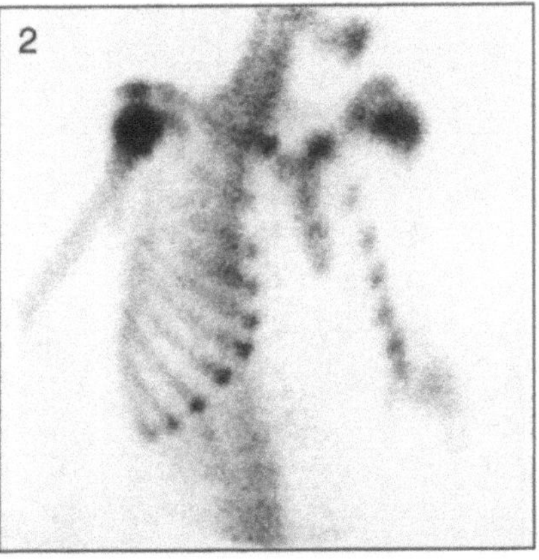

Technical Comment

– Note the different appearances of the sternum due to different degrees of obliquity

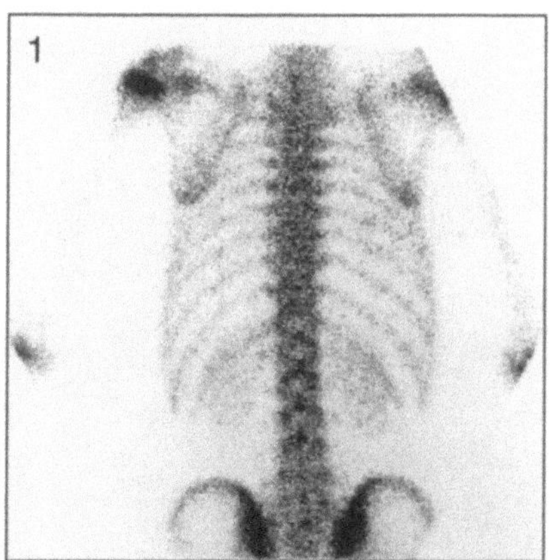

Fig. 1. Posterior view of thorax, spine and part pelvis

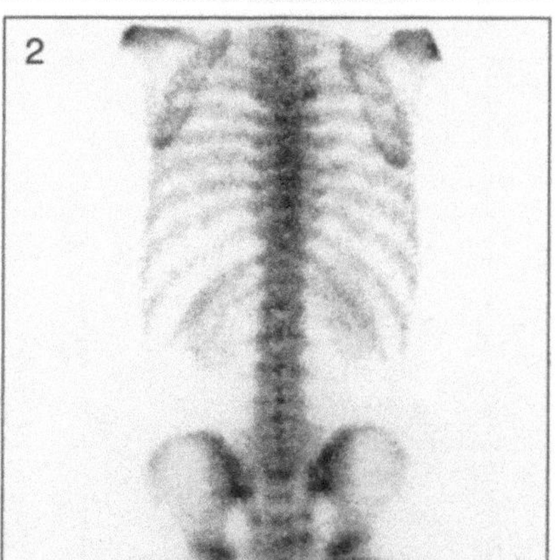

Fig. 2. Posterior view of thorax, spine and part pelvis

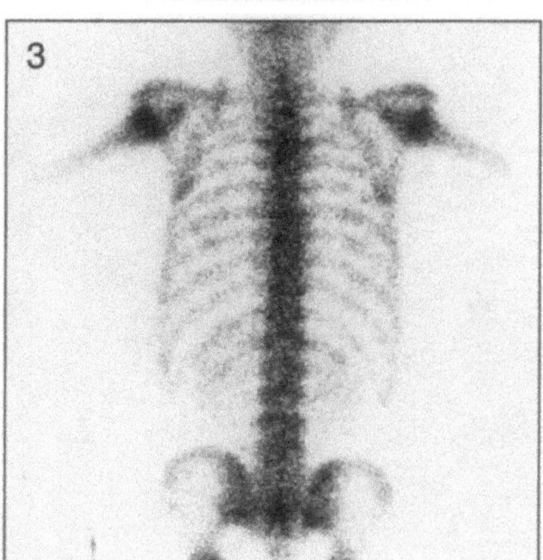

Fig. 3. Posterior view of thorax, spine and part pelvis

Fig. 1. Anterior view of pelvis
and femora

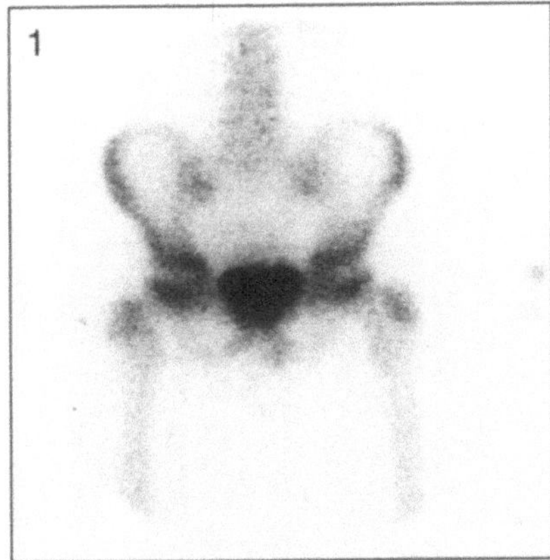

Fig. 2. Anterior view of pelvis
and femora

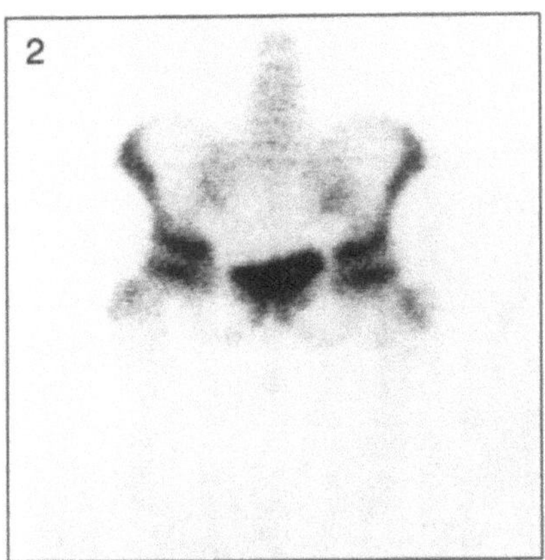

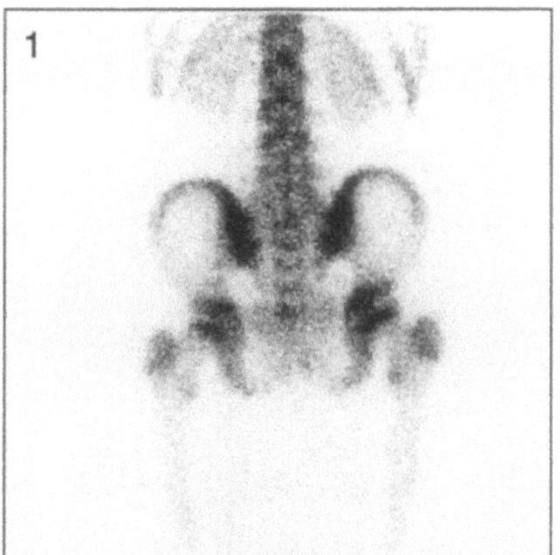

Fig. 1. Posterior view of spine, pelvis and femora

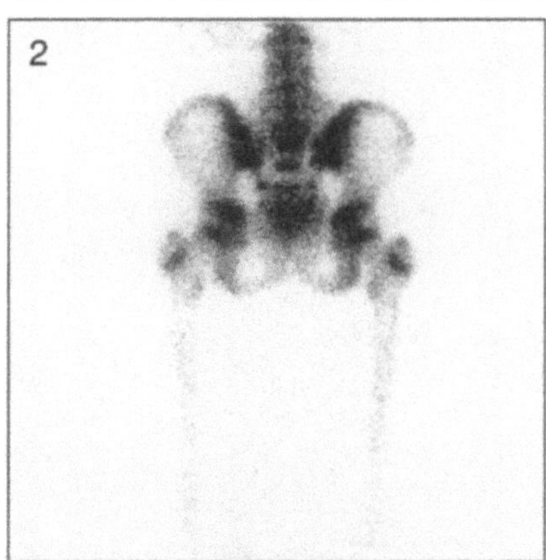

Fig. 2. Posterior view of pelvis and femora

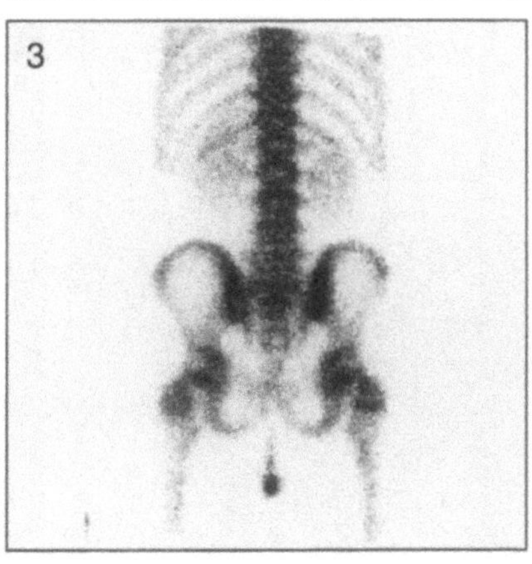

Fig. 3. Posterior view of spine, pelvis and femora

Technical Comment
– Urine contamination below the pelvis is seen in Fig. 3

Fig. 1. Pelvic inlet view

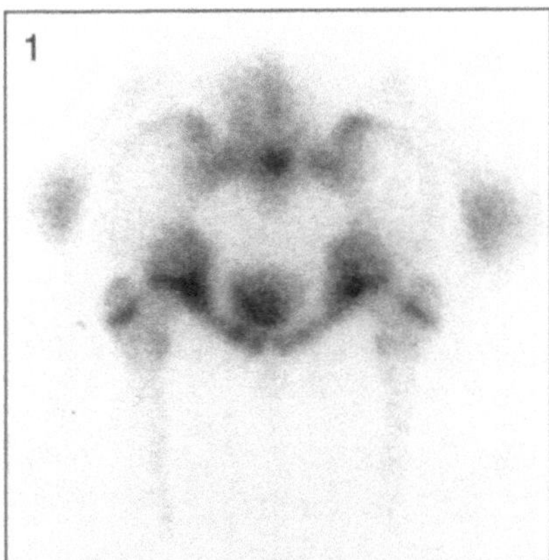

Fig. 2. Sitting view of spine and part pelvis

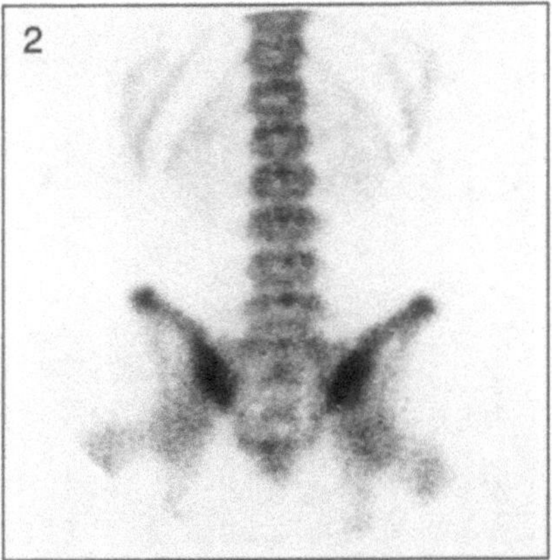

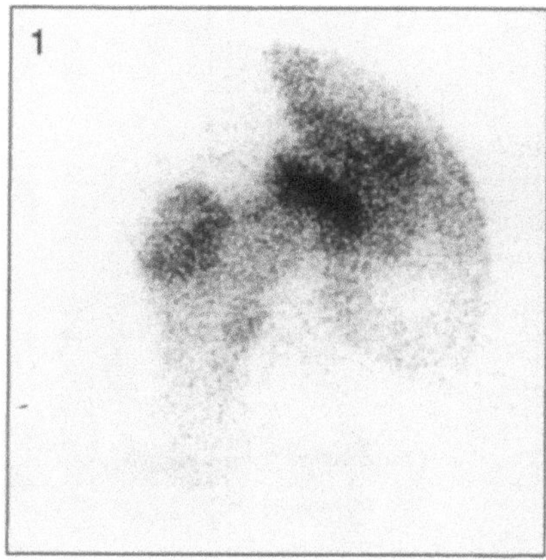 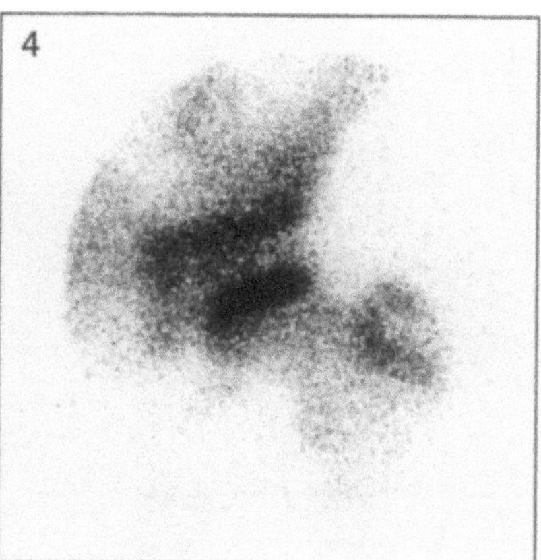

Fig. 1. Pinhole view of right hip

Fig. 4. Pinhole view of left hip

Fig. 1. Posterior view of femora

Fig. 4. Posterior view of femora and knees

Fig. 2. Posterior view of knees, tibia, fibula and ankles

Fig. 5. Posterior view of knees, tibia, fibula and ankles

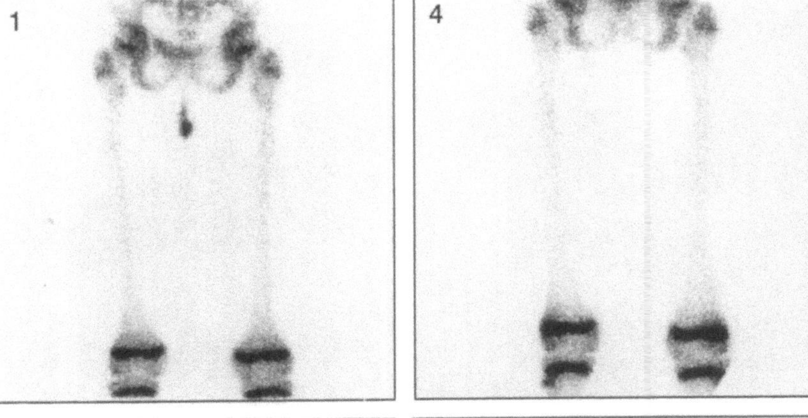

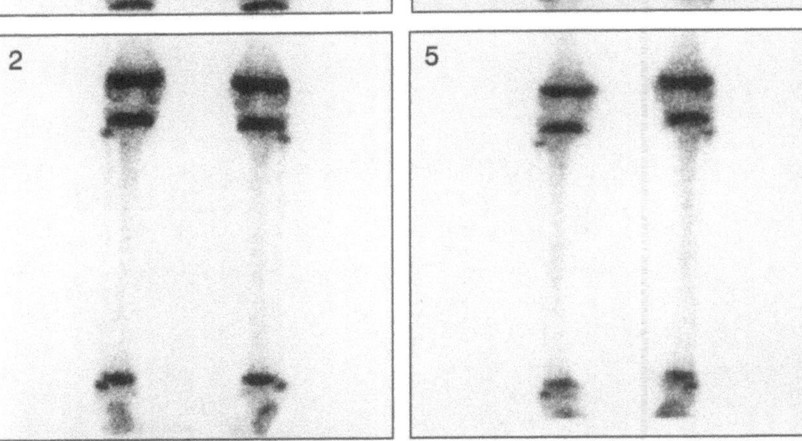

Technical Comment

– Urine contamination below the pelvis is seen in Fig. 1

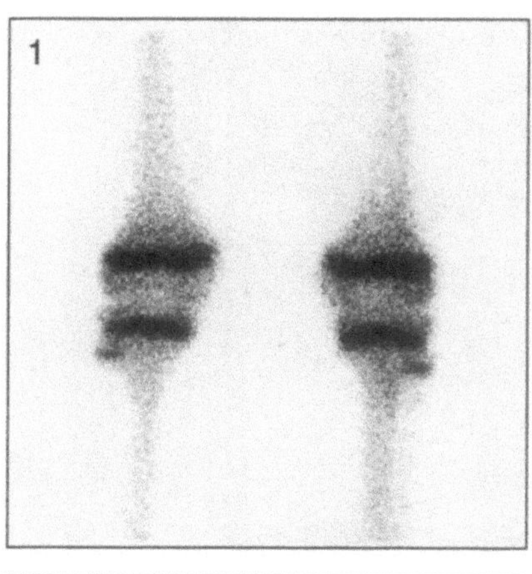

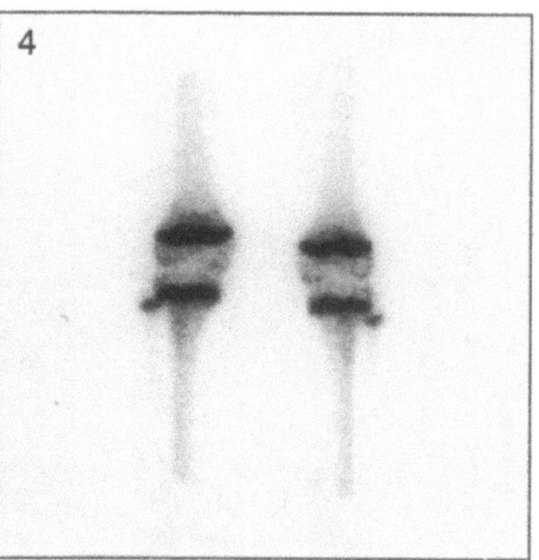

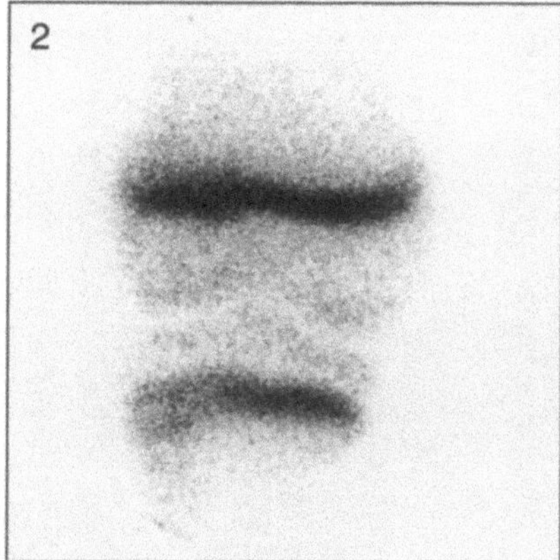

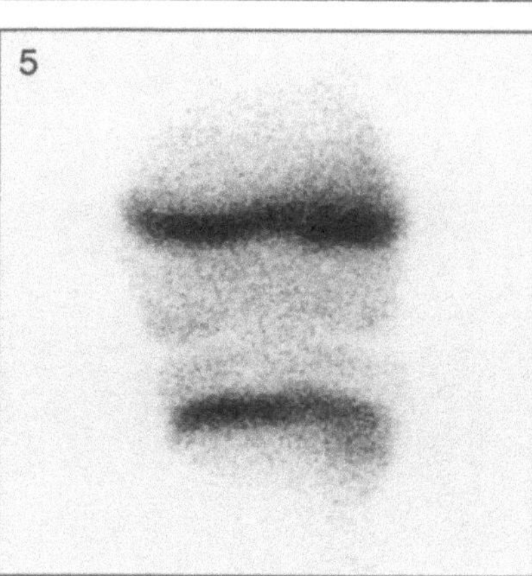

Fig. 1. Posterior magnified view of knees

Fig. 4. Posterior view of knees

Fig. 2. Anterior pinhole view of right knee

Fig. 5. Anterior pinhole view of left knee

Technical Comment

– Note that the fibula is better seen on the posterior view in Figs. 1 and 4 than on the anterior pinhole views in Figs. 2 and 5

Fig. 1. Posterior view of knees, tibia, fibula and ankles

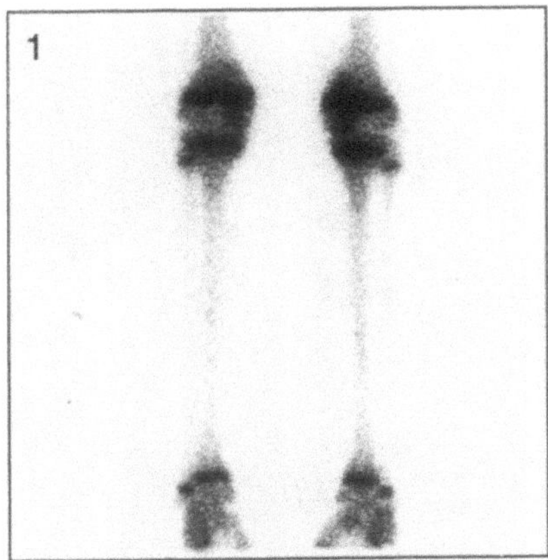

Fig. 2. Lateral view of feet

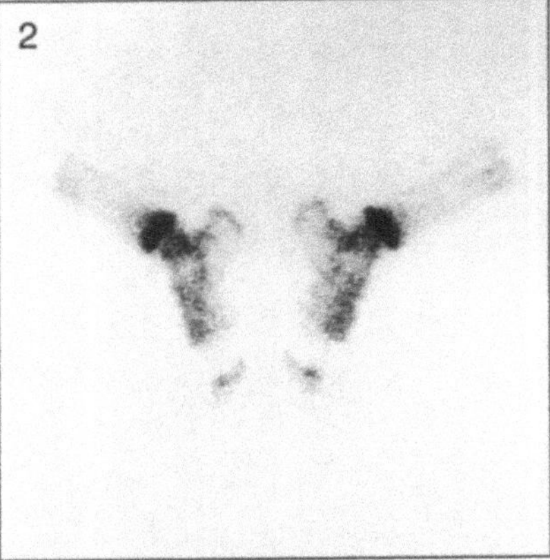

Fig. 3. Lateral lateral view of foot

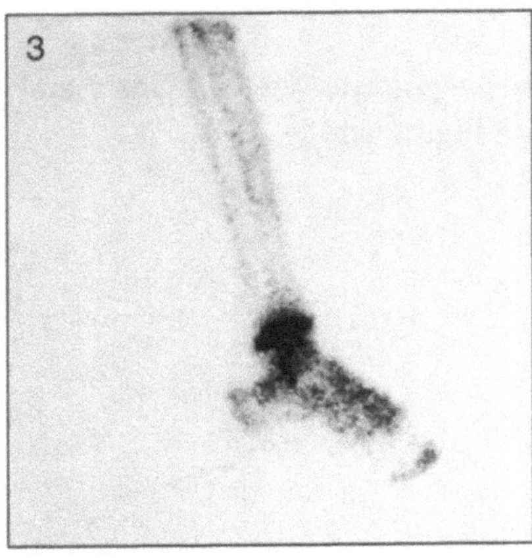

13: Age 11–12 Years

Fig. 1. Posterior view of thorax and spine

Fig. 4. Right lateral view of skull and posterior view of upper limbs

Fig. 2. Posterior view of pelvis

Fig. 5. Anterior view of pelvis

Fig. 3. Posterior view of knees

Fig. 6. Lateral view of feet

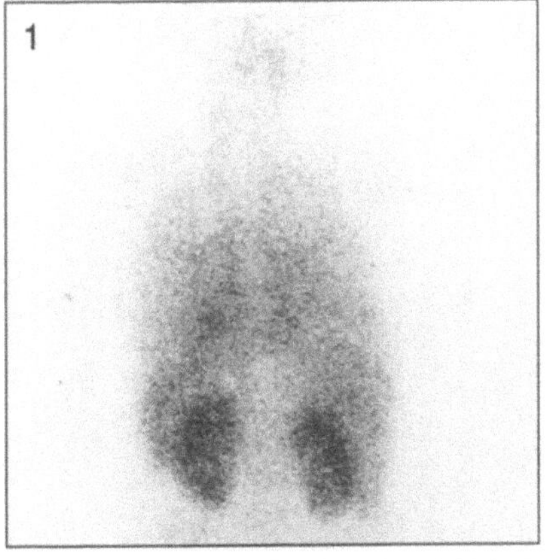
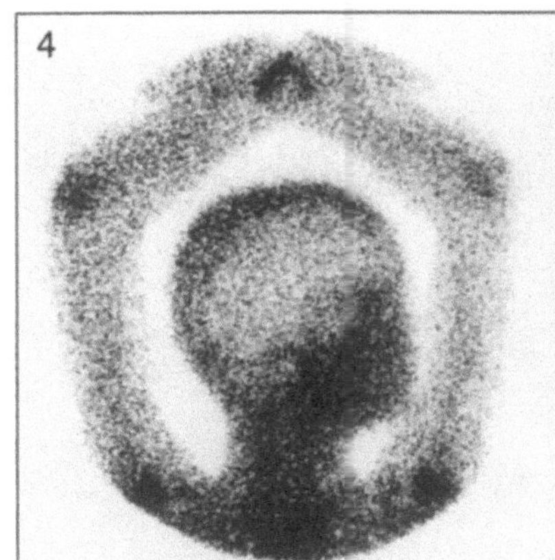
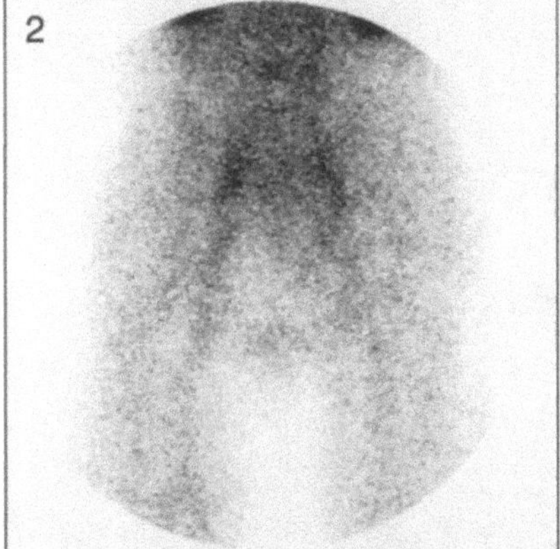
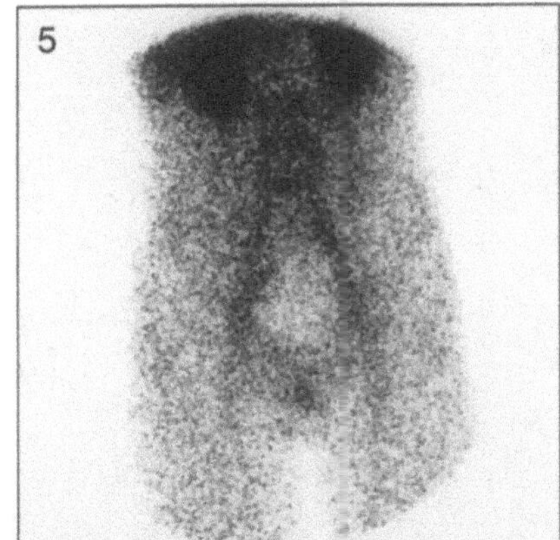
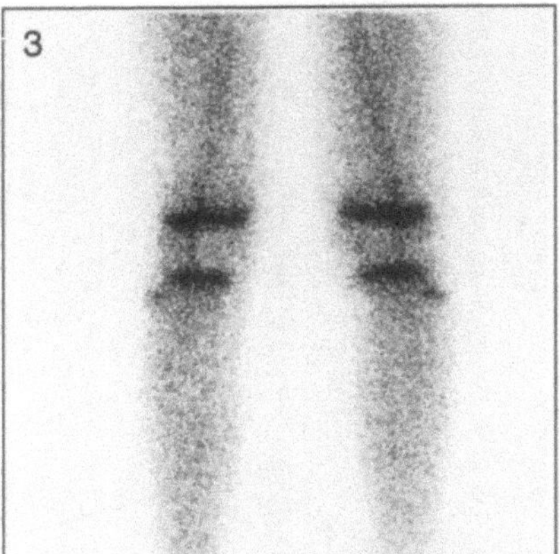
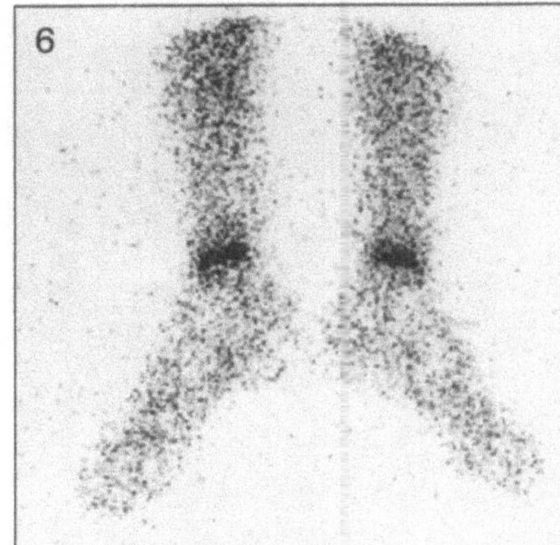

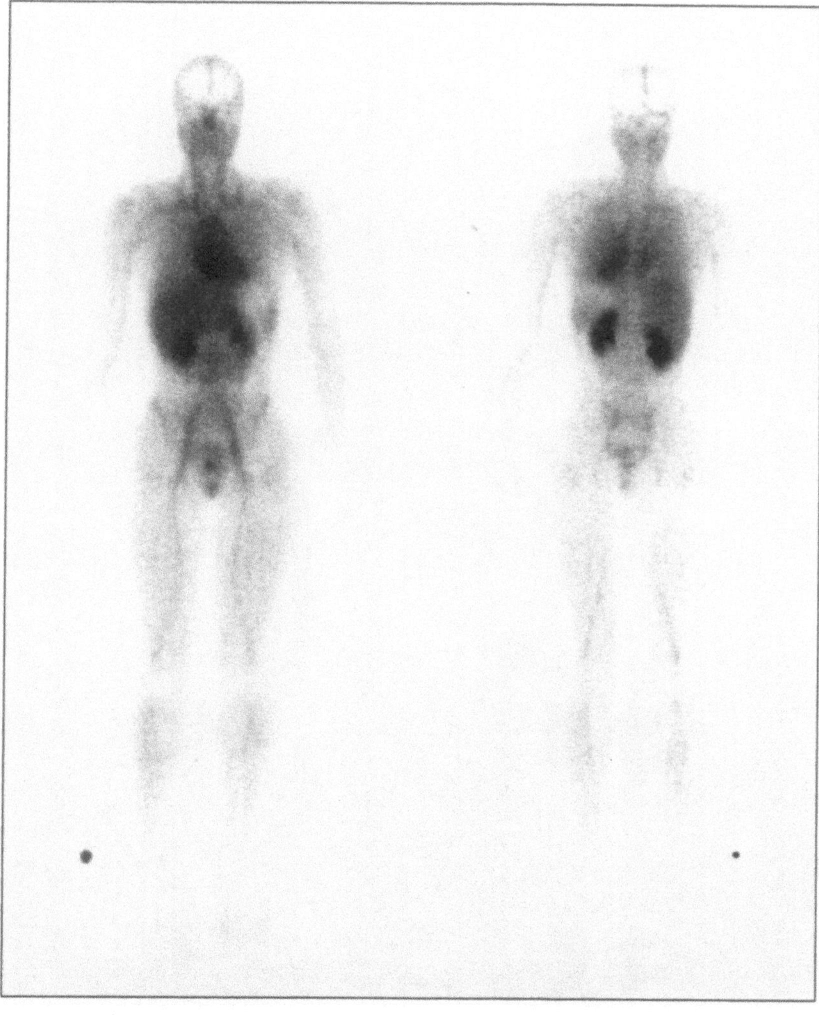

- A double headed whole body gamma camera was used
- Left image is the anterior view
- Right image is the posterior view

Technical Comments

- The photon deficient area between the heart, spleen and left kidney on the anterior view is due to a full stomach
- Marker on child's right side

- A double headed whole body
  gamma camera was used
- Left image is the anterior view
- Right image is the posterior
  view

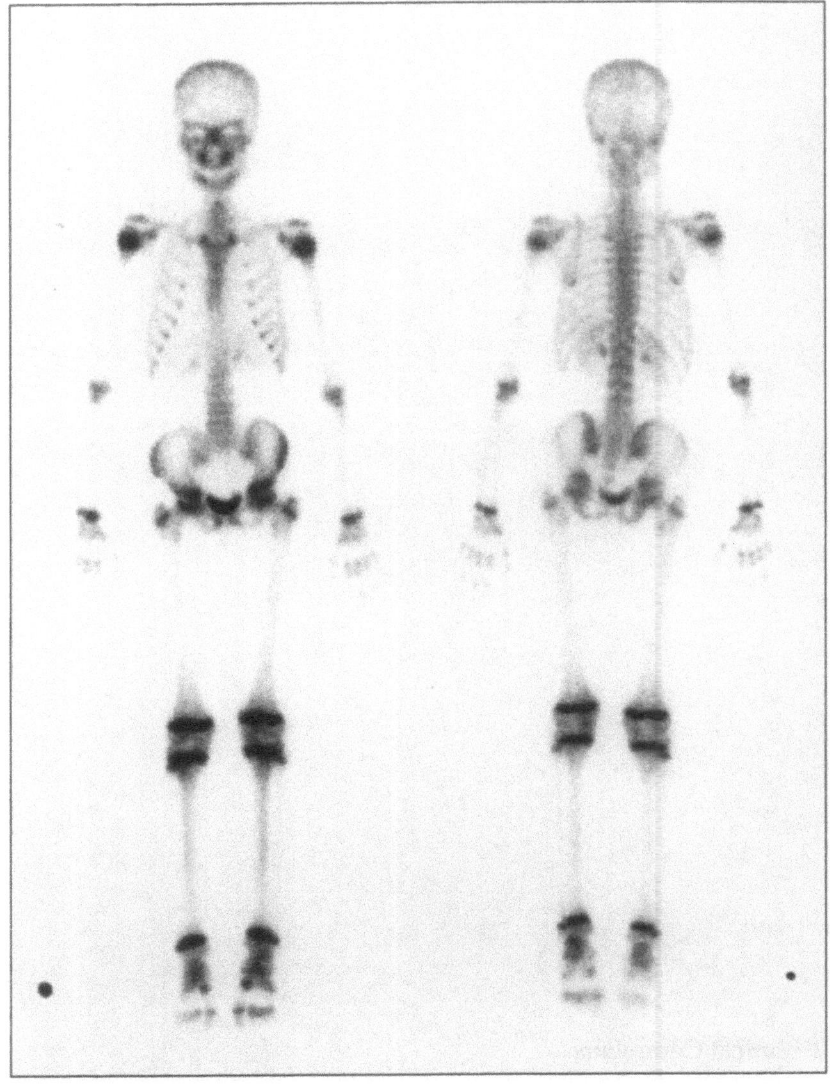

Technical Comment
- Marker on child's right side

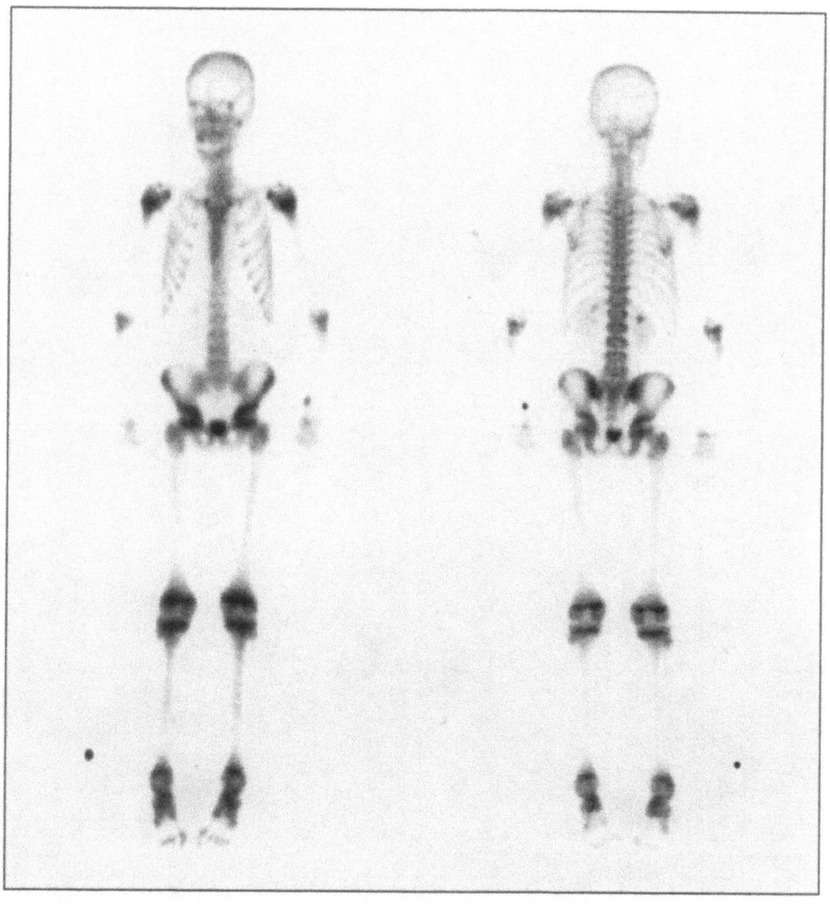

- A double headed whole body gamma camera was used
- Left image is the anterior view
- Right image is the posterior view

Technical Comments
- Note extravasation of isotope at the site of injection in the left hand
- Marker on child's right side

Fig. 1. Anterior view of skull and thorax

Fig. 4. Posterior view of skull and thorax

Fig. 2. Right lateral view of skull and right upper limb

Fig. 5. Left lateral view of skull and left upper limb

Fig. 3. Anterior view of skull and thorax

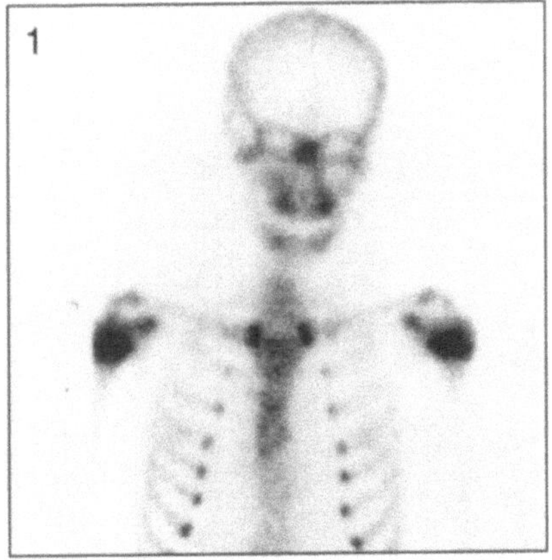

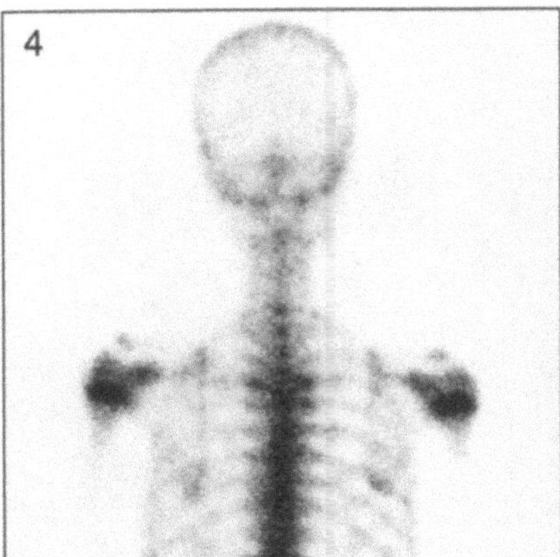

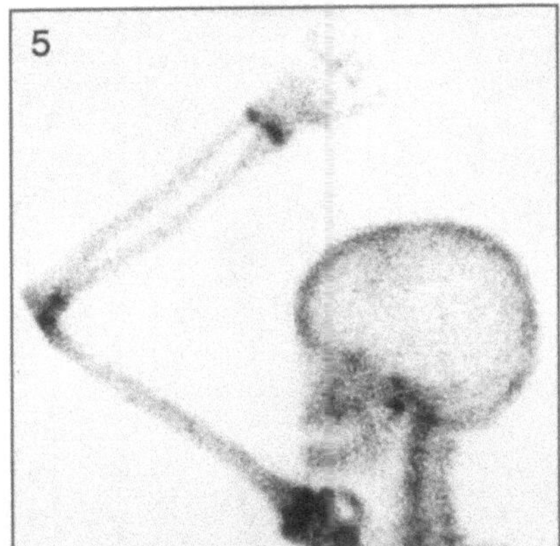

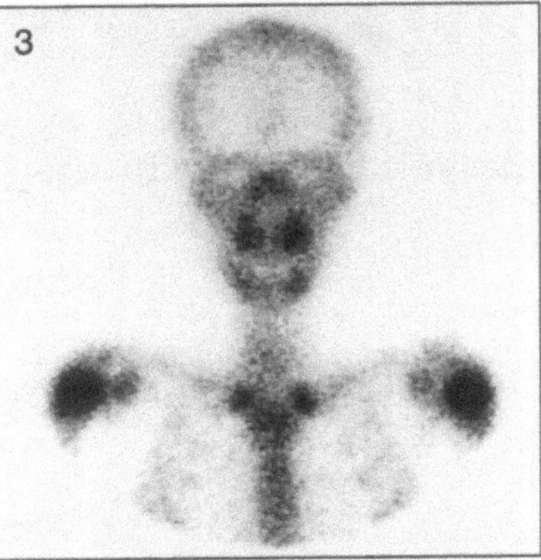

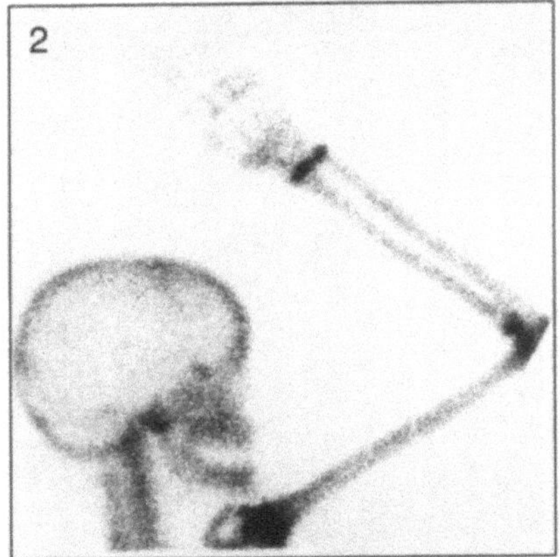

Technical Comments

- On the posterior view of thorax in Fig. 4 the angles of the scapulae show up as focal areas of increased activity especially on the right, this is normal
- Focal increased activity is noted to the left of the midline in the mandible in Fig. 3. The cause for this is uncertain but may be related to the teeth
- The distal end of the humerus has been excluded from the field of view in Figs. 2 and 5. This is due to the difficulty in positioning both the skull and upper limb at the same time in children at this age
- The lateral views of the skull (Figs. 2 and 5) were taken posteriorly

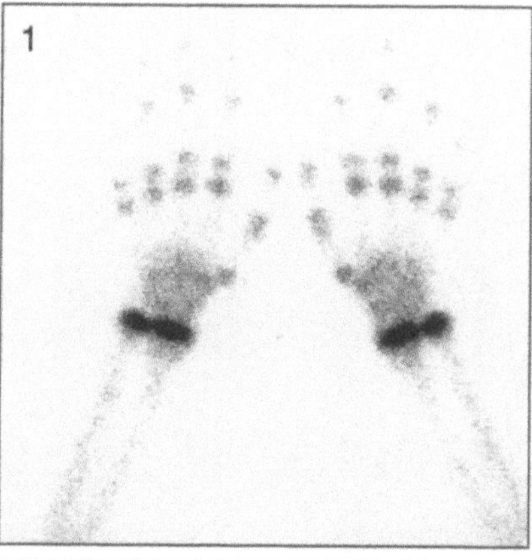

Fig. 1. Anterior view of hands

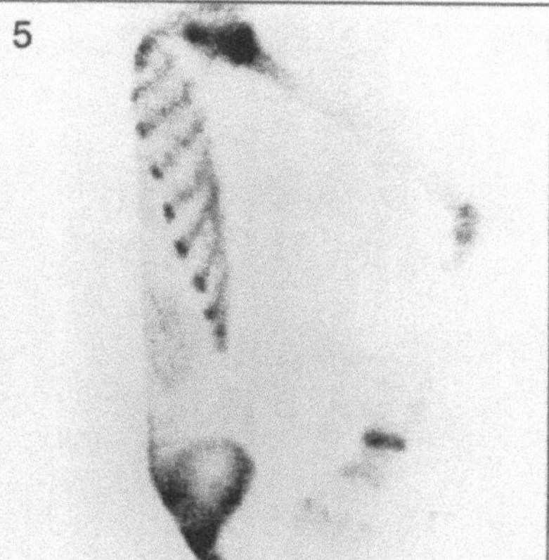

Fig. 2. Anterior view of part right thorax and right upper limb

Fig. 5. Anterior view of part left thorax and left upper limb

Technical Comment

– Figs. 2 and 5 show the difficulty of obtaining high quality images of the upper limbs adjacent to the thorax. Compare with p. 159

Fig. 1. Anterior view of thorax

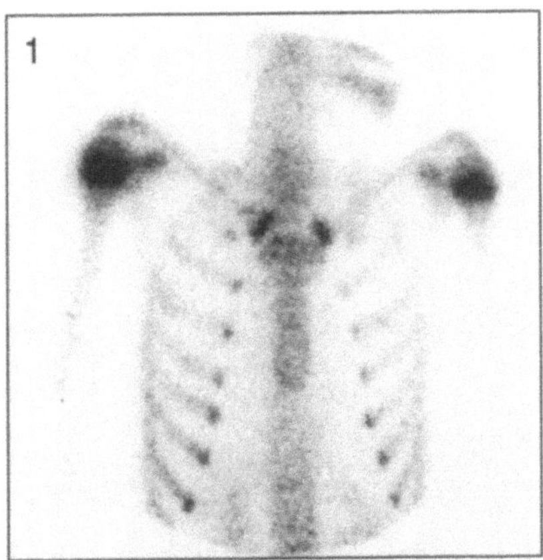

Fig. 2. Anterior view of thorax, spine and part pelvis

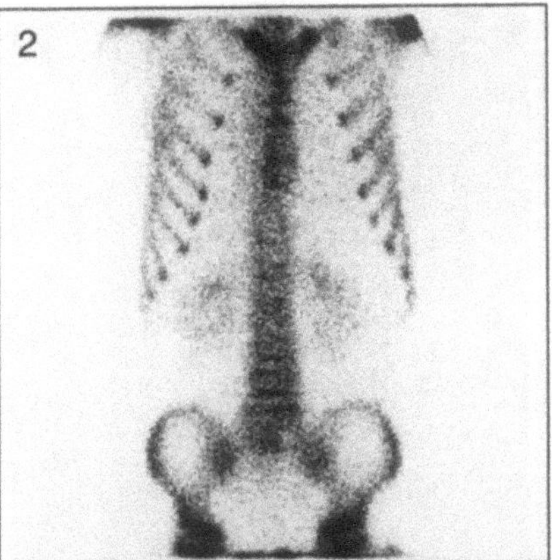

Fig. 3. Anterior view of thorax and spine

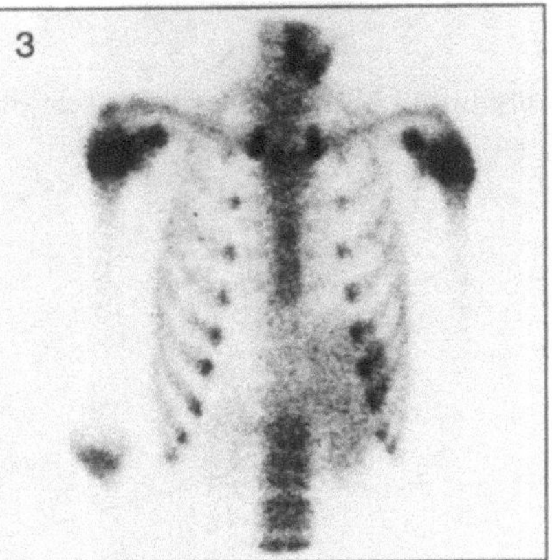

Technical Comments

– The unusual activity in the neck and in the left hypochondrium in Fig. 3 suggests 99m-Tc-pertechnetate in both the stomach and thyroid gland
– The clarity of the lower lumbar spine on the anterior views in Figs. 2 and 3 is again noted

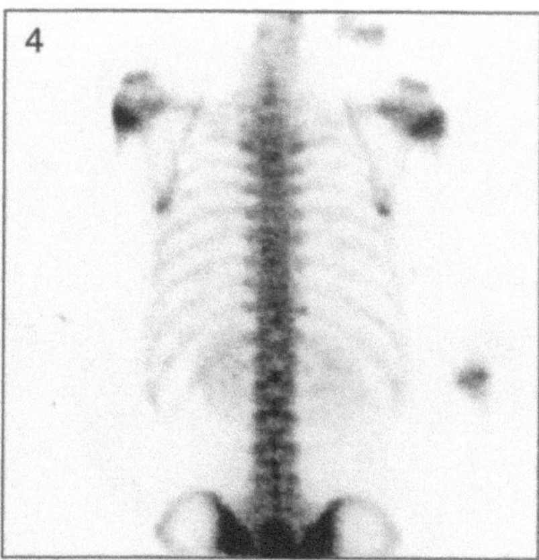

Fig. 1. Posterior view of thorax and spine

Fig. 4. Posterior view of thorax and spine

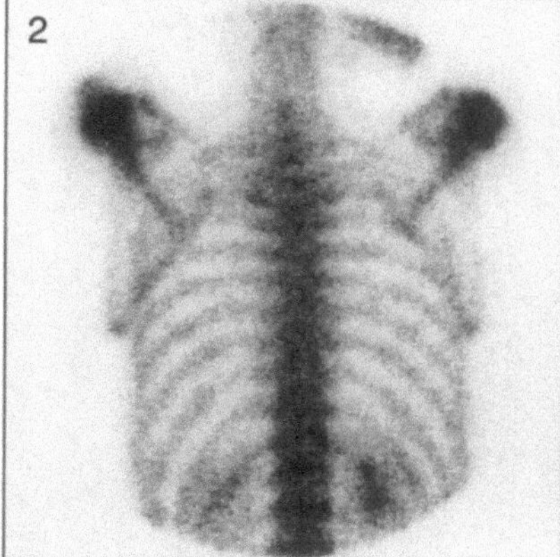

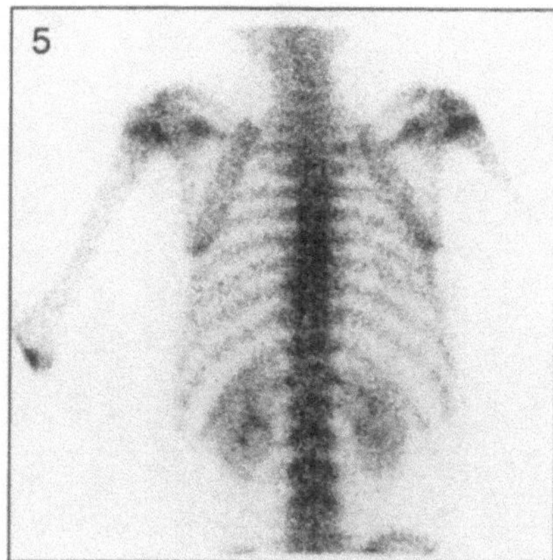

Fig. 2. Posterior view of thorax and spine

Fig. 5. Posterior view of thorax and spine

Technical Comments
- The scapulae appear different due to the varying positions of the hume- ri in these images
- There is retention of isotope in the right kidney in Fig. 2

Fig. 1. Right anterior oblique view of thorax

Fig. 4. Left anterior oblique view of thorax

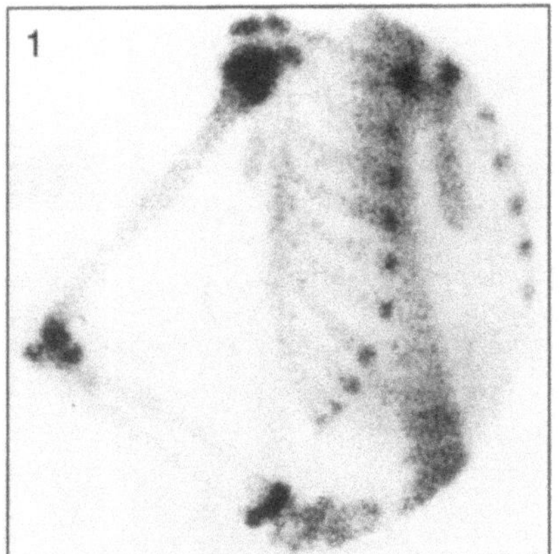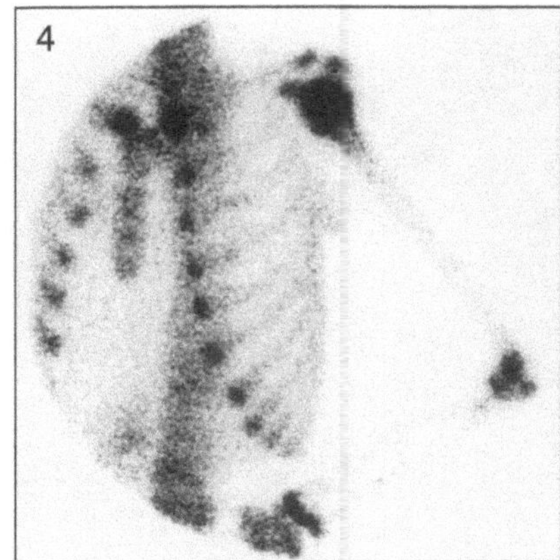

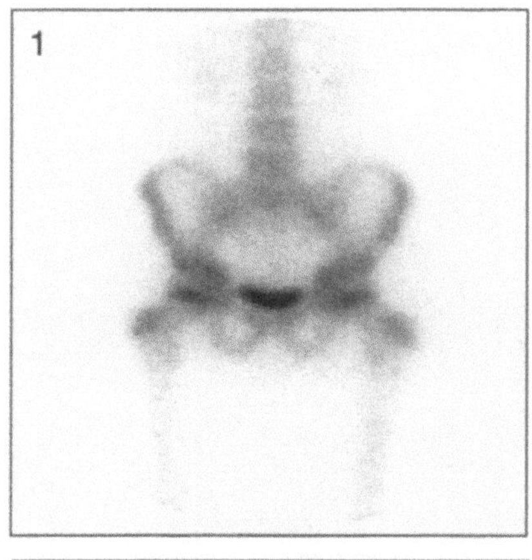

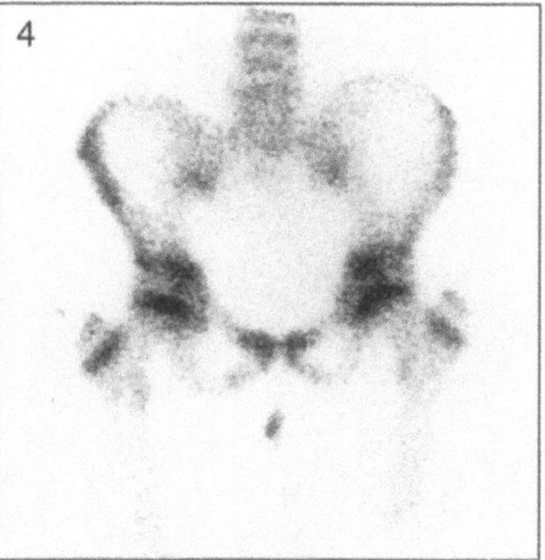

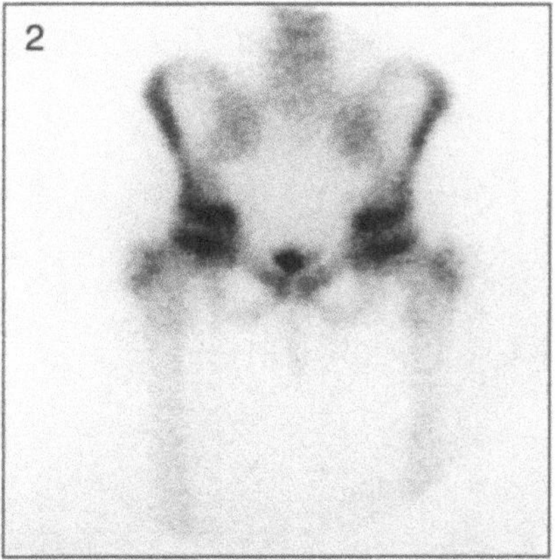

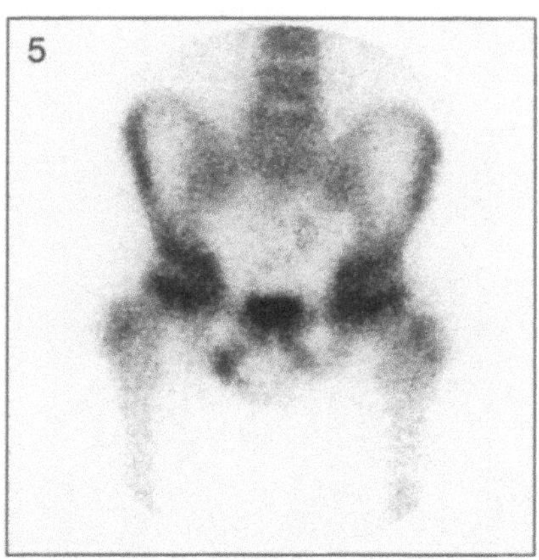

Fig. 1. Anterior view of spine, pelvis and femora

Fig. 4. Anterior view of spine, pelvis and femora

Fig. 2. Anterior view of pelvis and femora

Fig. 5. Anterior view of spine, pelvis and femora

Technical Comment

– Urine contamination below the pelvis is seen in Fig. 4

▶ **Potential Pitfall**

– Focal increased activity is noted in the region of the right synchondrosis in Fig. 5. This represents urine contamination and does not represent pathology in the pubic bone

Fig. 1. Posterior view of spine, pelvis and femora

Fig. 4. Posterior view of spine, pelvis and femora

Fig. 2. Posterior view of spine, pelvis and femora

Fig. 5. Posterior view of spine, pelvis and femora

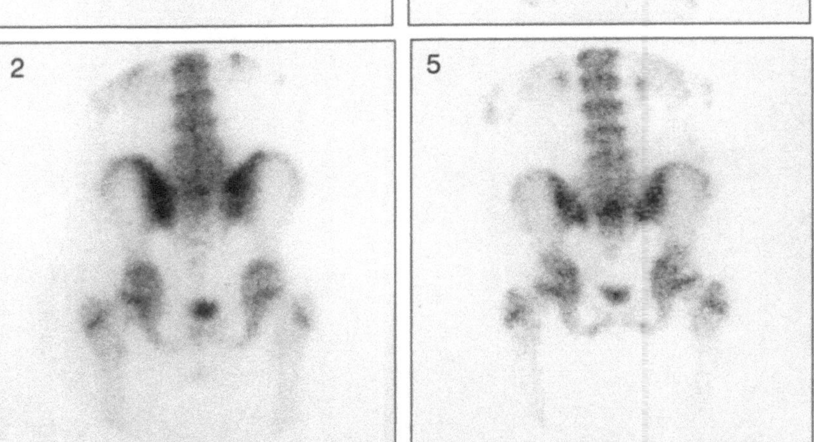

Technical Comment

– Urine contamination below the pelvis is seen in Figs. 1, 2 and 4

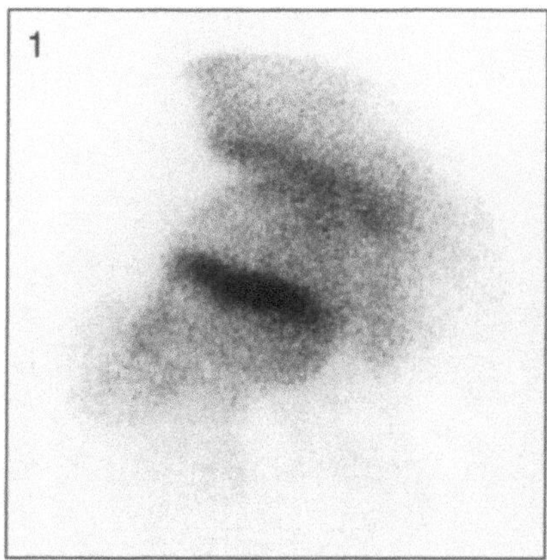 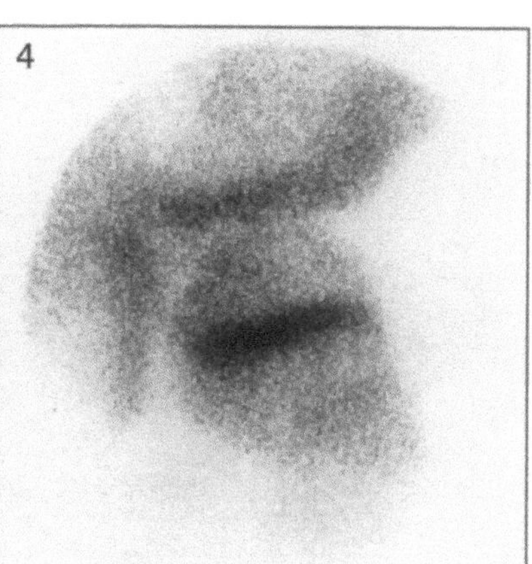

Fig. 1. Pinhole view of right hip

Fig. 4. Pinhole view of left hip

Fig. 1. Posterior view of femora and knees

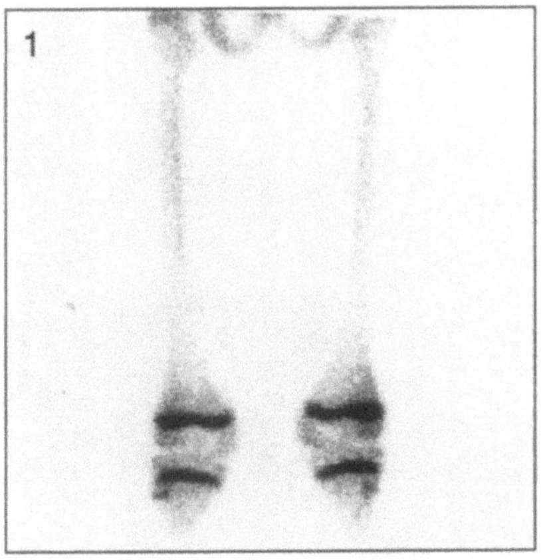

Fig. 4. Posterior view of femora and knees

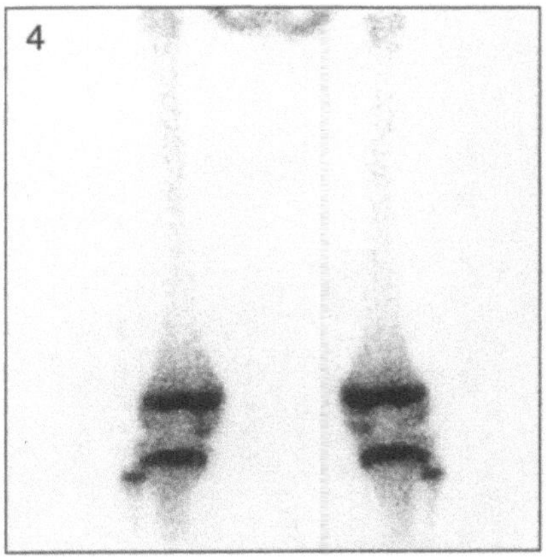

Fig. 2. Posterior view of tibia, fibula and ankles

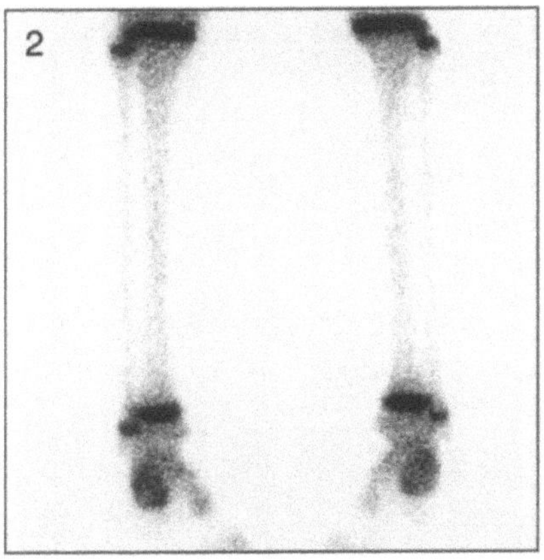

Fig. 5. Posterior view of tibia, fibula and ankles

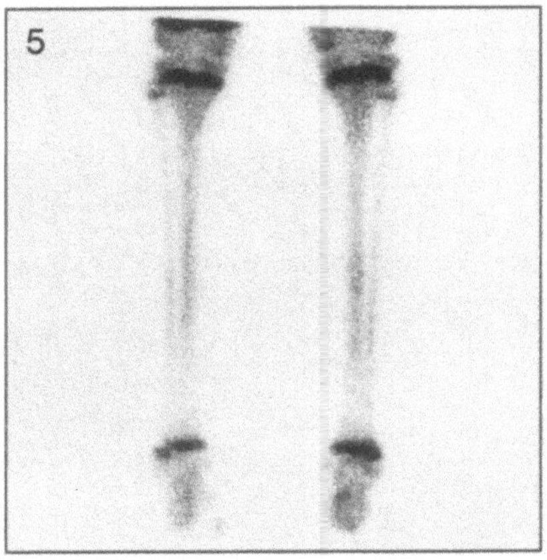

Technical Comment

– Note the rather patchy asymmetrical distribution of the isotope in the mid-portions of the tibia in Fig. 5. Although the authors are unable to offer a physiological explanation, nevertheless the observation has been made frequently with no clinical consequence of this observation

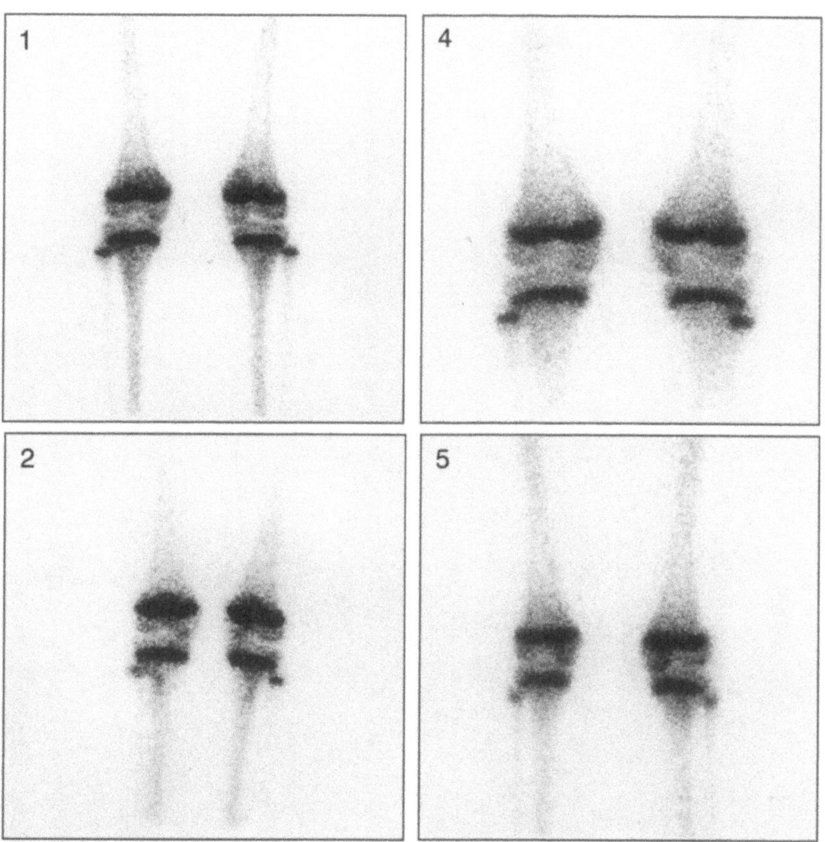

Fig. 1. Posterior view of knees

Fig. 4. Posterior view of knees

Fig. 2. Anterior view of knees

Fig. 5. Posterior view of knees

Technical Comment

– The epiphyseal plate of the femur is not well defined in Fig. 2, the anterior view, because the patella overlies the epiphyseal plate

Fig. 1. Posterior view of tibia, fibula and ankles (Same figure as Fig. 2, p. 198)

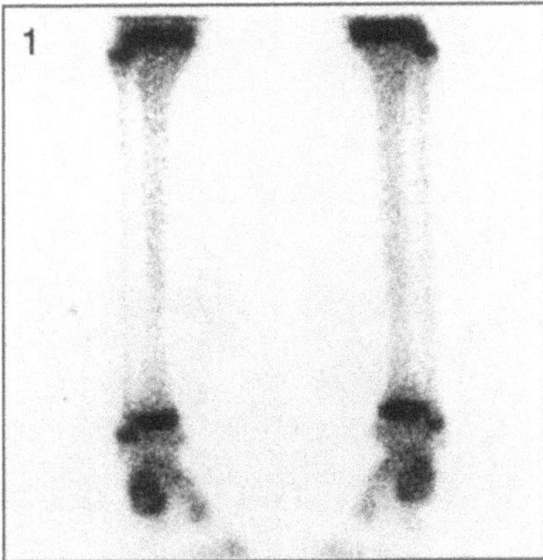

Fig. 2. Anterior view of ankles and feet

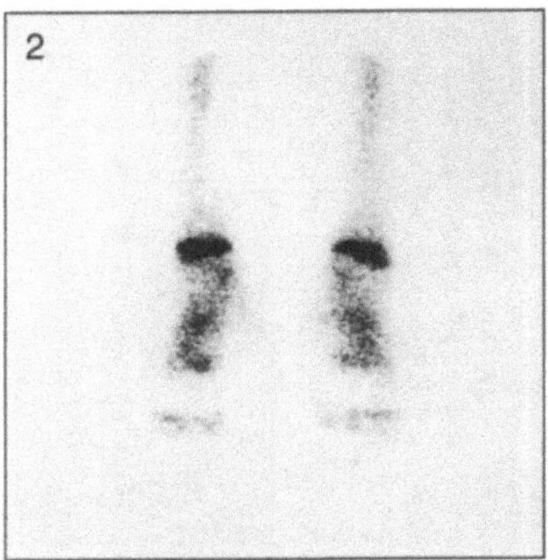

14: Age 12–13 Years

Fig. 1. Posterior view of thorax and spine

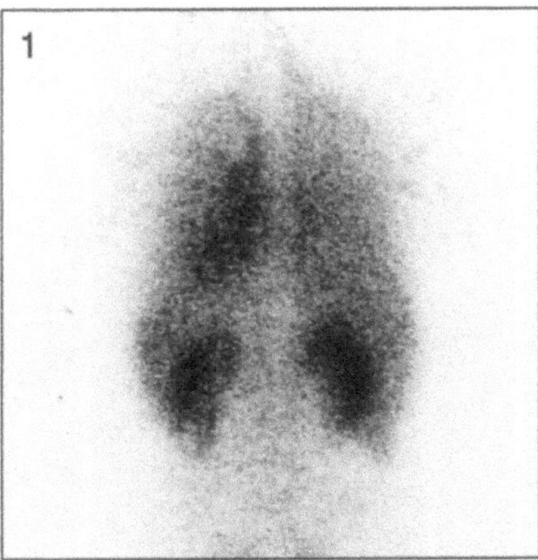

Fig. 2. Anterior view of knees

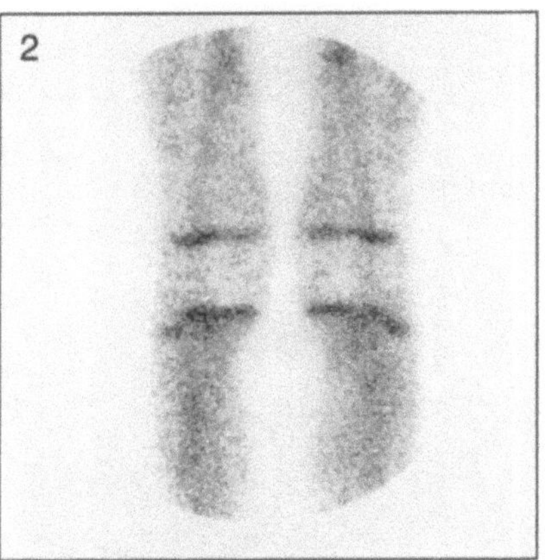

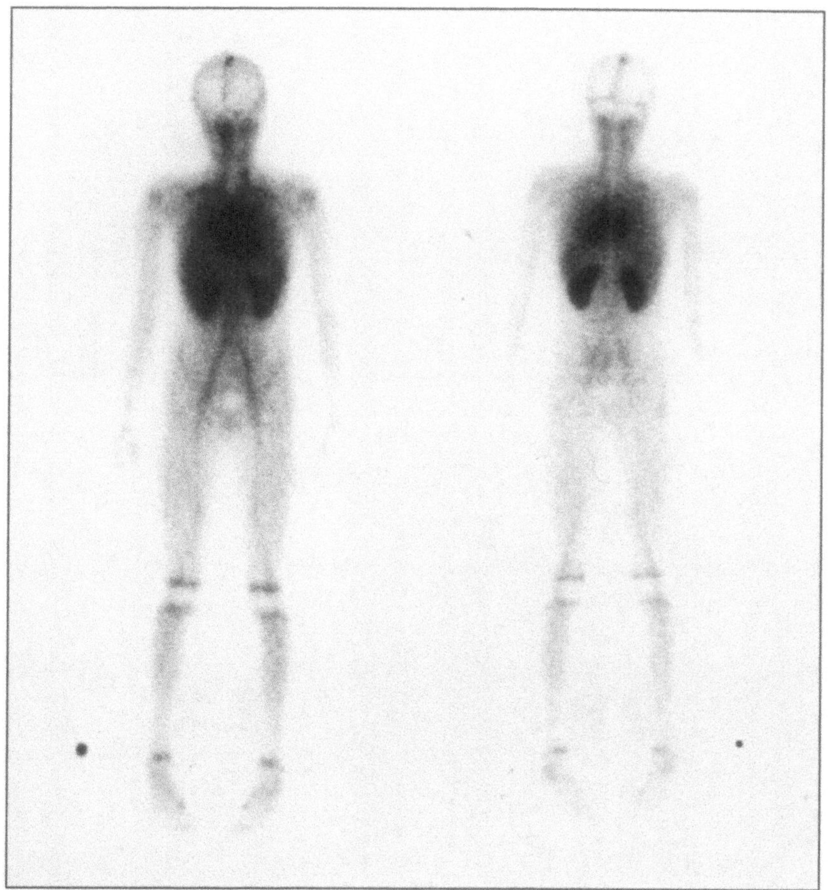

- A double headed whole body
  gamma camera was used
- Left image is the anterior view
- Right image is the posterior
  view

Technical Comment
- Marker on child's right side

– A double headed whole body
  gamma camera was used
– Left image is the anterior view
– Right image is the posterior
  view

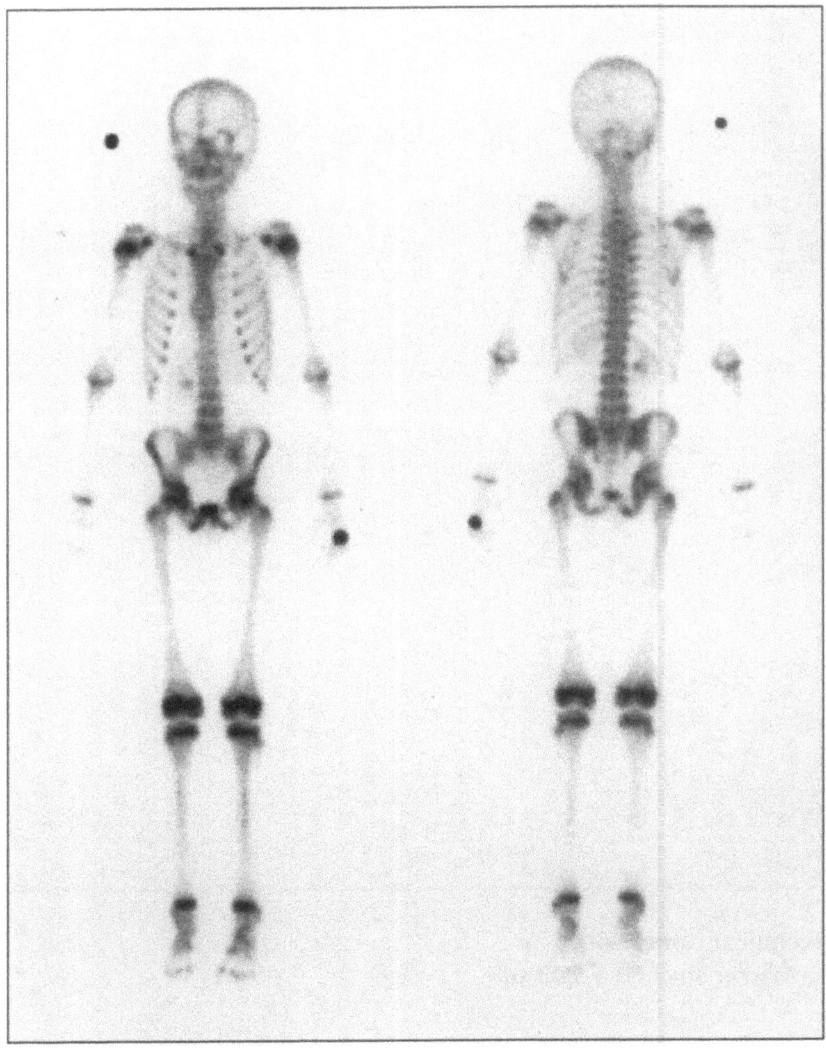

Technical Comments
– Poor positioning of the right foot is noted
– Note extravasation of isotope at the site of injection in the left hand
– Marker on child's right side

▶ **Potential Pitfall**
– Increased activity is noted in the inferior pubic ramus due to the synchondrosis, this should not be mistaken for pathology

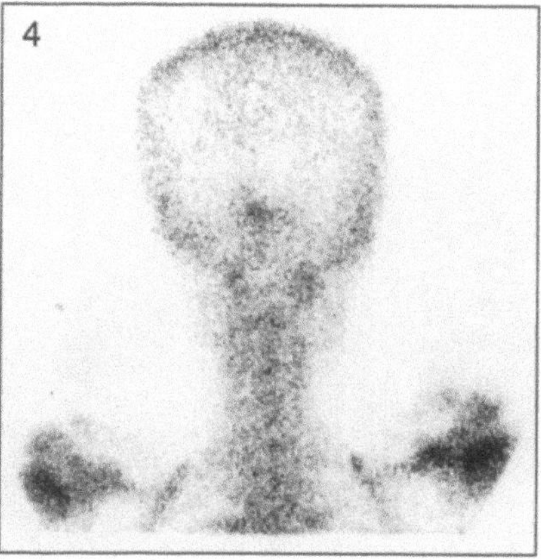

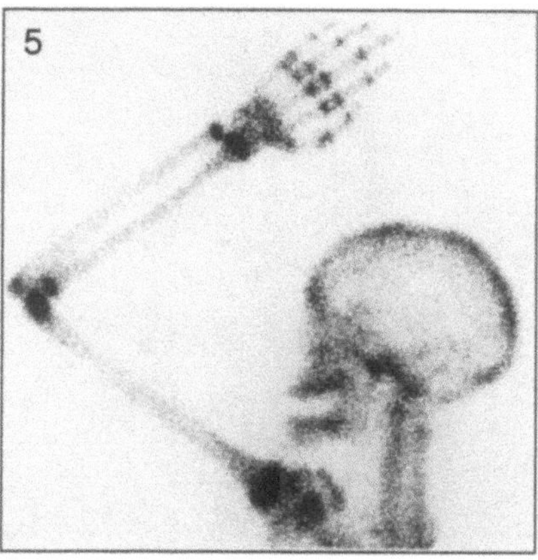

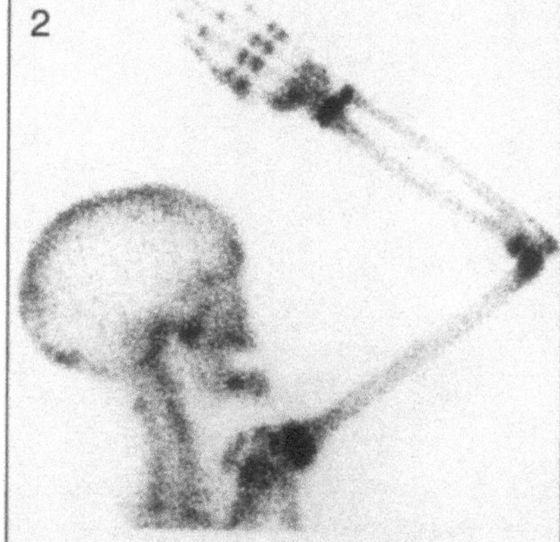

Fig. 1. Anterior view of skull and thorax

Fig. 4. Posterior view of skull

Fig. 2. Right lateral view of skull and right upper limb

Fig. 5. Left lateral view of skull and left upper limb

Technical Comment
– The lateral views of the skull (Figs. 2 and 5) were taken posteriorly

Fig. 1. Anterior view of skull and thorax

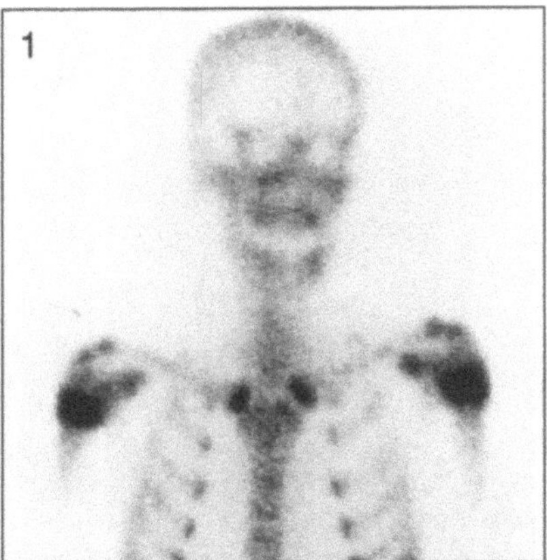

Fig. 2. Right lateral view of skull and posterior view of thorax

Fig. 5. Left lateral view of skull and posterior view of thorax

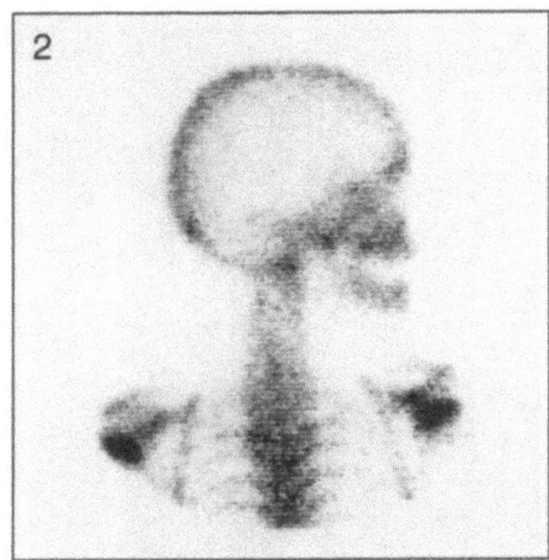

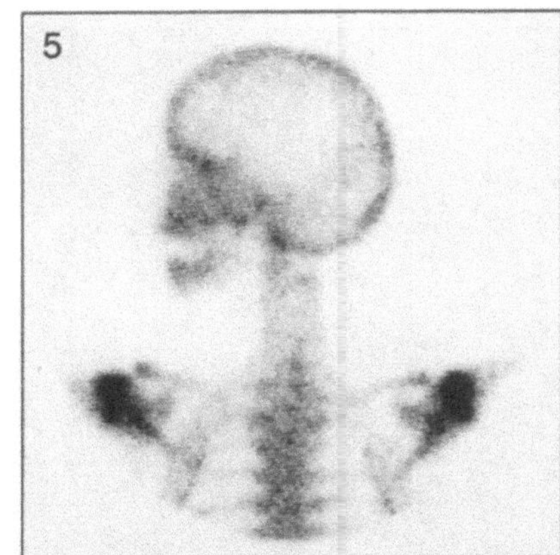

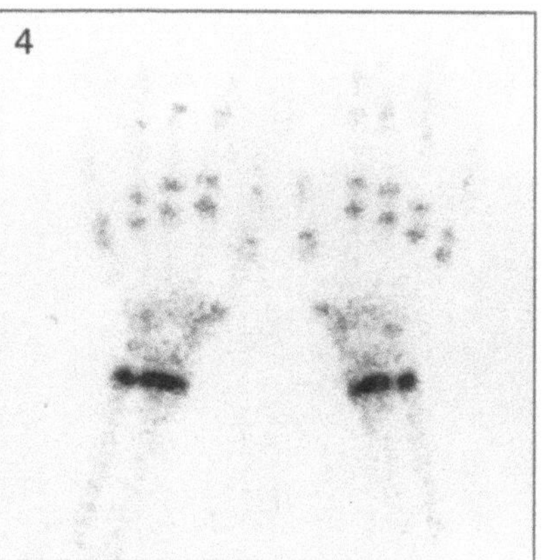

Fig. 1. Anterior view of hands

Fig. 4. Anterior view of hands

Fig. 1. Anterior view of thorax and spine

Fig. 4. Anterior view of thorax and spine

Fig. 2. Anterior view of thorax and spine

Fig. 5. Anterior view of thorax

Fig. 3. Left anterior oblique view of thorax

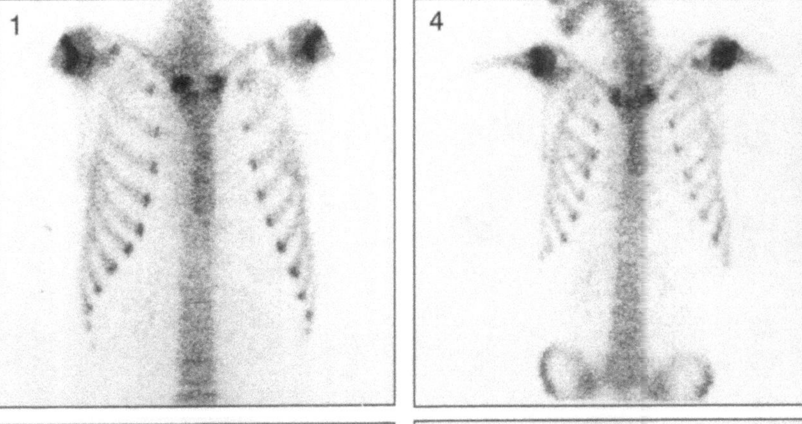

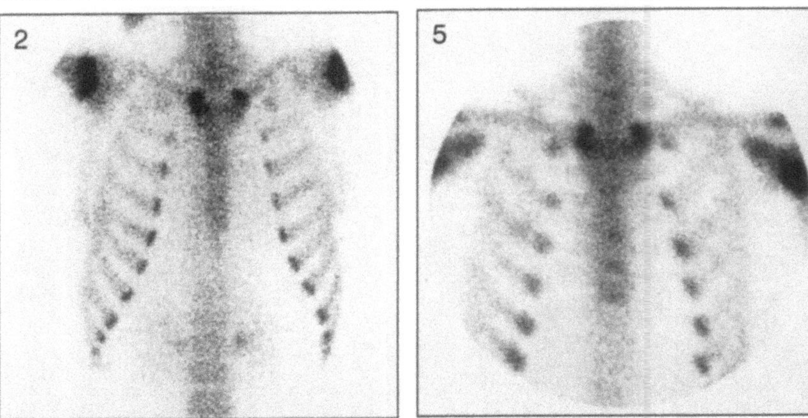

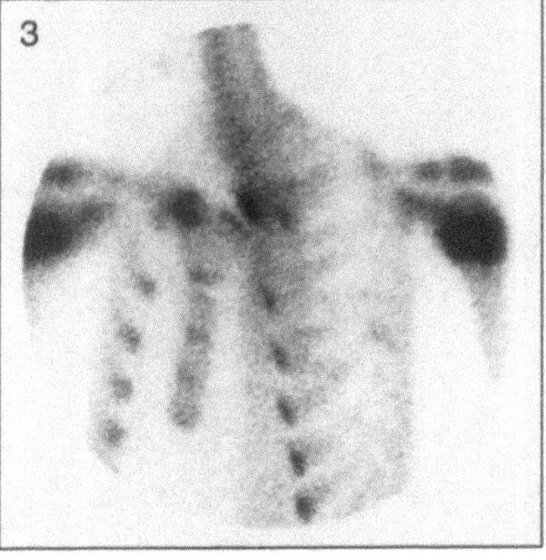

Technical Comment

– Note the variation of the sternum in these different anterior and anterior oblique projections

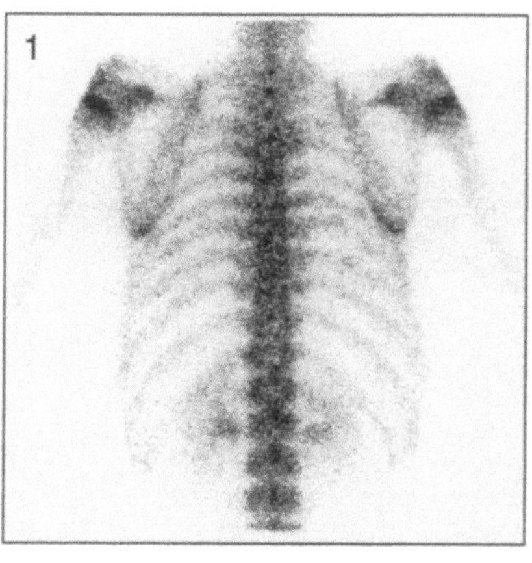

Fig. 1. Posterior view of thorax and spine

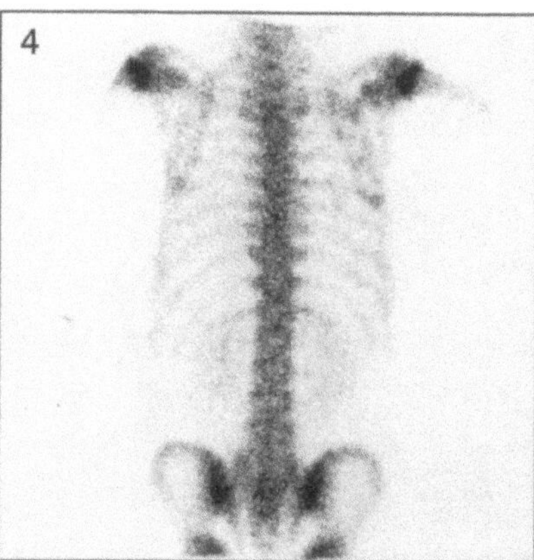

Fig. 4. Posterior view of thorax, spine and part pelvis

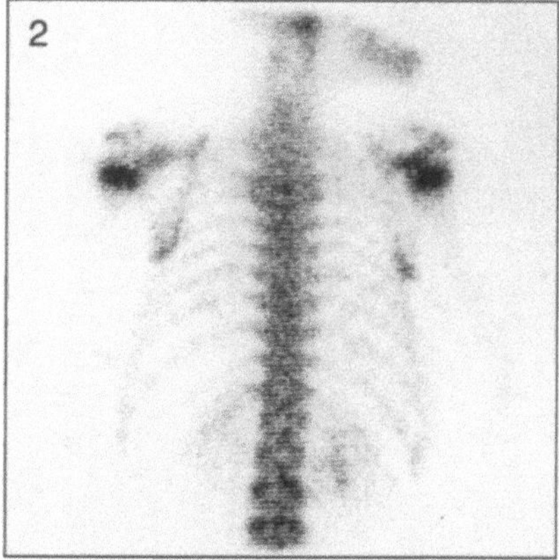

Fig. 2. Posterior view of thorax and spine

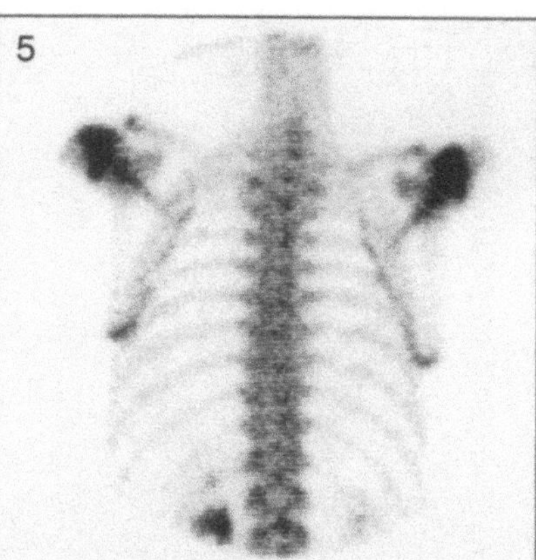

Fig. 5. Posterior view of thorax and spine

Technical Comments

- Asymmetrical activity is noted in the renal pelvis in Fig. 5, this is within normal limits
- The appearances of the scapulae vary with the positioning of the upper limbs

Fig. 1. Anterior view of spine and pelvis

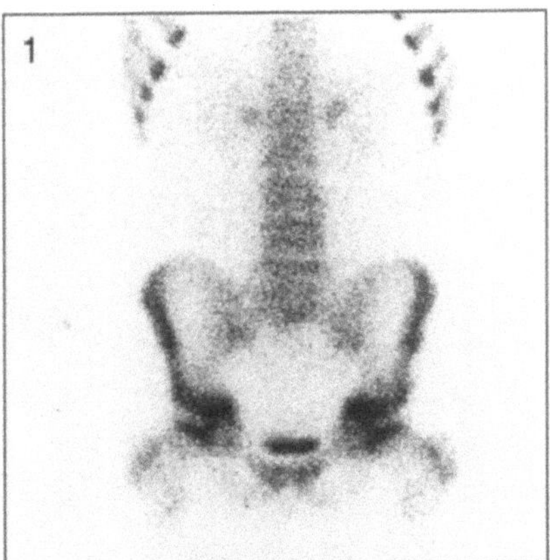

Fig. 2. Anterior view of spine and pelvis

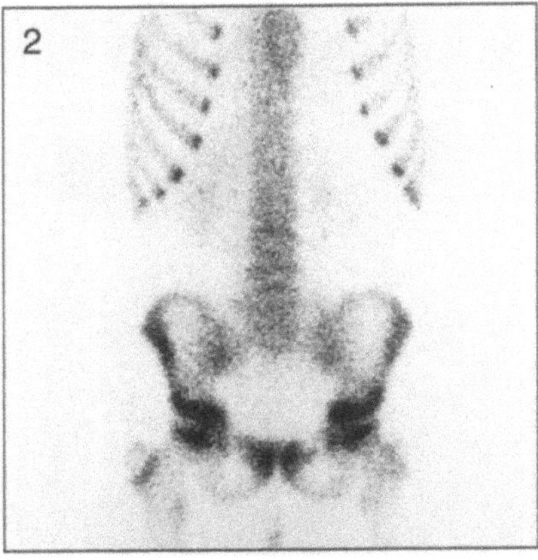

Fig. 3. Anterior view of spine and pelvis

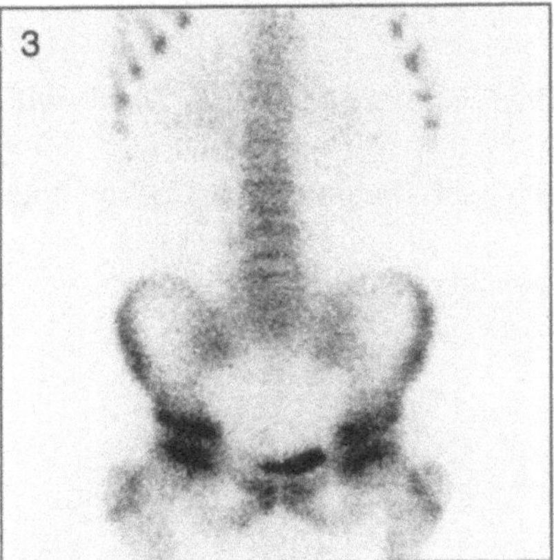

Technical Comments
- Note the clarity of the lower lumbar spine on these anterior views
- Urine contamination below the pelvis is seen in Fig. 2

▶ **Potential Pitfall**
- The pubic rami show normal increased uptake of tracer in all images

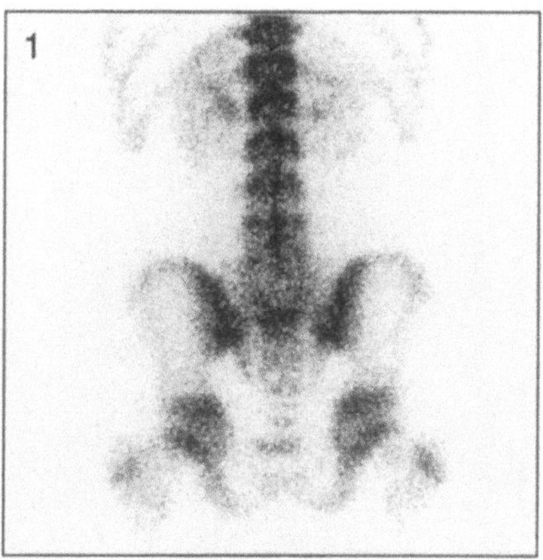

Fig. 1. Posterior view of spine and pelvis

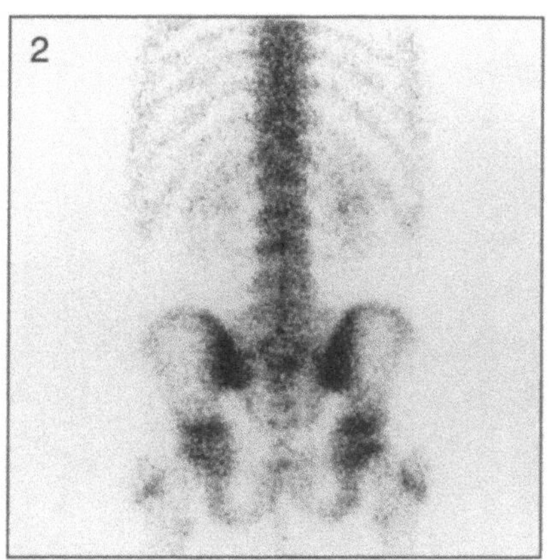

Fig. 2. Posterior view of spine and pelvis

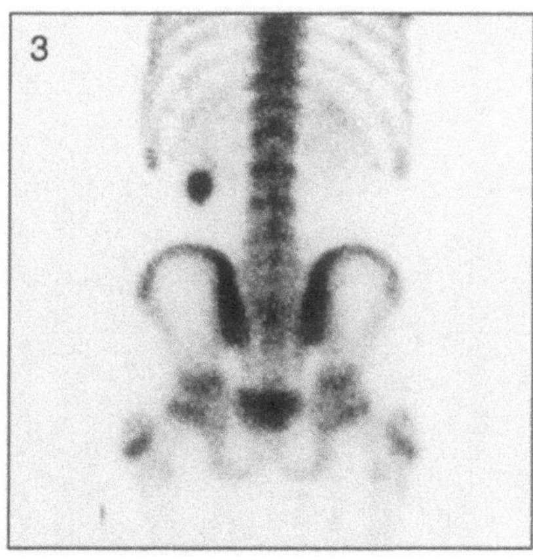

Fig. 3. Posterior view of spine and pelvis

Technical Comments

- Note the isotope in the left renal pelvis seen in Fig. 3 suggesting dilation of the renal pelvis
- Urine contamination below the pelvis is seen in Fig. 2

Fig. 1. Pinhole view of right hip

Fig. 4. Pinhole view of left hip

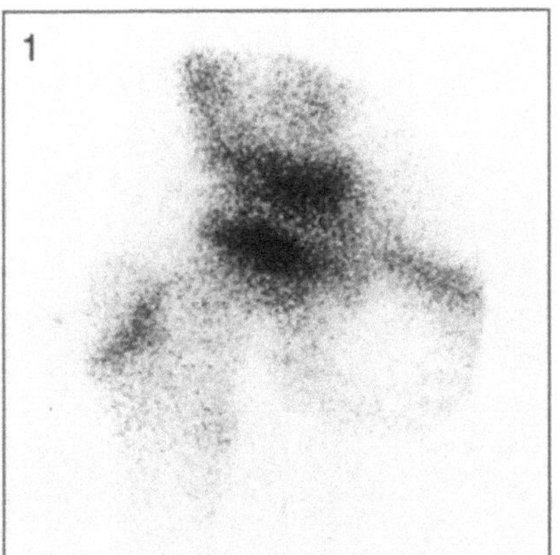
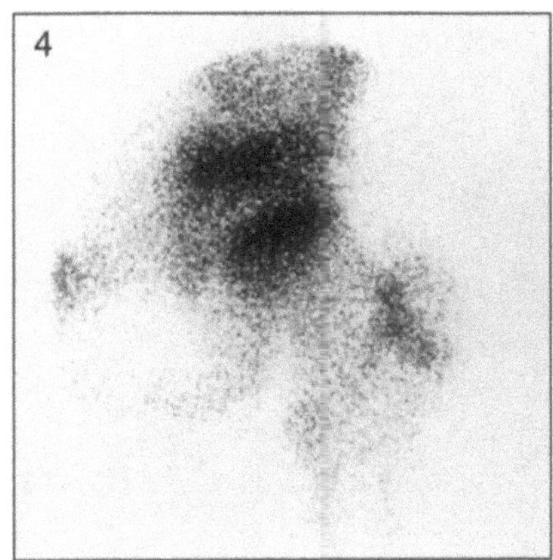

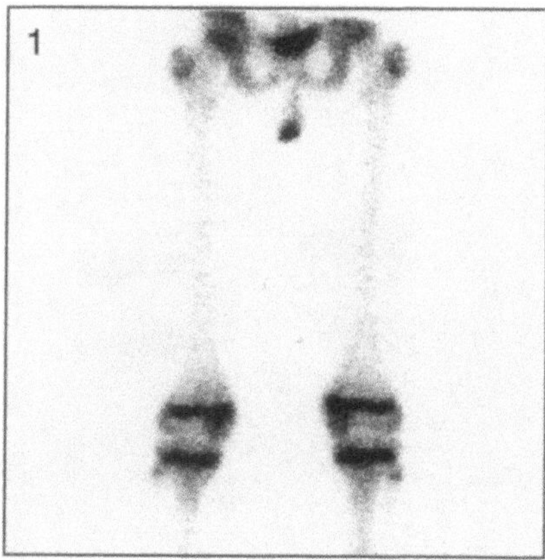

Fig. 1. Posterior view of femora and knees

Fig. 2. Posterior view of knees, tibia, fibula and ankles

Technical Comments
- Slight increased activity is noted in the left mid tibia in Fig. 2, similar to the appearances in Fig. 5, p. 198
- Urine contamination below the pelvis is seen in Fig. 1

Fig. 1. Anterior view of knees

Fig. 4. Anterior view of knees

Fig. 2. Anterior pinhole view of right knee

Fig. 5. Anterior pinhole view of left knee

Fig. 3. Lateral view of left knee

Fig. 6. Lateral view of right knee

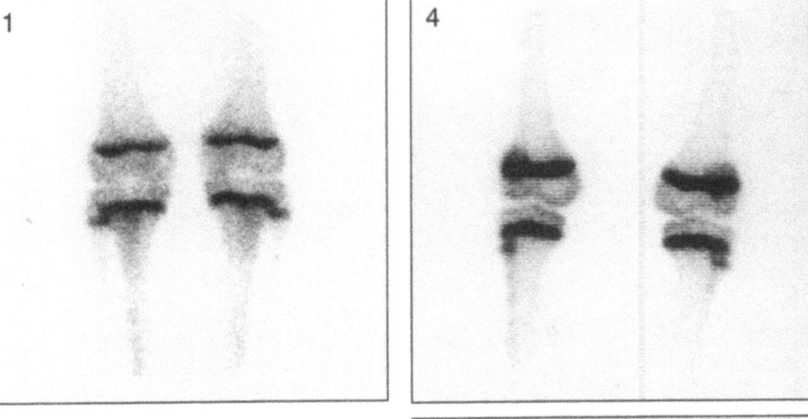

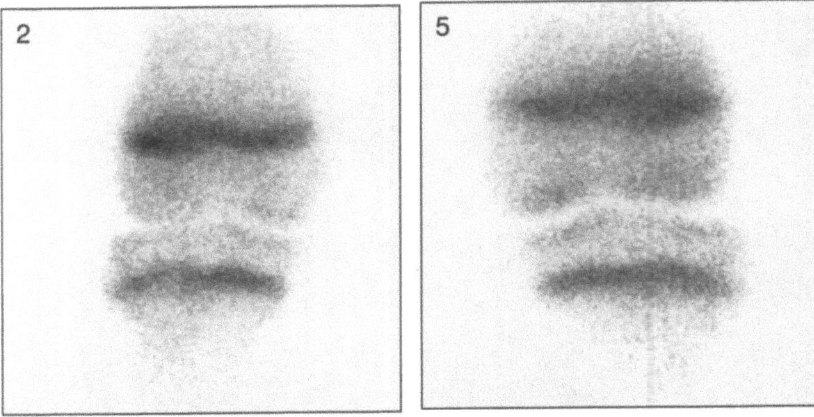

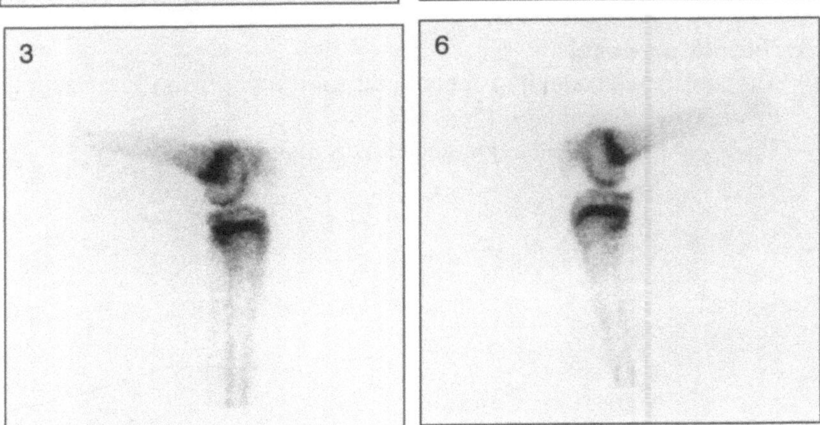

Technical Comment

– The fibula is not seen in Figs. 3 and 6 because the images are medial lateral views

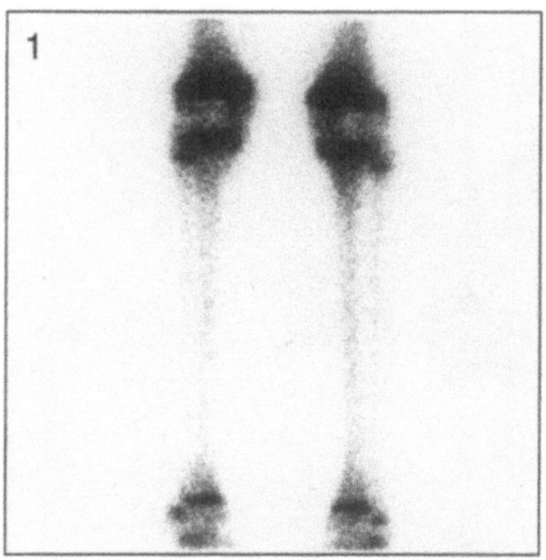

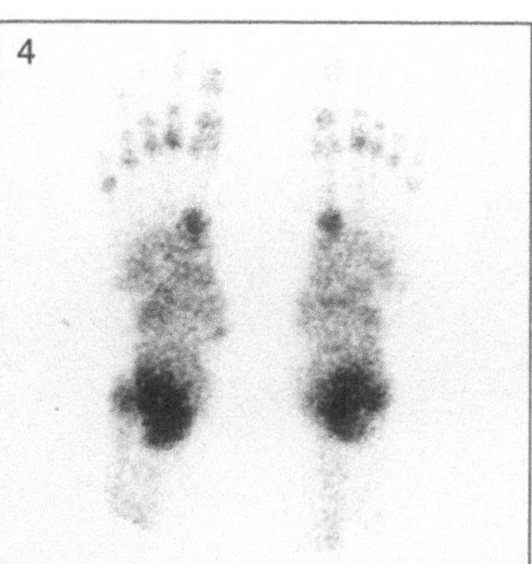

Fig. 1. Posterior view of knees, tibia, fibula and ankles

Fig. 4. Anterior view of feet

► **Potential Pitfall**
  - The lack of clarity of the epiphyseal plates around the knees in Fig. 1 is due to overexposure during the acquisition

15: Age 13–14 Years

Fig. 1. Posterior view of skull and upper limbs

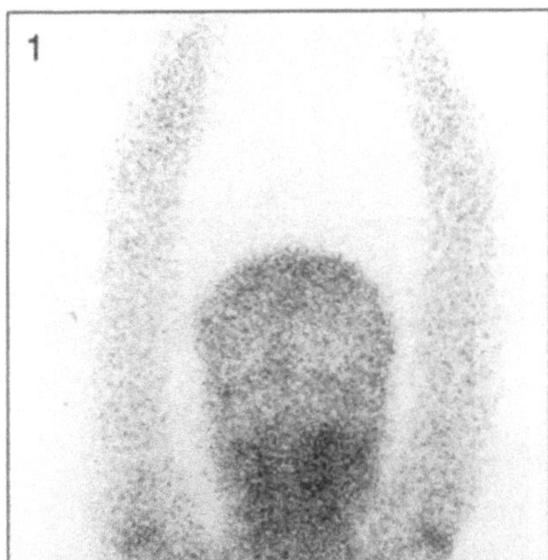

Fig. 2. Anterior view of spine and pelvis

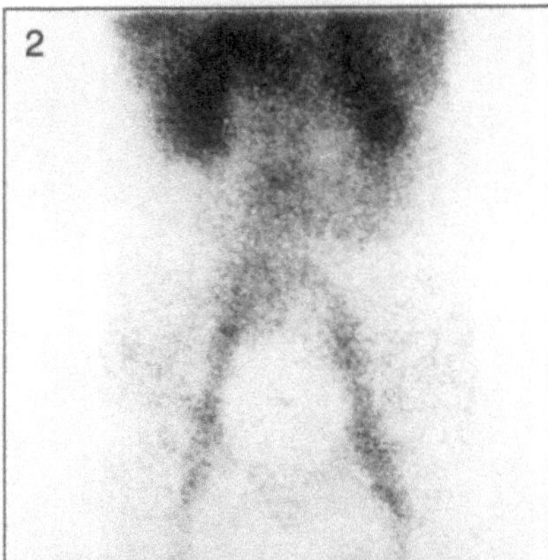

Fig. 3. Anterior view of knees

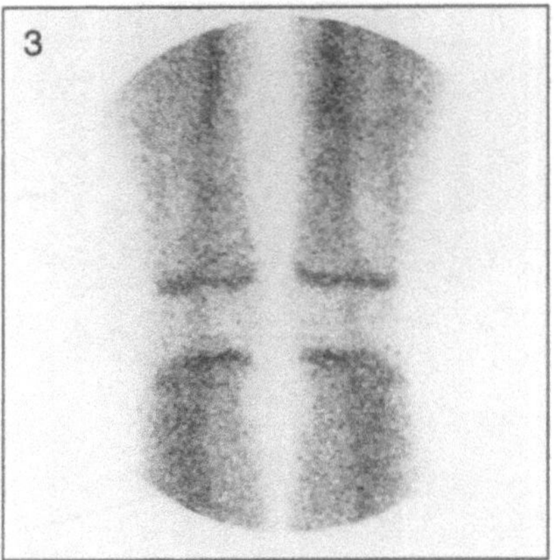

Technical Comment
– The full bladder is seen as a photon deficient area in Fig. 2

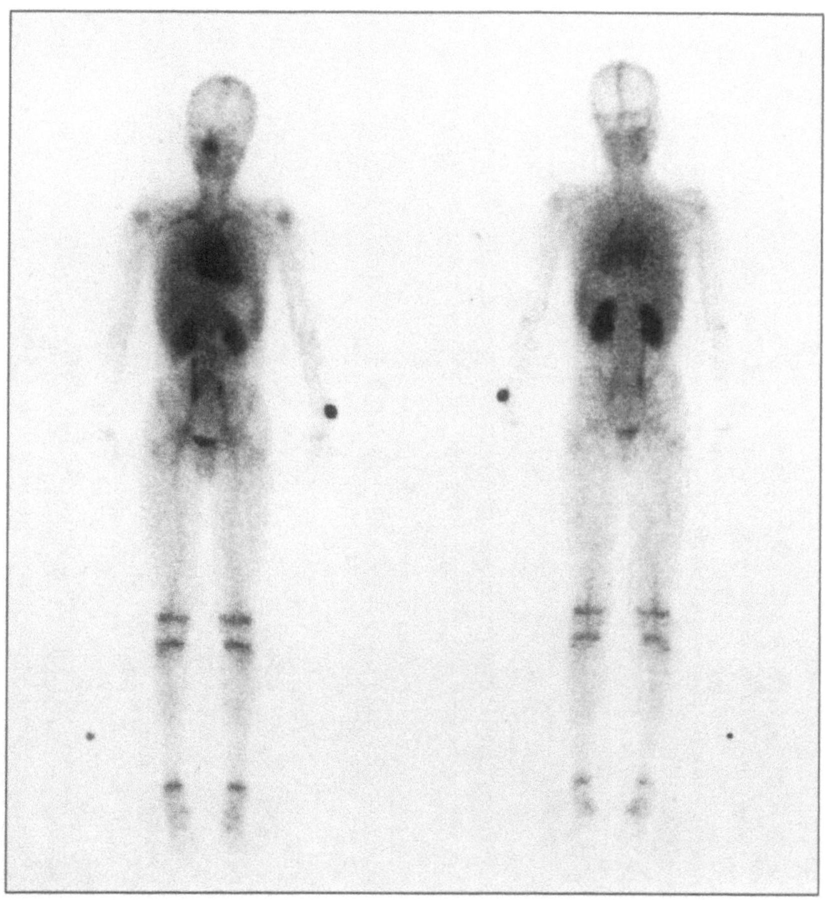

- A double headed whole body gamma camera was used
- Left image is the anterior view
- Right image is the posterior view

Technical Comments
- Note extravasation of isotope at the site of injection in the left hand
- Marker on child's right side

– A double headed whole body
  gamma camera was used
– Left image is the anterior view
– Right image is the posterior
  view

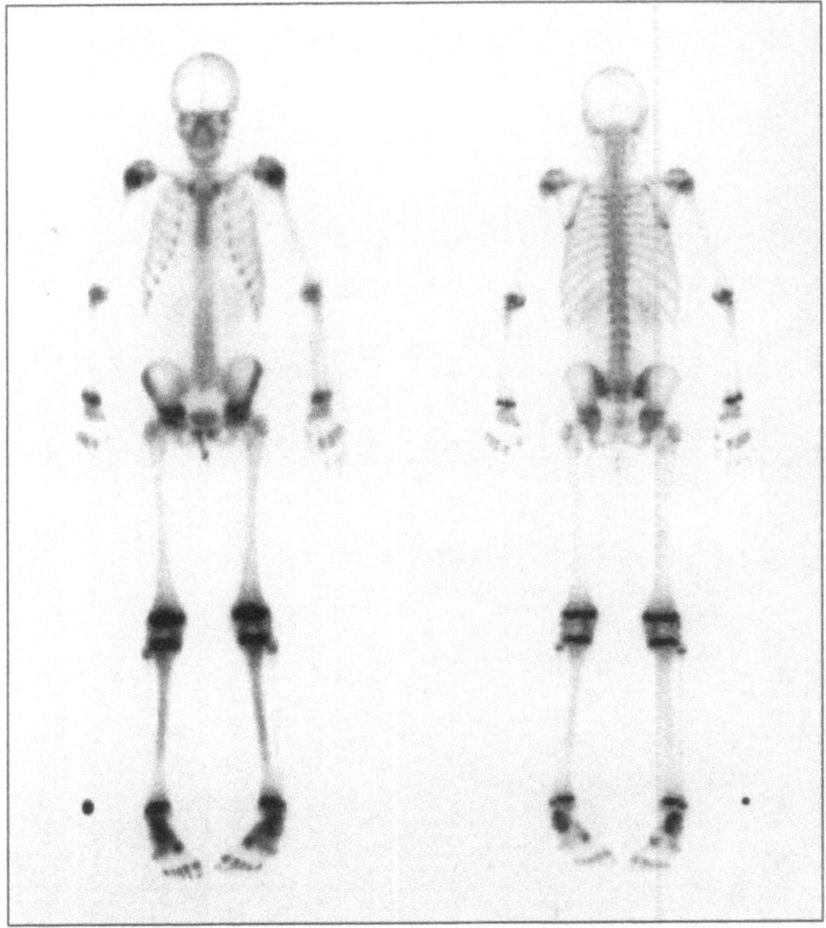

Technical Comments
– Urine contamination below the pelvis is seen
– Marker on child's right side

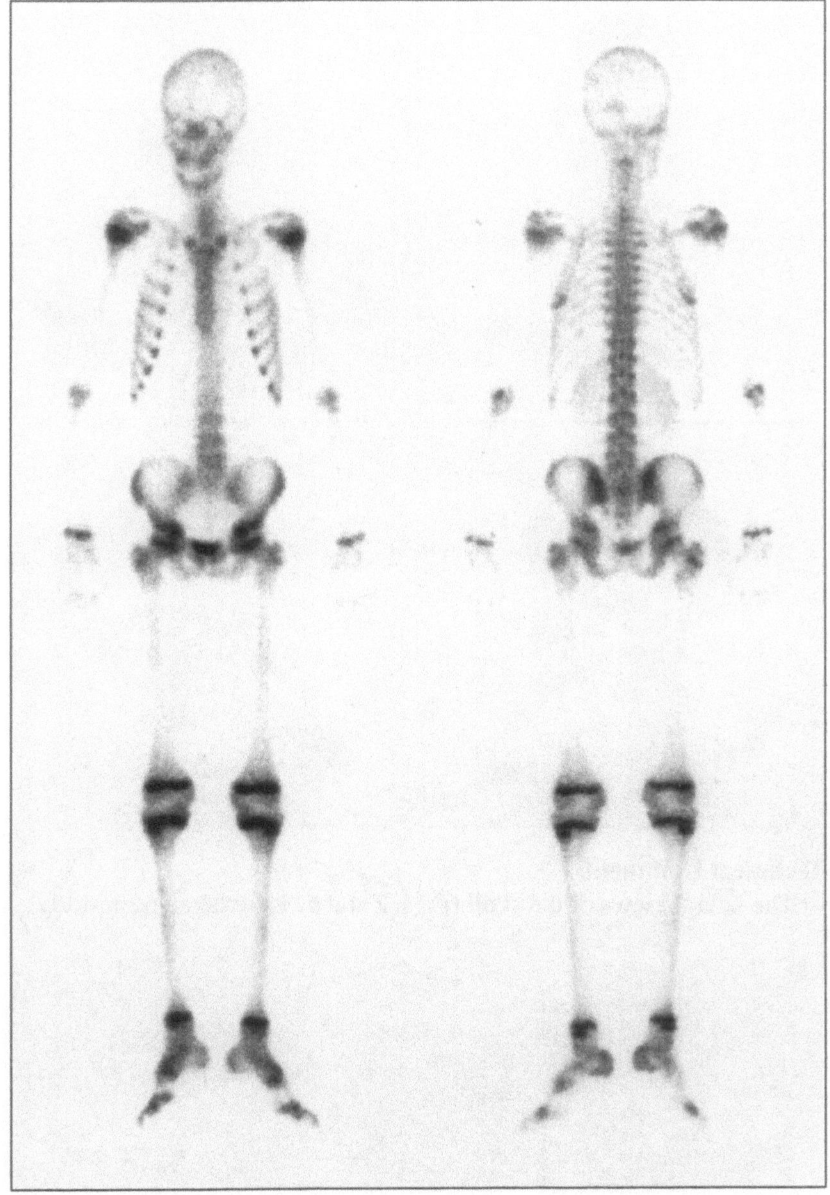

- A double headed whole body
  gamma camera was used
- Left image is the anterior view
- Right image is the posterior
  view

Technical Comment
- Both feet are poorly positioned

Fig. 1. Anterior view of skull

Fig. 4. Posterior view of skull

Fig. 2. Right lateral view of skull

Fig. 5. Left lateral view of skull

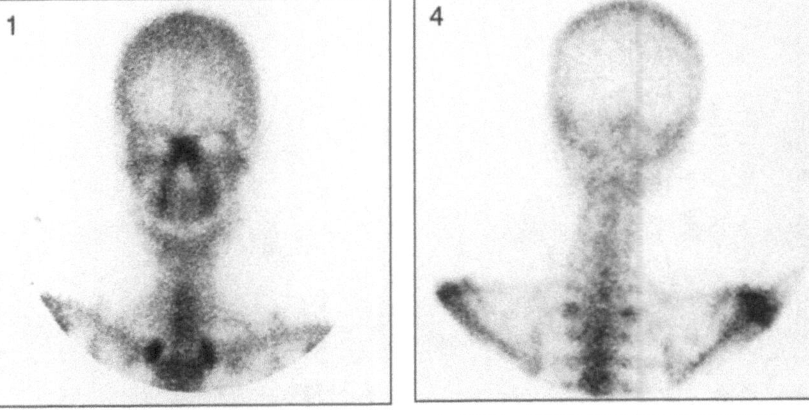

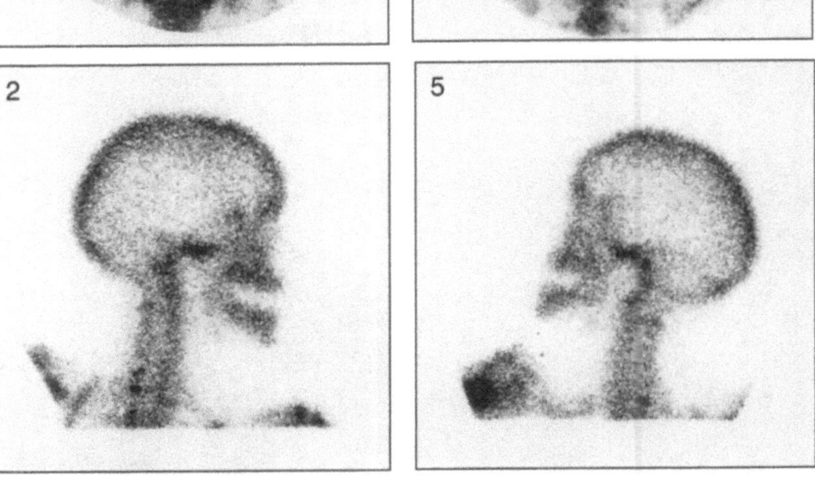

Technical Comment
– The lateral views of the skull (Figs. 2 and 5) were taken posteriorly

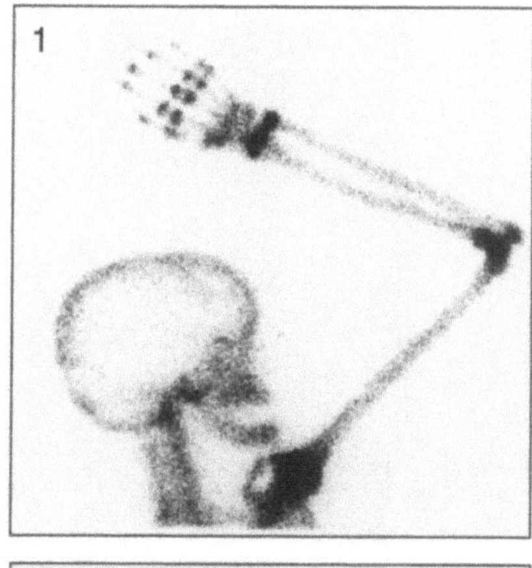

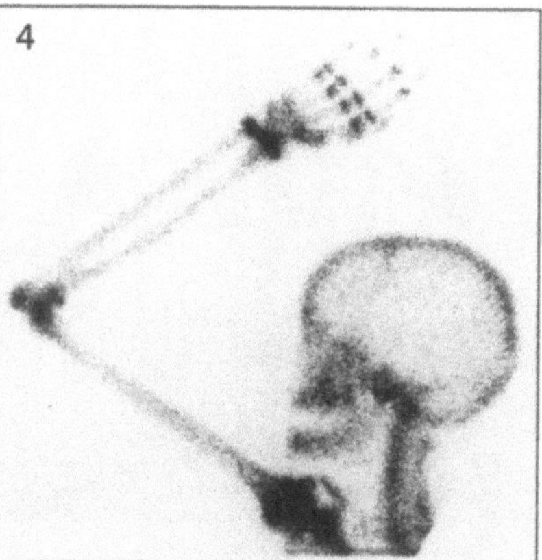

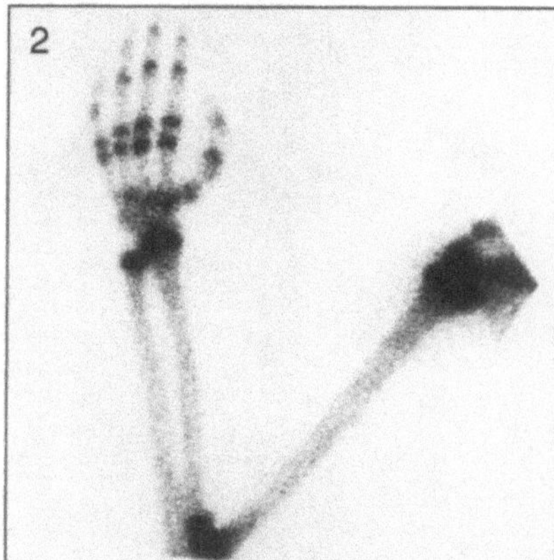

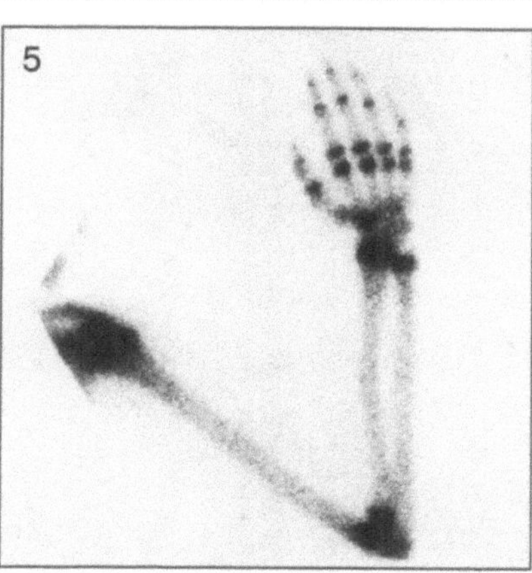

Fig. 1. Right lateral view of skull and right upper limb

Fig. 4. Left lateral view of skull and left upper limb

Fig. 2. Anterior view of right upper limb

Fig. 5. Anterior view of left upper limb

Technical Comment
– The lateral views of the skull (Figs. 1 and 4) were taken posteriorly

Fig. 1. Anterior view of thorax

Fig. 4. Anterior view of thorax and spine

Fig. 2. Anterior view of thorax and spine

Fig. 5. Anterior view of thorax

Fig. 3. Left anterior oblique view of thorax

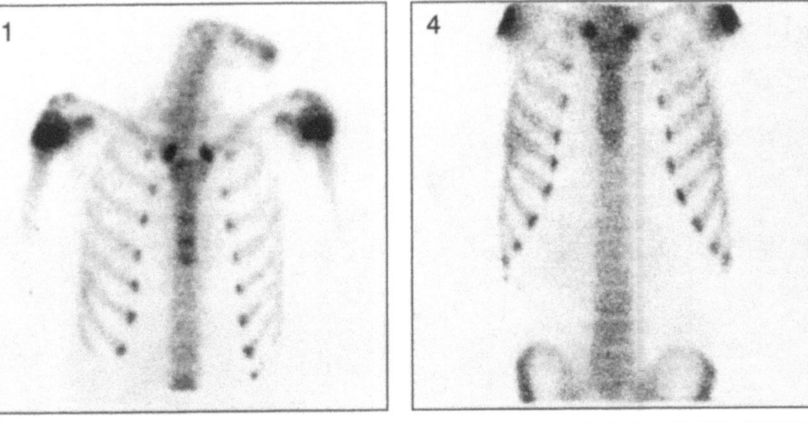

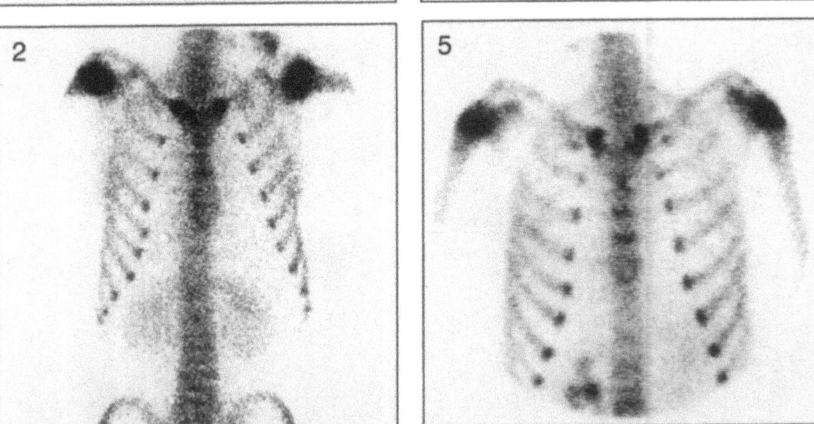

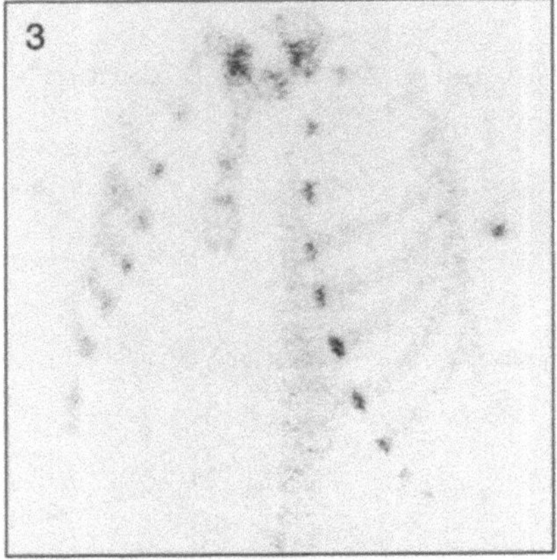

Technical Comments
- Note the differences in the sternum on both the anterior projections and also the anterior oblique
- Note the clarity of the lower lumbar spine in Figs. 2 and 4

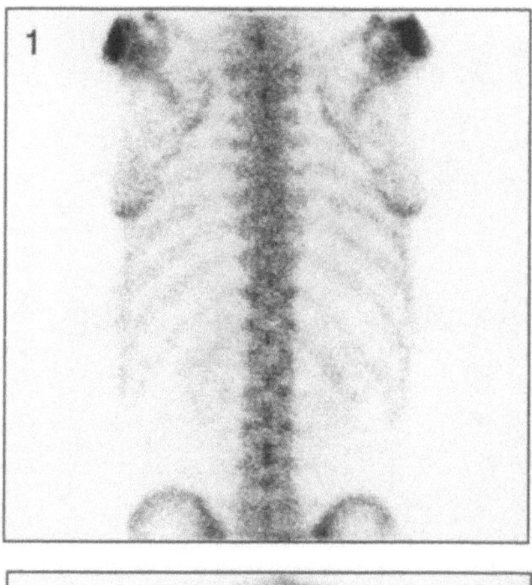

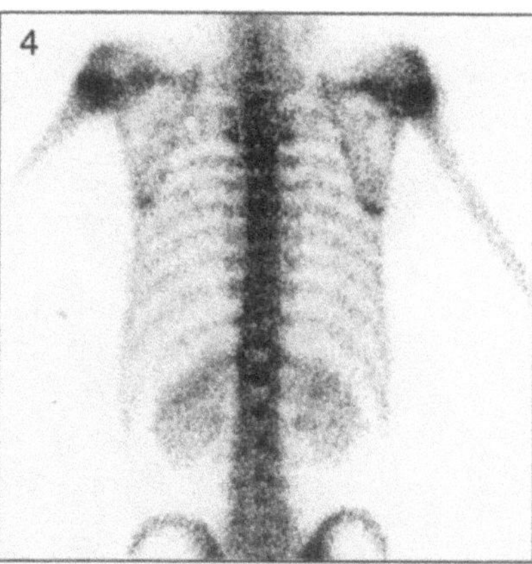

Fig. 1. Posterior view of thorax and spine

Fig. 4. Posterior view of thorax and spine

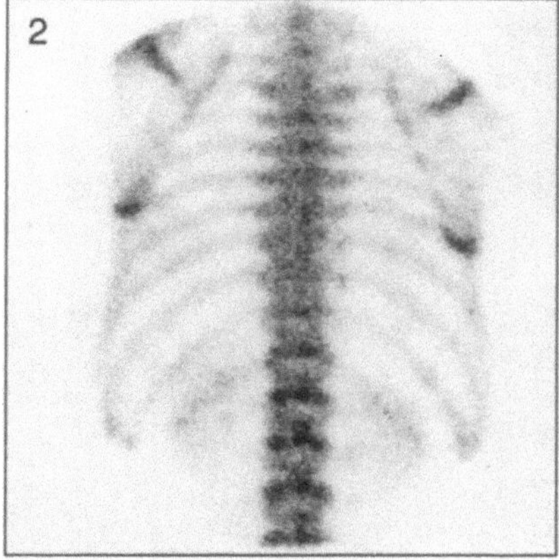

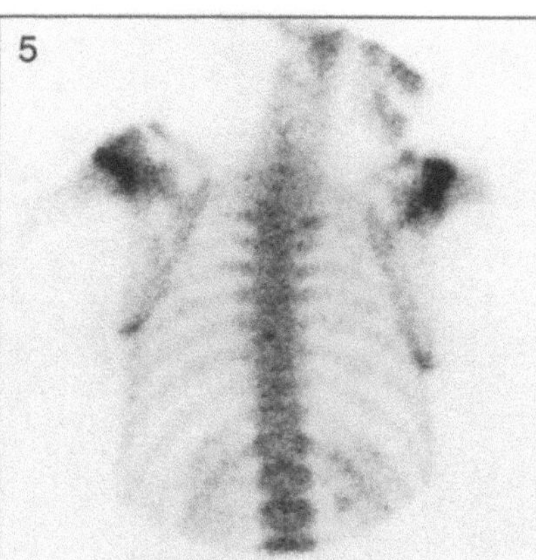

Fig. 2. Posterior view of thorax and spine

Fig. 5. Posterior view of thorax and spine

Technical Comment

– Note the difference in the positions of the scapulae in Fig. 1, compared to Fig. 4, this is due to the different position of the upper limbs

Fig. 1. Anterior view of spine, pelvis and femora

Fig. 4. Anterior view of spine, pelvis and femora

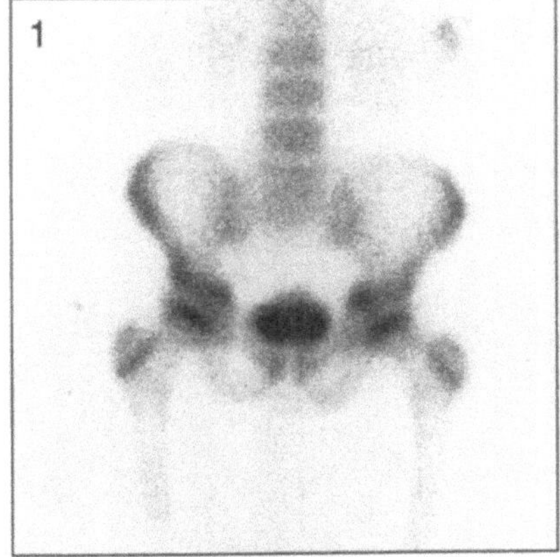

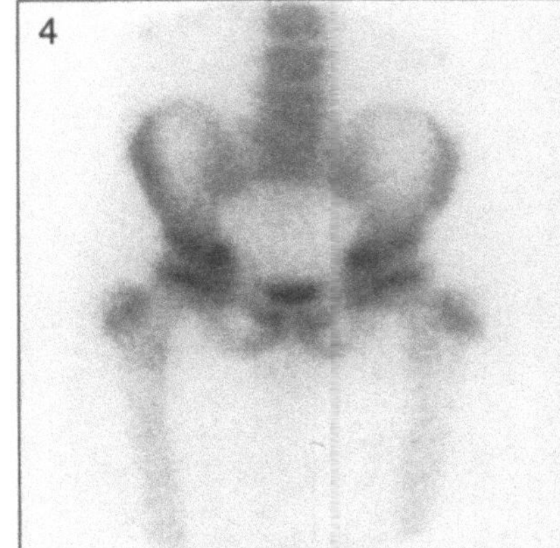

Fig. 2. Anterior view of spine, pelvis and femora

Fig. 5. Anterior view of pelvis and femora

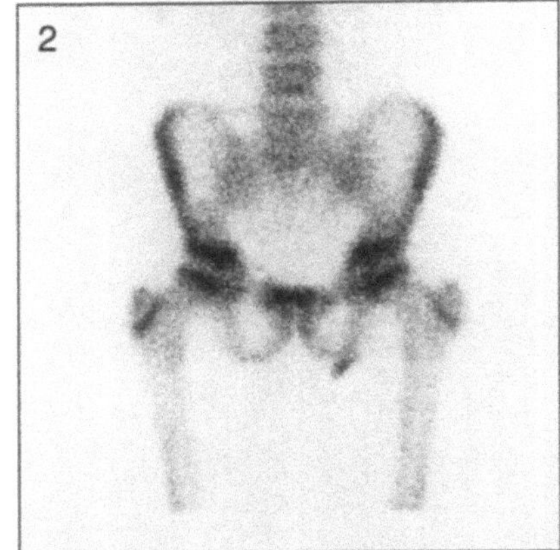

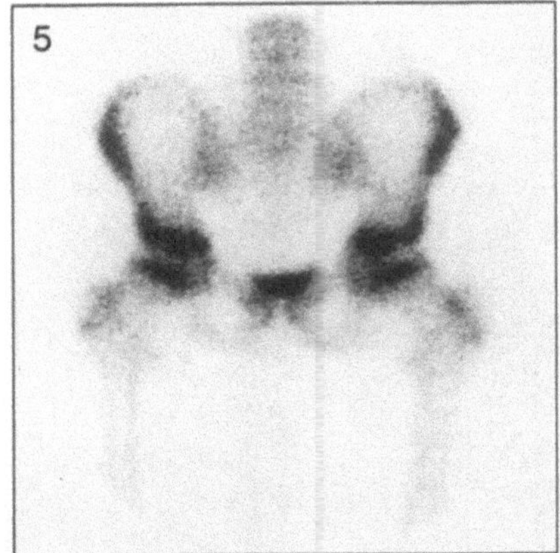

Technical Comments
- The lower lumbar spine is again clearly seen on the anterior views in Figs. 1, 2 and 4
- Urine contamination adjacent to the left ischium is seen in Fig. 2

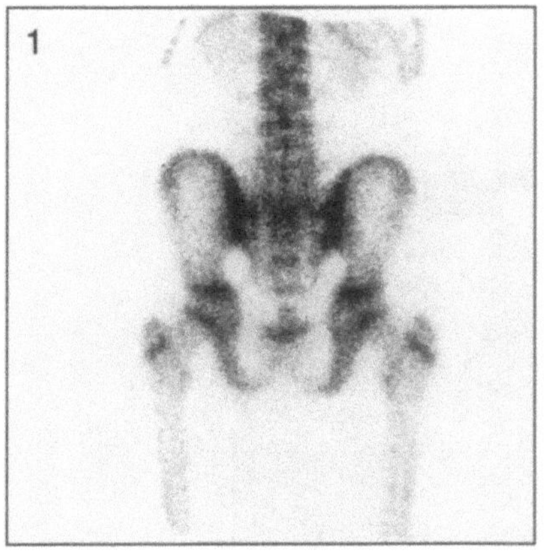

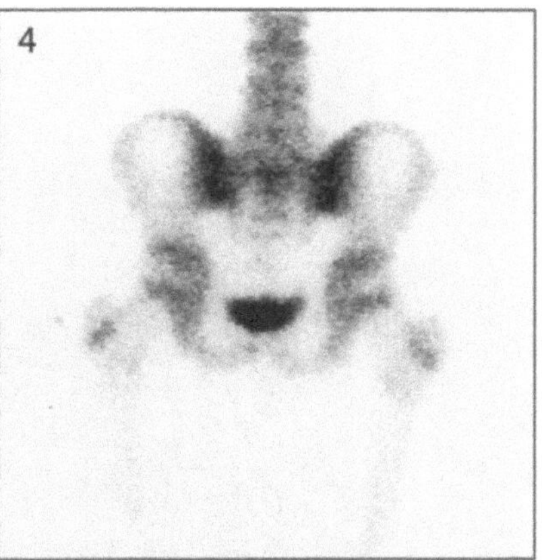

Fig. 1. Posterior view of spine, pelvis and femora

Fig. 4. Posterior view of spine, pelvis and femora

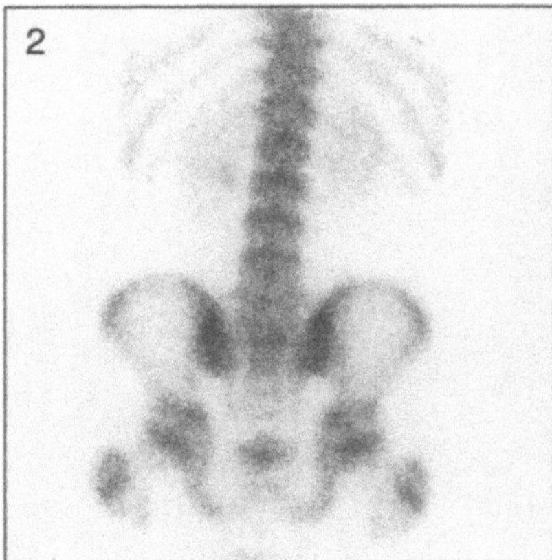

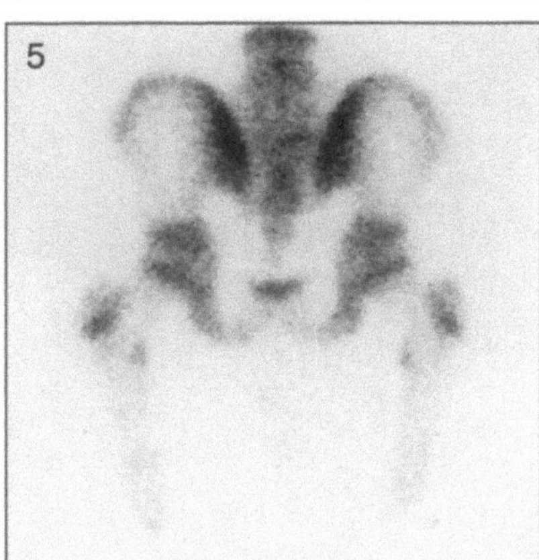

Fig. 2. Posterior view of spine and pelvis

Fig. 5. Posterior view of pelvis and femora

15: Age 13–14 Years

Fig. 1. Pinhole view of right hip

Fig. 4. Pinhole view of left hip

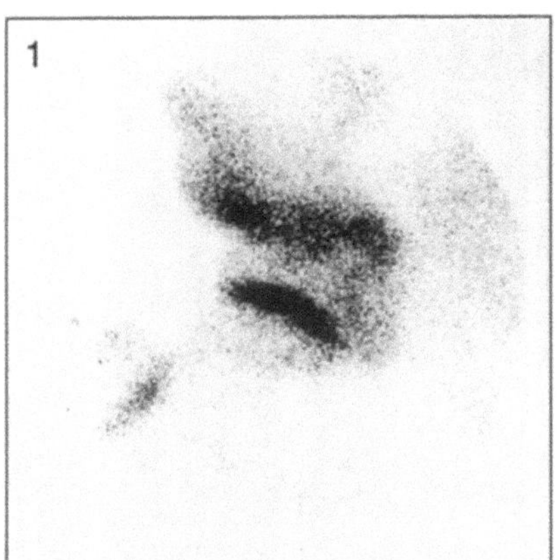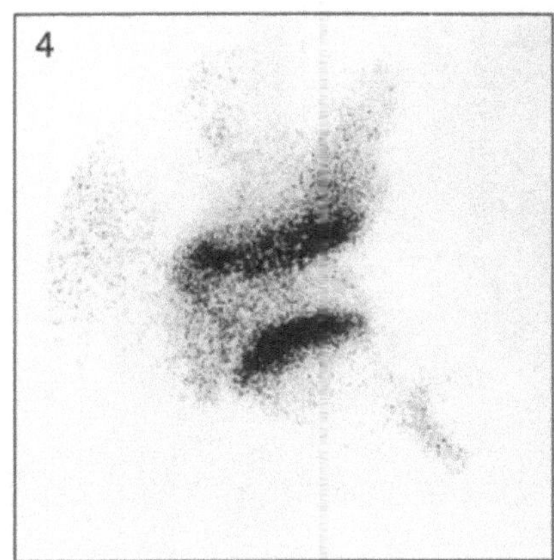

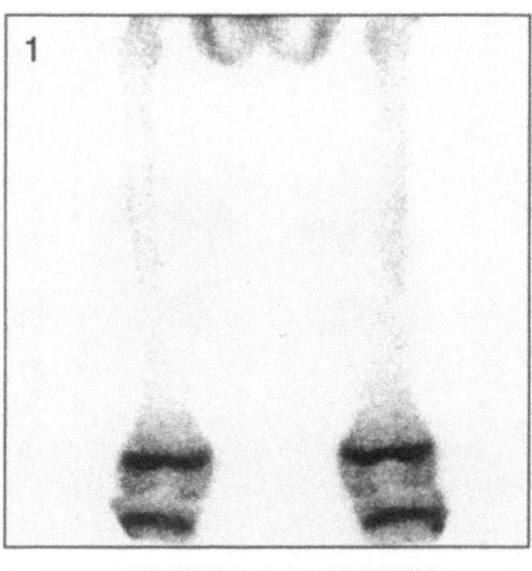
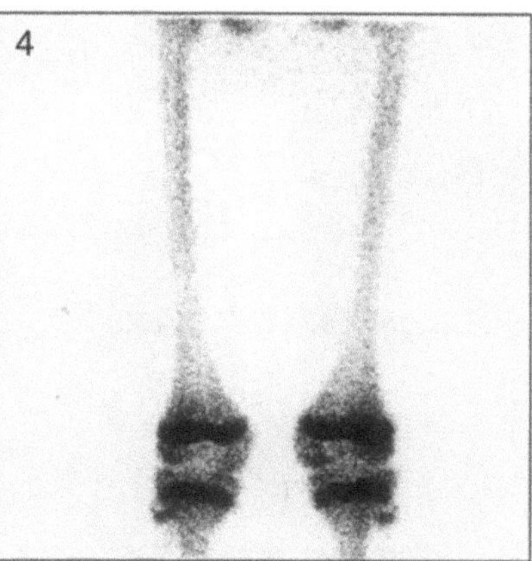
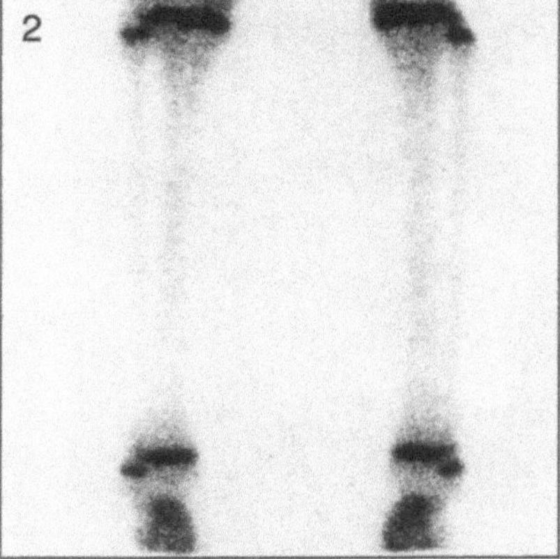
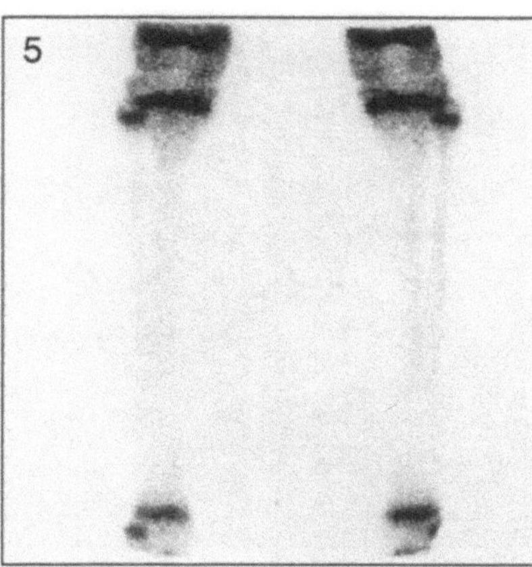

Fig. 1. Posterior view of femora and knees

Fig. 4. Posterior view of femora and knees

Fig. 2. Posterior view of tibia, fibula and ankles

Fig. 5. Posterior view of tibia and fibula

Fig. 1. Anterior view of knees

Fig. 4. Posterior view of knees

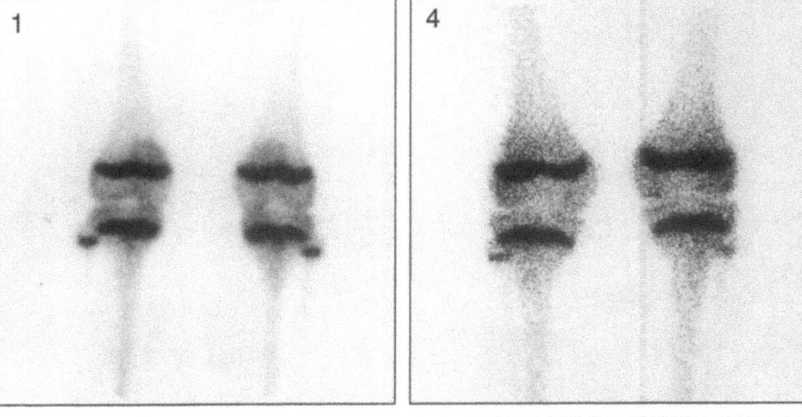

Fig. 2. Anterior pinhole view of right knee

Fig. 5. Anterior pinhole view of left knee

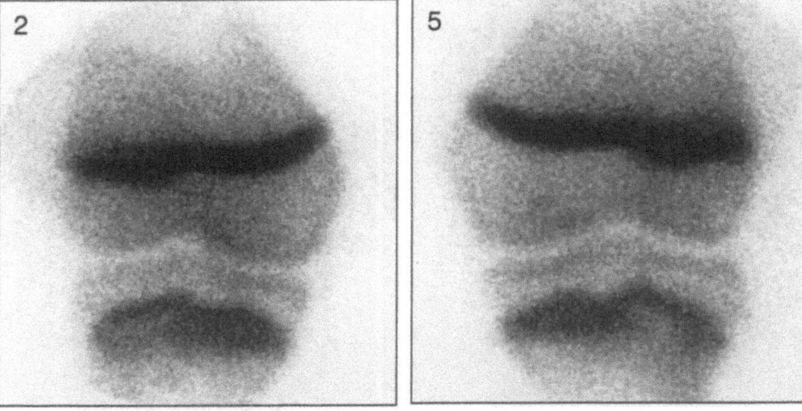

Fig. 3. Lateral view of knees

Fig. 6. Lateral view of knees

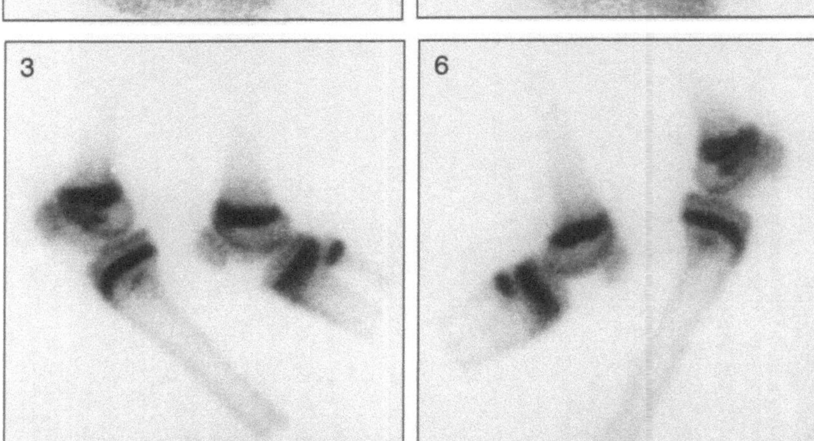

Technical Comment

– Figs. 3 and 6 represent lateral images of the knees, on the left there is a medial lateral and a lateral lateral while on the right it is the reverse order. The lateral lateral images show the head of the fibula clearly while the medial lateral show the head of the fibula within the tibia but below the epiphyseal plate

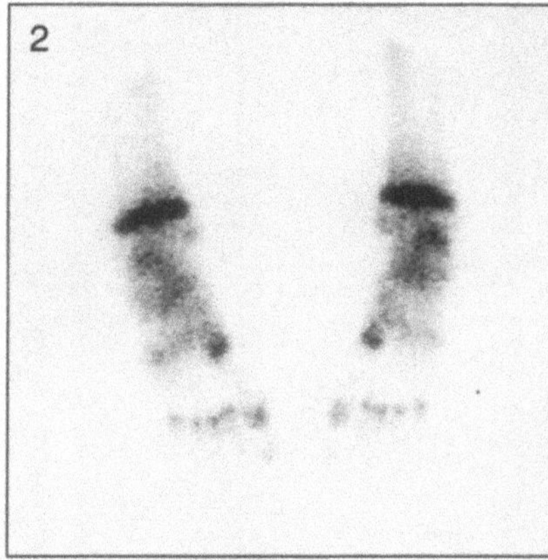

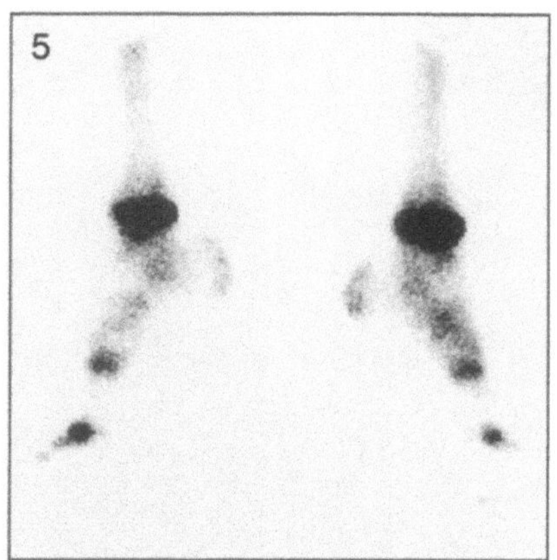

Fig. 1. Lateral view of tibia and ankles

Fig. 2. Anterior view of feet

Fig. 5. Lateral view of feet

Technical Comment

– Note external rotation of the feet in Fig. 1, resulting in non visualization of the epiphyseal plate at the distal end of the fibula

16: Age 14–15 Years

– A double headed whole body
  gamma camera was used
– Left image is the anterior view
– Right image is the posterior
  view

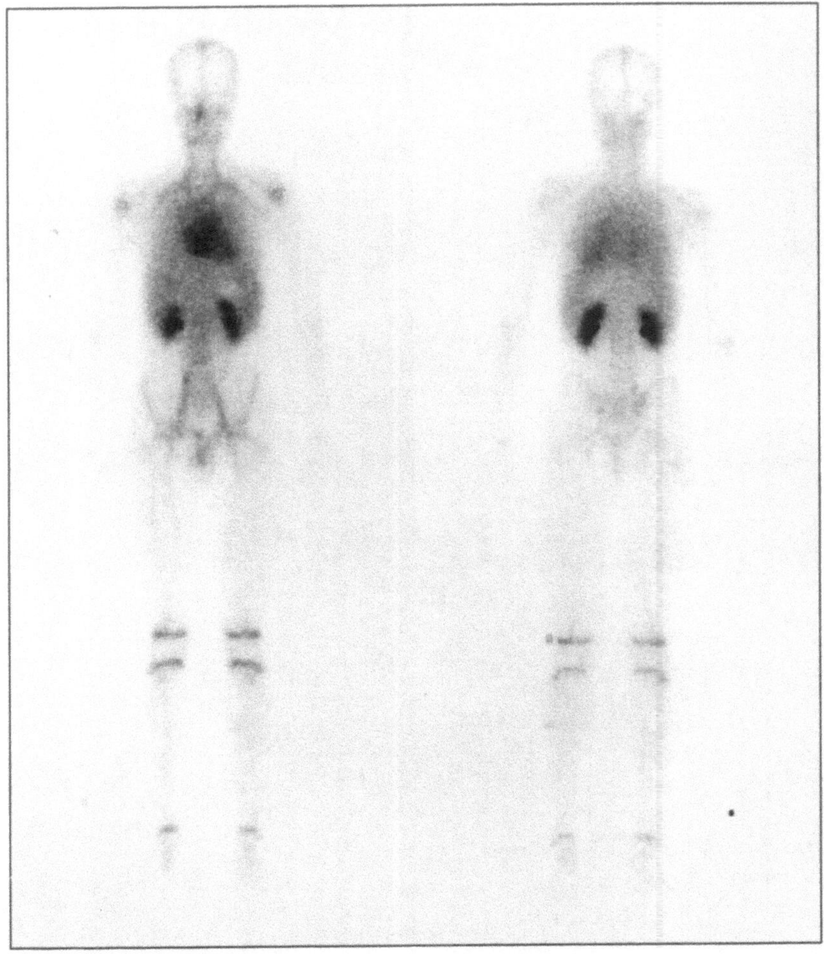

Technical Comment
– Marker on child's right side

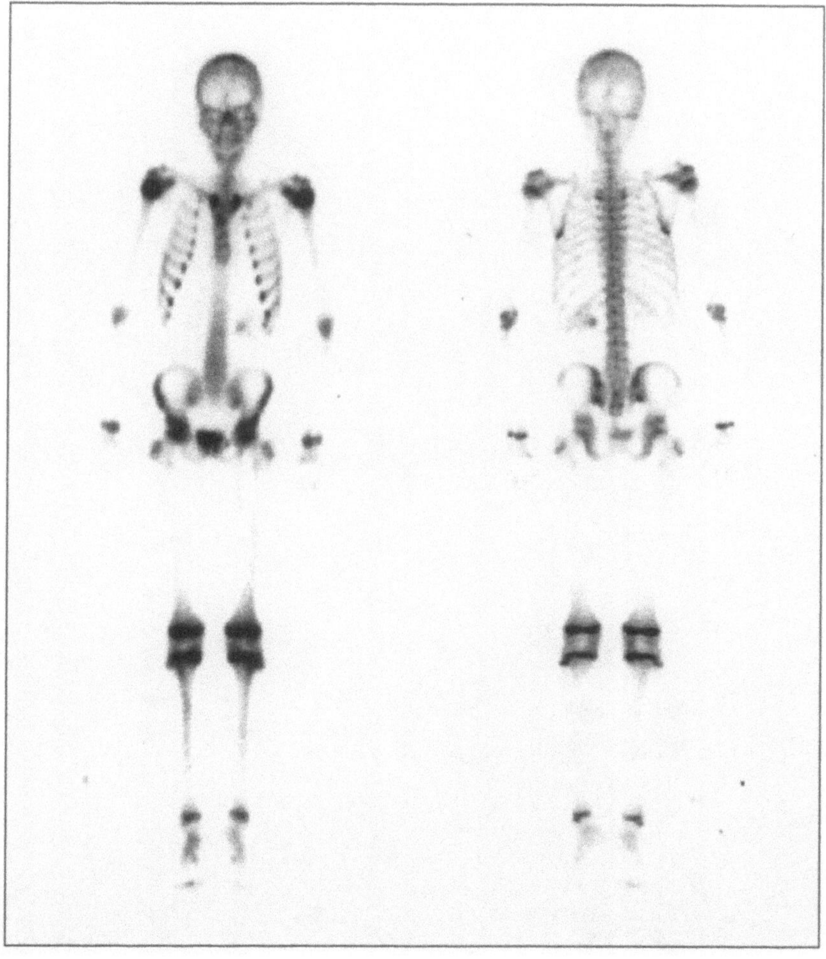

- A double headed whole body gamma camera was used
- Left image is the anterior view
- Right image is the posterior view

Technical Comments

- Note on the anterior projection (on the left) that the epiphyseal plates at the elbow, the wrists and knees are not as distinct as on the posterior view (on the right). This is due to the greater distance of the patient from the detector on the anterior view
- Marker on child's right side

- A double headed whole body
  gamma camera was used
- Left image is the anterior view
- Right image is the posterior
  view

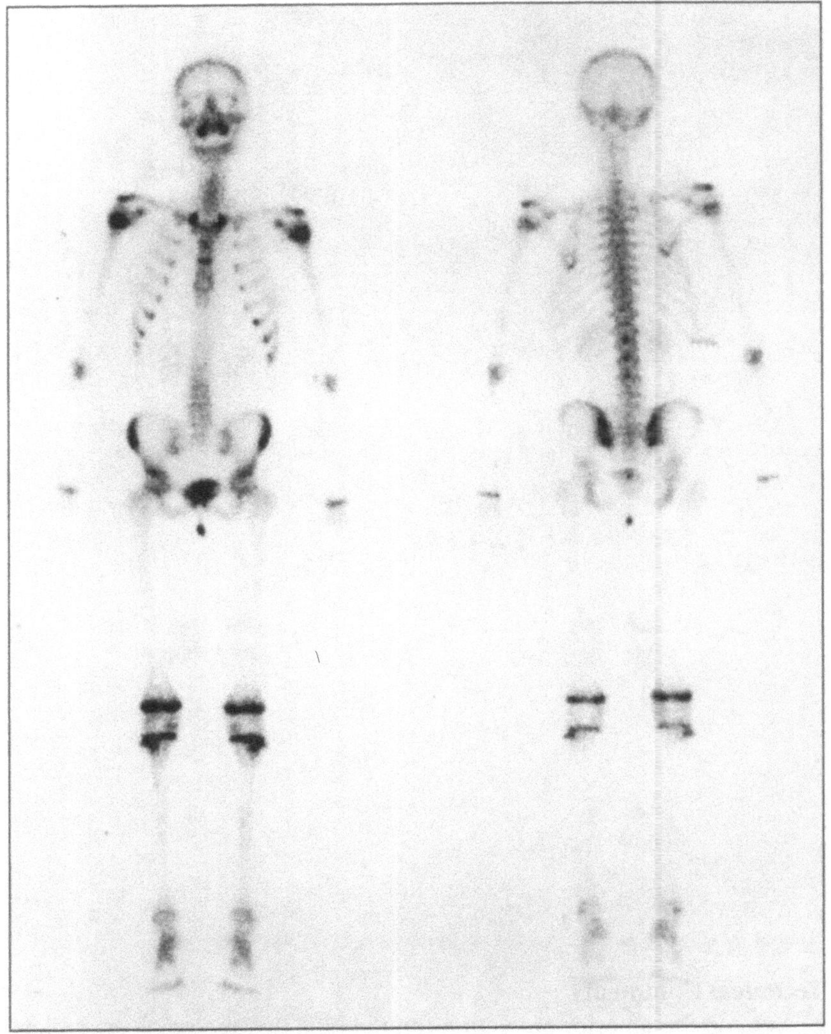

Technical Comments

- Note the increased activity in the sternum on the anterior view, this is
  within the normal range. Sternal fusion has not yet occurred
- Urine contamination below the pelvis is seen
- Marker on child's right side

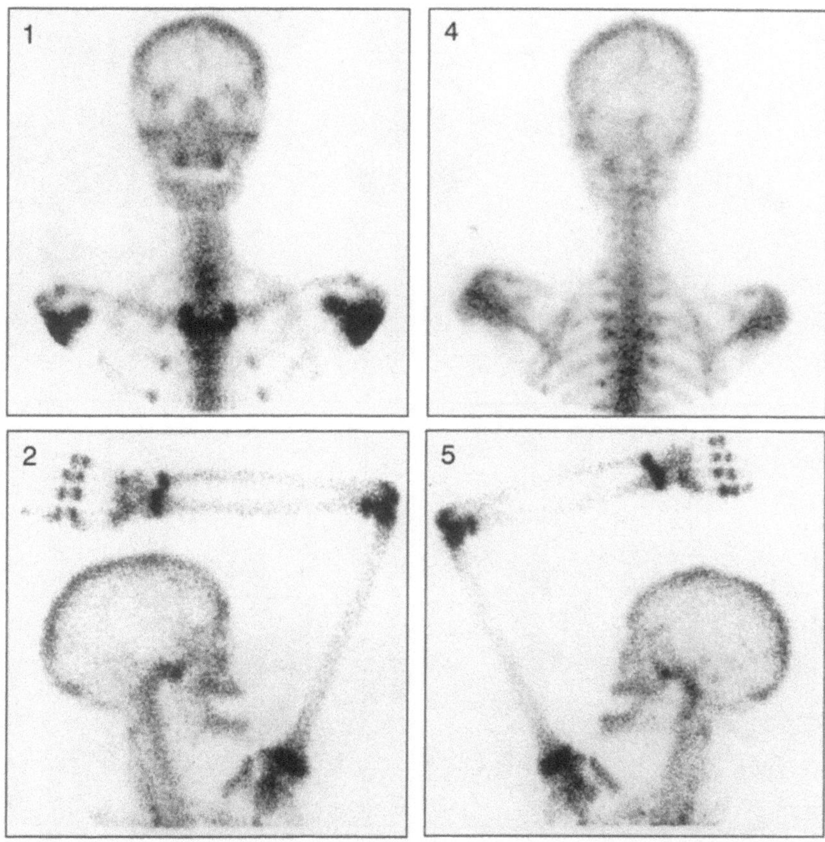

Fig. 1. Anterior view of skull and thorax

Fig. 4. Posterior view of skull and thorax

Fig. 2. Right lateral view of skull and right upper limb

Fig. 5. Left lateral view of skull and left upper limb

Technical Comments
- The fingers of the hands in Figs. 2 and 5 have been excluded from the field of view because of the child's size
- The lateral views of the skull (Figs. 2 and 5) were taken posteriorly

Fig. 1. Right lateral view of skull and right upper limb

Fig. 4. Left lateral view of skull and left upper limb

Fig. 2. Anterior view of hands

Fig. 5. Anterior view of hands

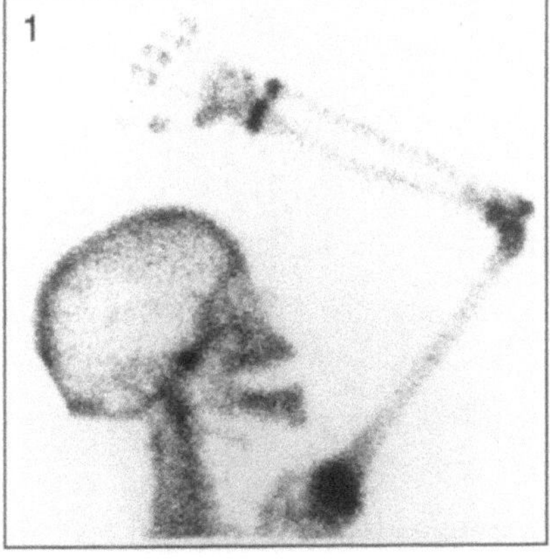

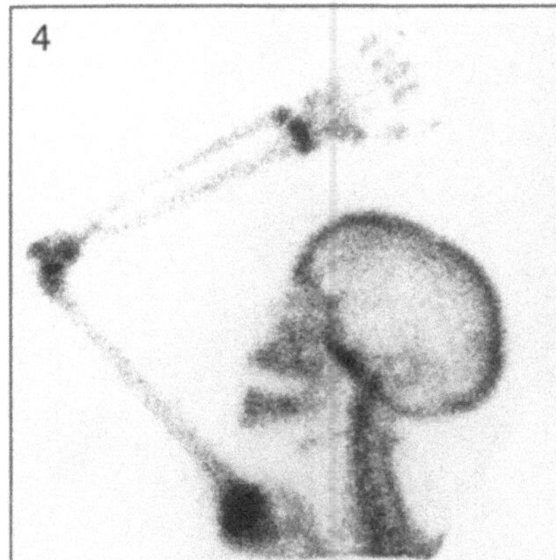

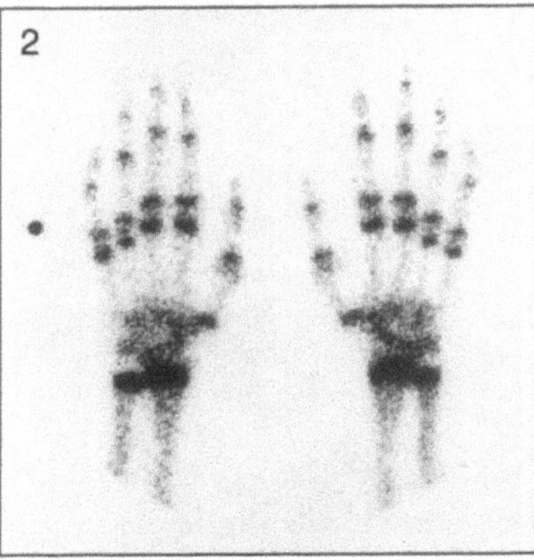

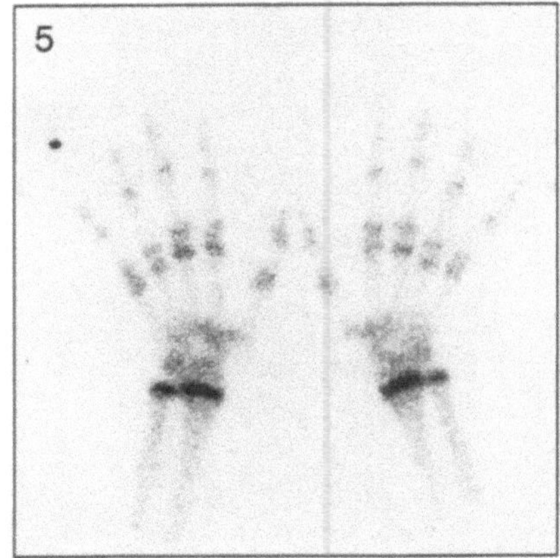

Technical Comments
- Note that part of the occipital bone in Fig. 1 has been excluded from the field of view
- The lateral views of the skull (Figs. 1 and 4) were taken posteriorly
- There is a marker adjacent to the right hand in Figs. 2 and 5

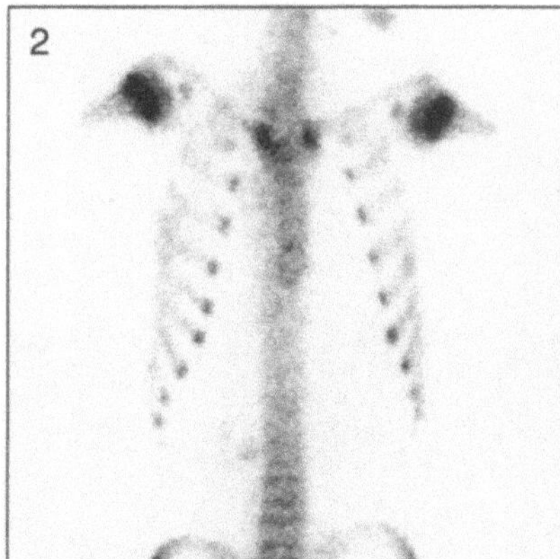

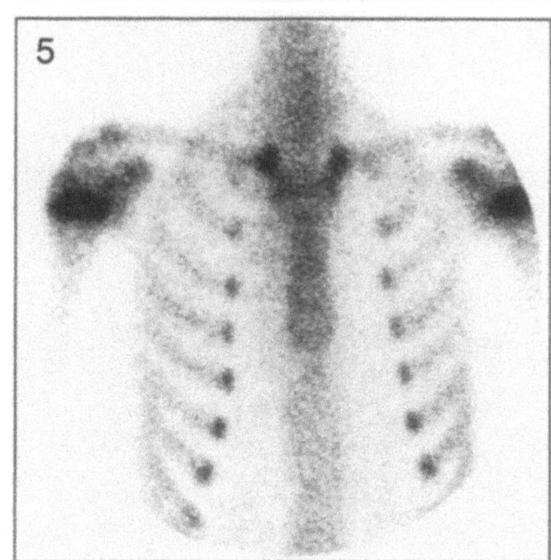

Fig. 1. Anterior view of thorax

Fig. 4. Anterior view of thorax and spine

Fig. 2. Anterior view of thorax and spine

Fig. 5. Anterior view of thorax

Technical Comment
– Note the variation of the sternum in these four different children, all within the normal range

Fig. 1. Right anterior oblique view of thorax

Fig. 4. Left anterior oblique view of thorax

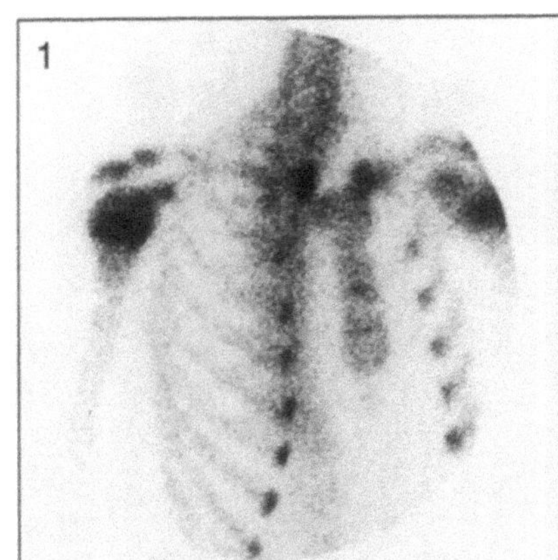
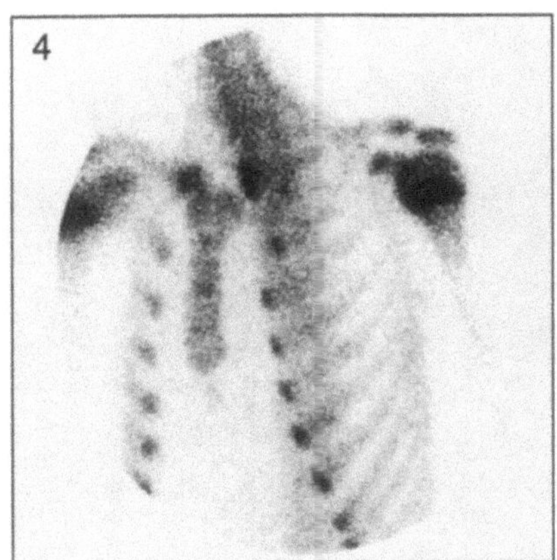

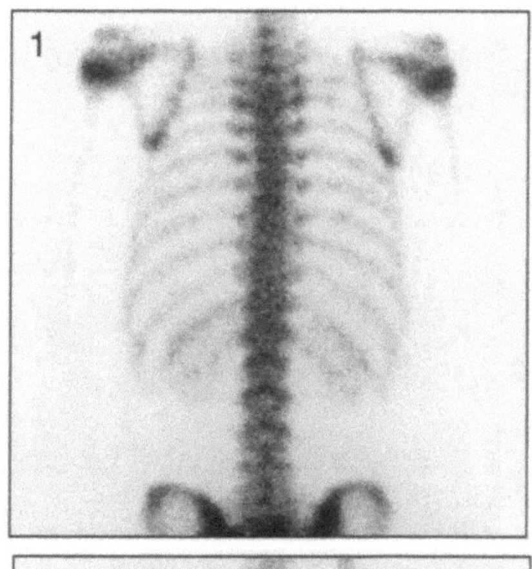

Fig. 1. Posterior view of thorax and spine

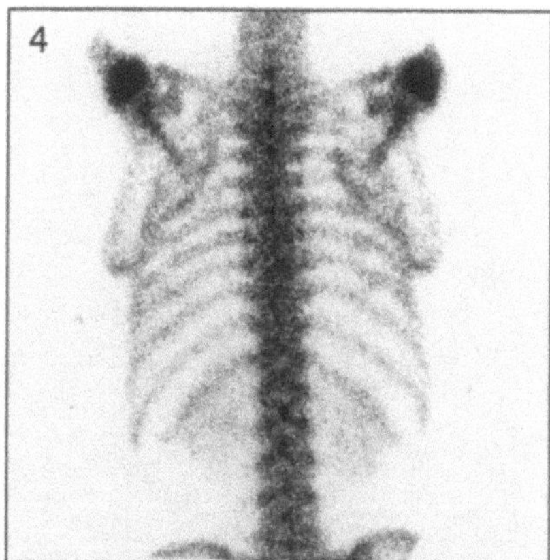

Fig. 4. Posterior view of thorax and spine

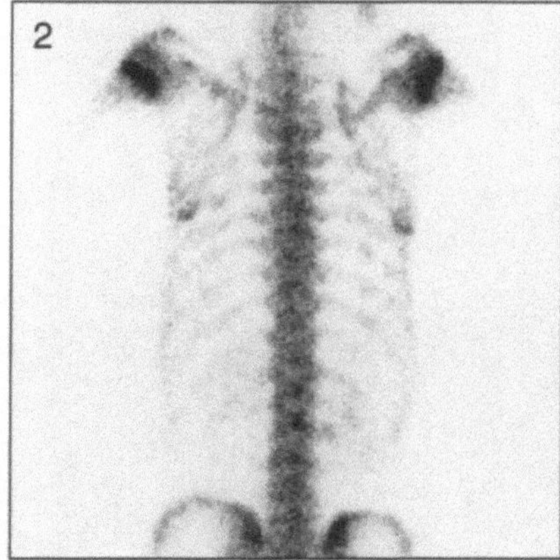

Fig. 2. Posterior view of thorax and spine

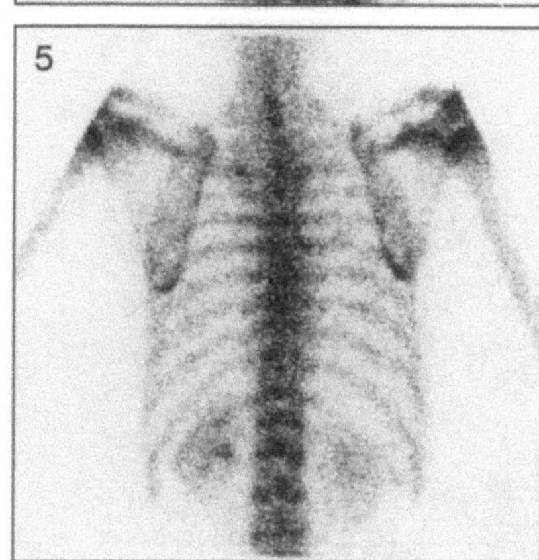

Fig. 5. Posterior view of thorax and spine

Technical Comment
– Note the variation of the appearances of the scapulae

▶ **Potential Pitfall**
– Focal increased activity is noted in Fig. 5 in the region of the mid-portion of the left third rib, this represents activity from the sterno-clavicular joint not a fracture of the rib

Fig. 1. Anterior view of spine and pelvis

Fig. 4. Anterior view of spine, pelvis and femora

Fig. 2. Anterior view of spine, pelvis and femora

Fig. 5. Anterior view of spine and pelvis

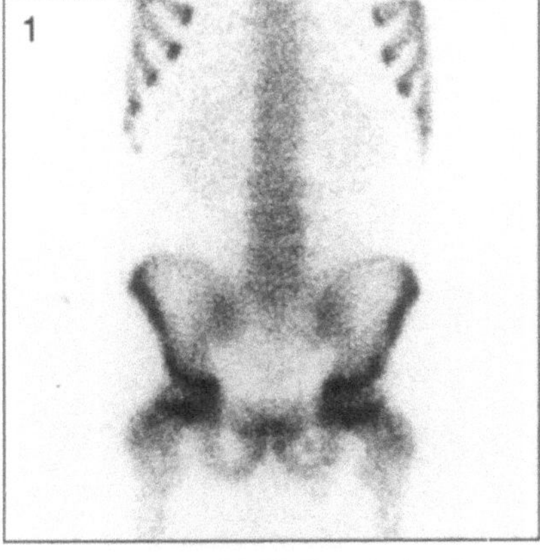

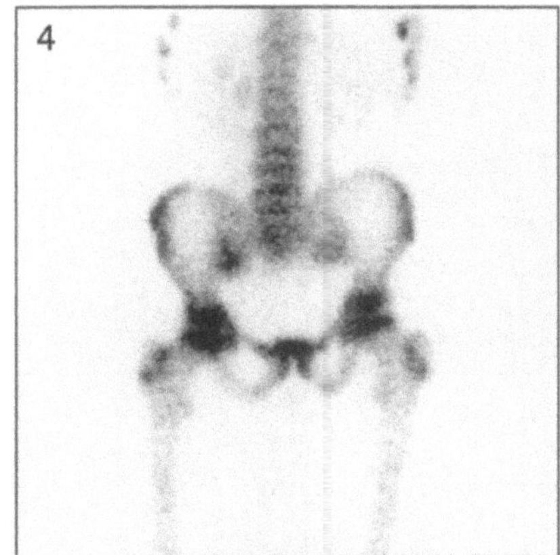

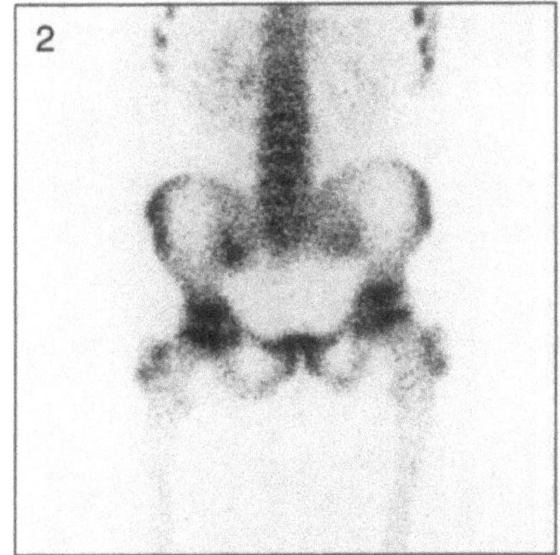

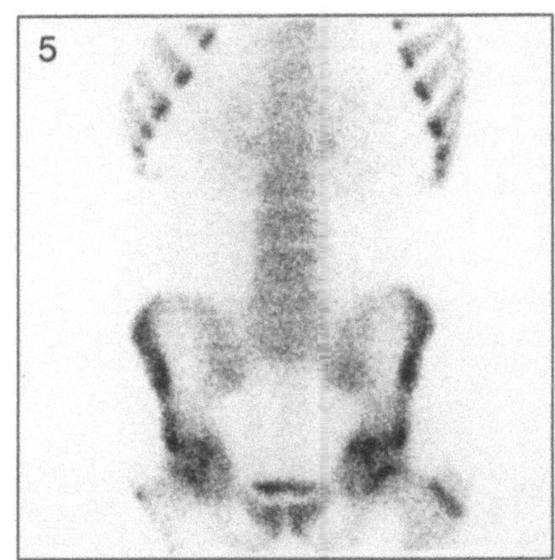

Technical Comment
– Urine contamination to the left of the pubic bone is seen in Fig. 1

► **Potential Pitfall**
– The apparent increased activity in the right sacro-iliac joint and also the asymmetry between the anterior superior iliac spines in Figs. 2 and 4 is due to rotation and tilting of the pelvis and is not pathological

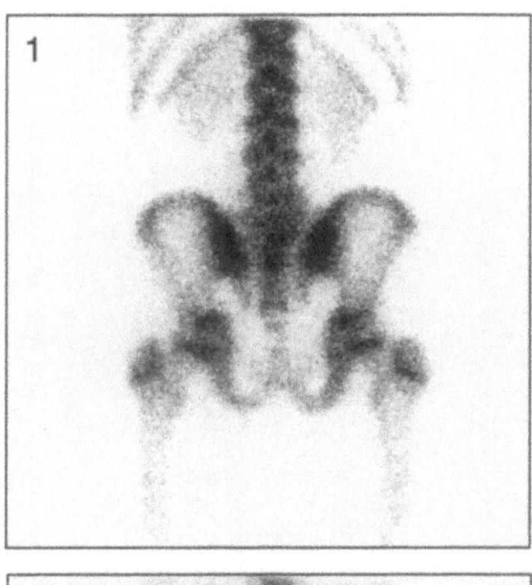

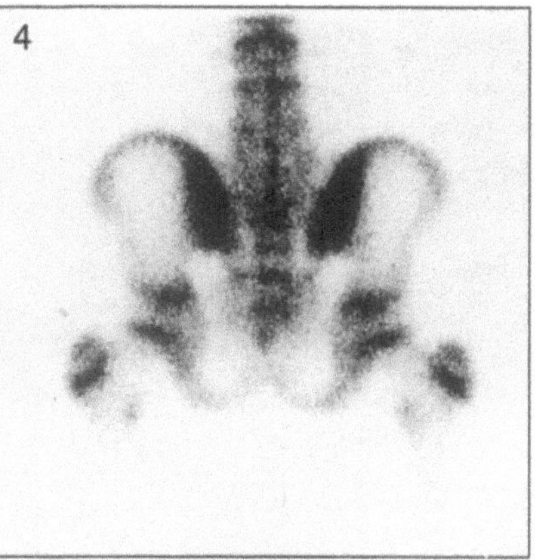

Fig. 1. Posterior view of spine, pelvis and femora

Fig. 4. Posterior view of spine, pelvis and femora

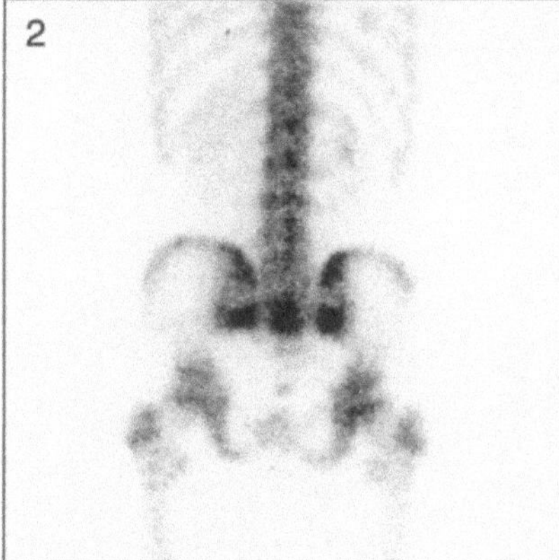

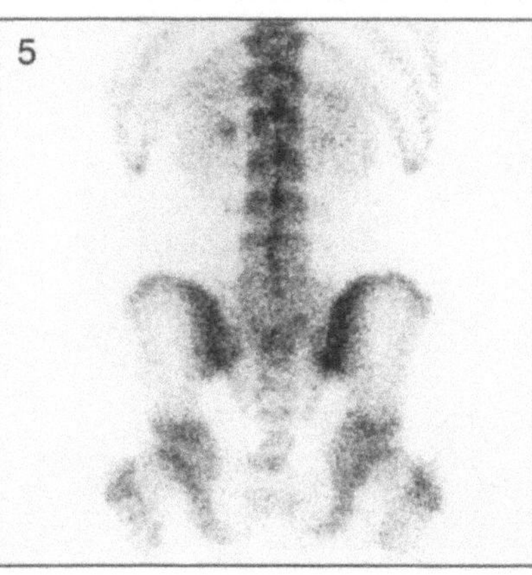

Fig. 2. Posterior view of spine, pelvis and femora

Fig. 5. Posterior view of spine and pelvis

Technical Comment

– Note in Fig. 2 the non-uniformity of activity in both sacro-iliac joints with a relatively cool area in the mid-portion of the joints, this is within the normal range

Fig. 1. Pinhole view of right hip

Fig. 4. Pinhole view of left hip

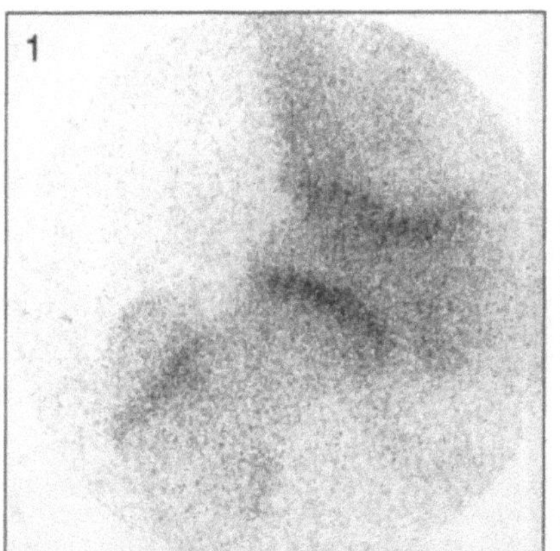
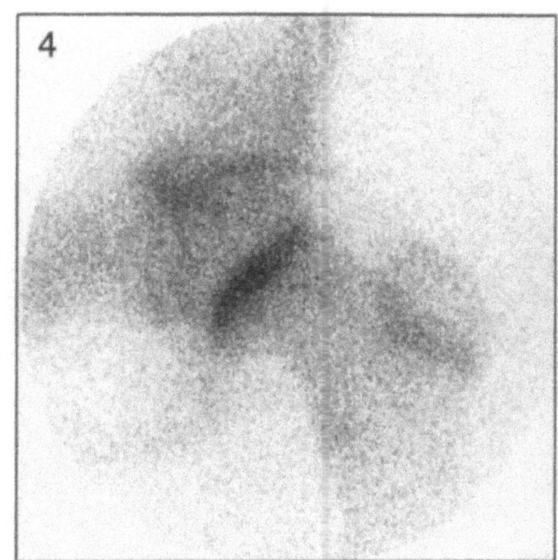

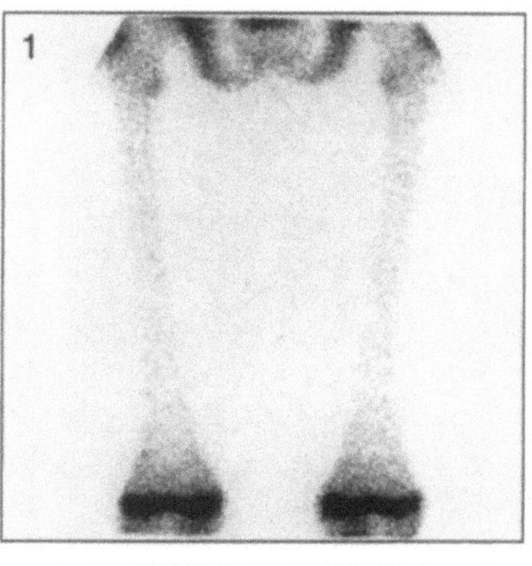

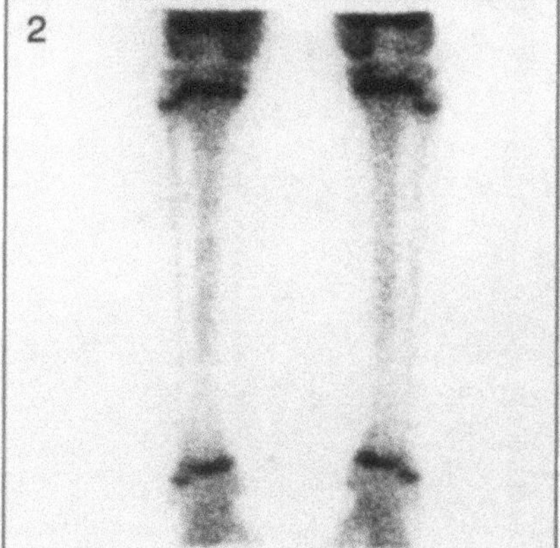

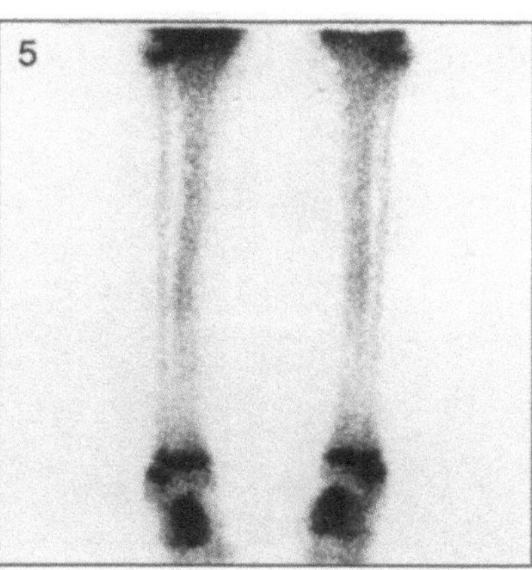

Fig. 1. Posterior view of femora

Fig. 4. Posterior view of femora
and knees

Fig. 2. Posterior view of tibia,
fibula and ankles

Fig. 5. Posterior view of tibia,
fibula and ankles

Technical Comment

– Note the non-homogeneous distribution of isotope in the tibia with
  apparent slight increase in the mid-portion in Figs. 2 and 5. Both these
  children underwent 10 years follow up and no abnormality has been
  detected (see also pp. 198 and 213)

Fig. 1. Posterior view of knees

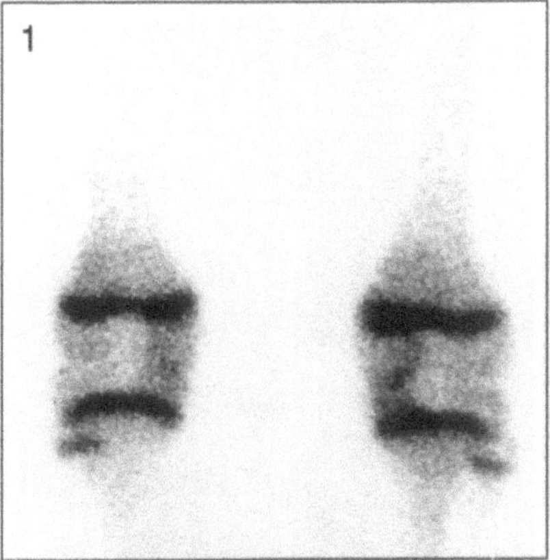

Fig. 2. Anterior pinhole view of
right knee

Fig. 5. Anterior pinhole view of
left knee

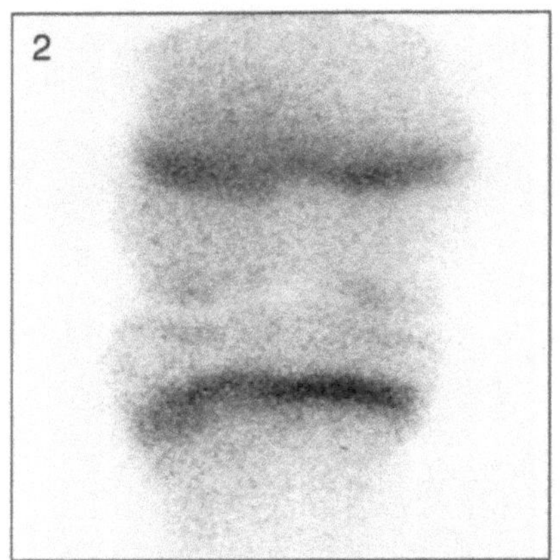
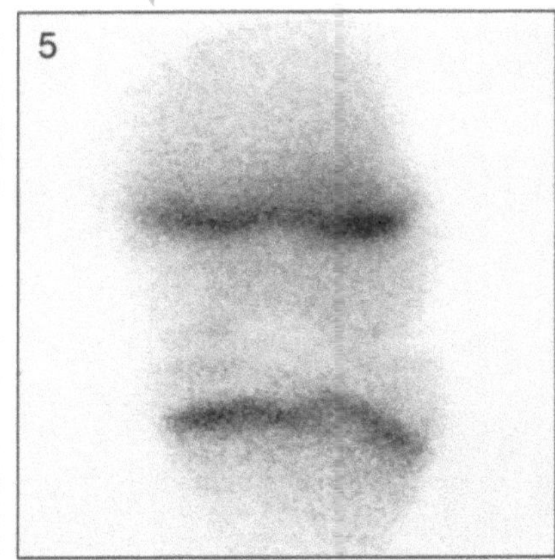

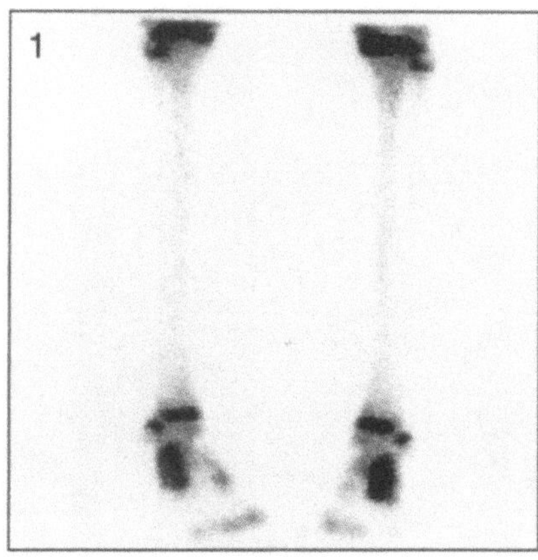 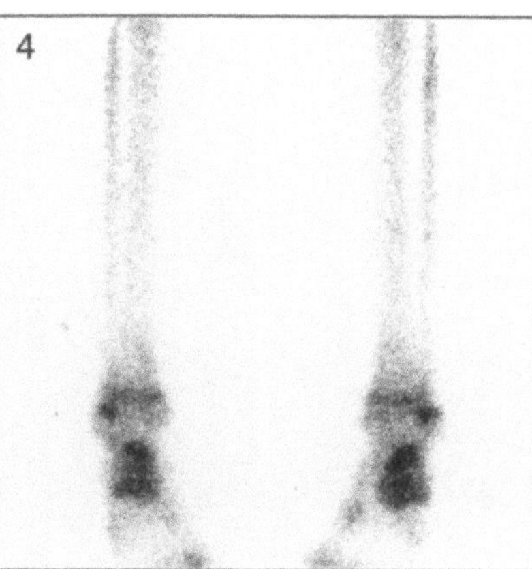

Fig. 1. Posterior view of tibia, fibula and ankles

Fig. 4. Posterior view of tibia, fibula and ankles

Technical Comment
– Note the patchy distribution of the isotope in the fibulae in Fig. 4

17: Age 15–17 Years

Fig. 1. Posterior view of pelvis

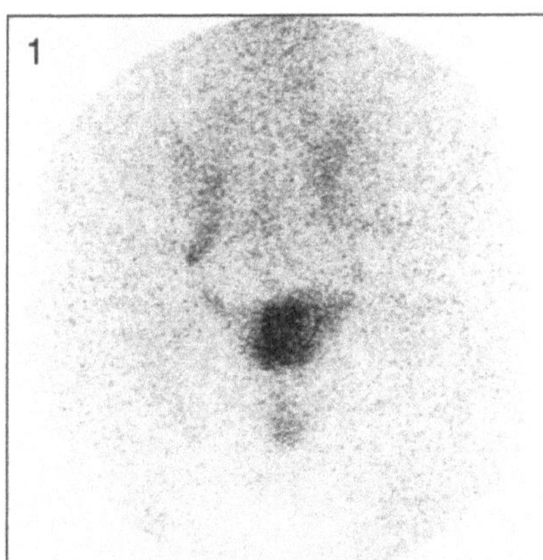

Fig. 2. Anterior view of knees

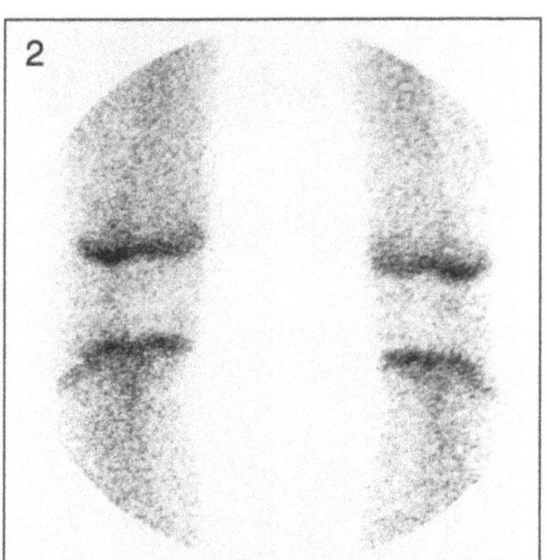

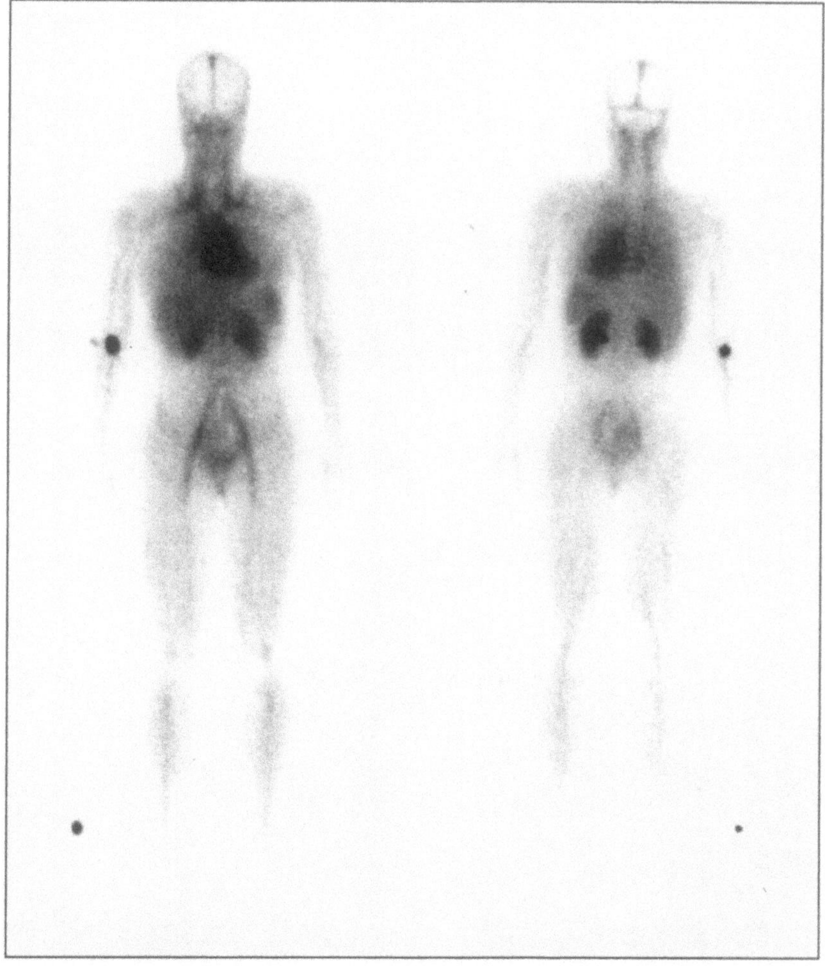

- A double headed whole body gamma camera was used
- Left image is the anterior view
- Right image is the posterior view

Technical Comments
- Note extravasation of isotope at the site of injection in the right elbow
- Marker on child's right side

– A double headed whole body
  gamma camera was used
– Left image is the anterior view
– Right image is the posterior
  view

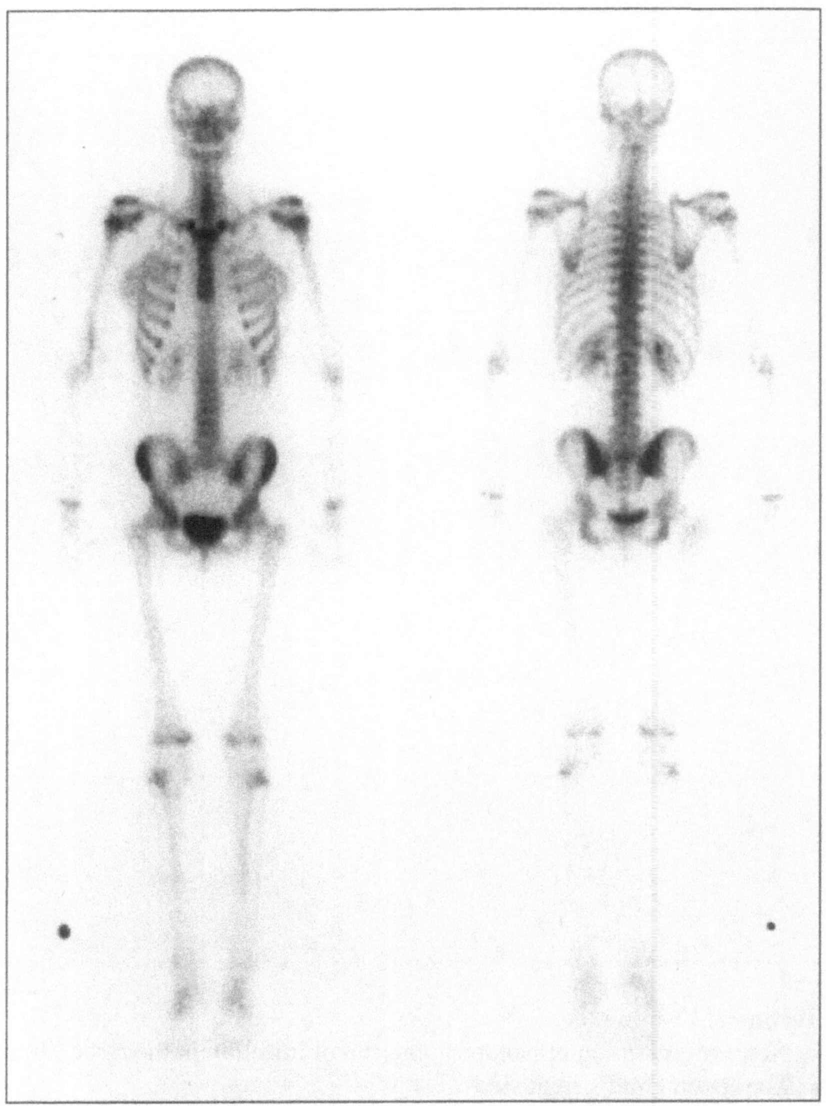

Technical Comments

– Isotope in the breasts causes indistinctness of the ribs on the anterior
  view
– Marker on child's right side

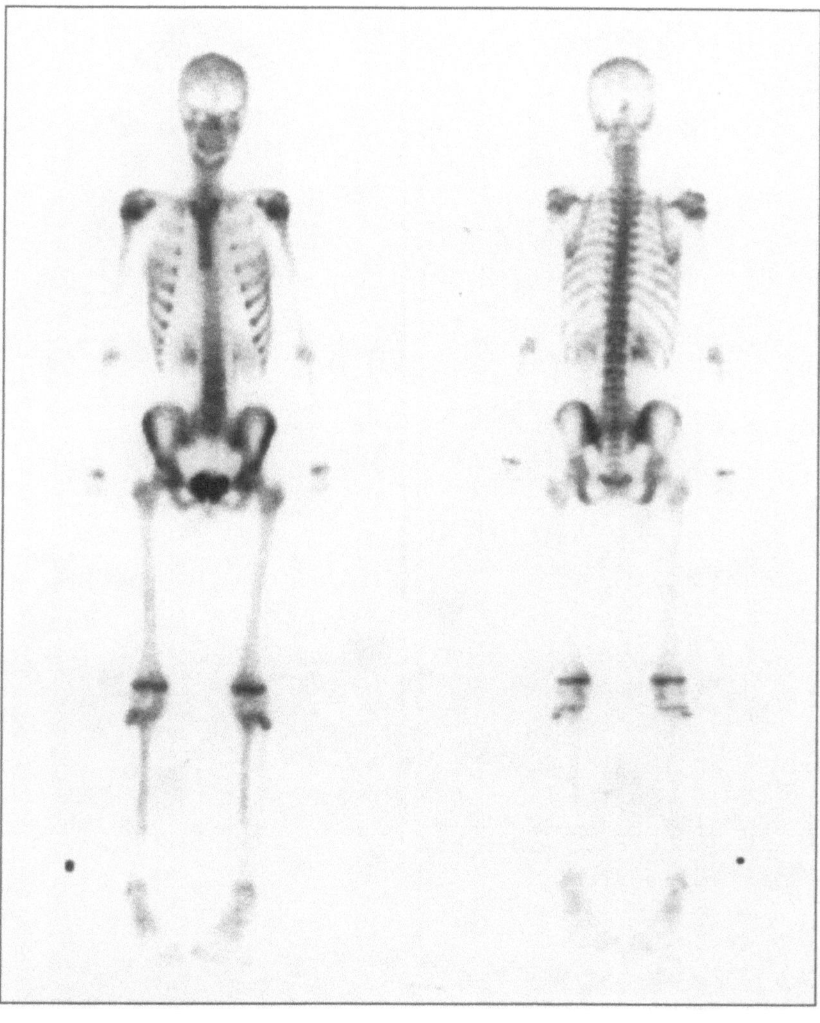

- A double headed whole body gamma camera was used
- Left image is the anterior view
- Right image is the posterior view

Technical Comments
- Isotope in the breasts causes indistinctness of the ribs on the anterior view
- Marker on child's right side

Fig. 1. Anterior view of skull

Fig. 4. Posterior view of skull

Fig. 2. Right lateral view of skull
and right upper limb

Fig. 5. Left lateral view of skull
and left upper limb

Fig. 3. Anterior view of hands

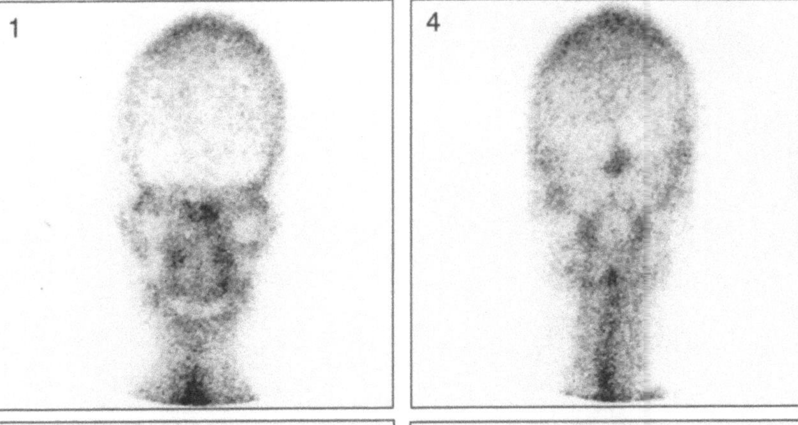

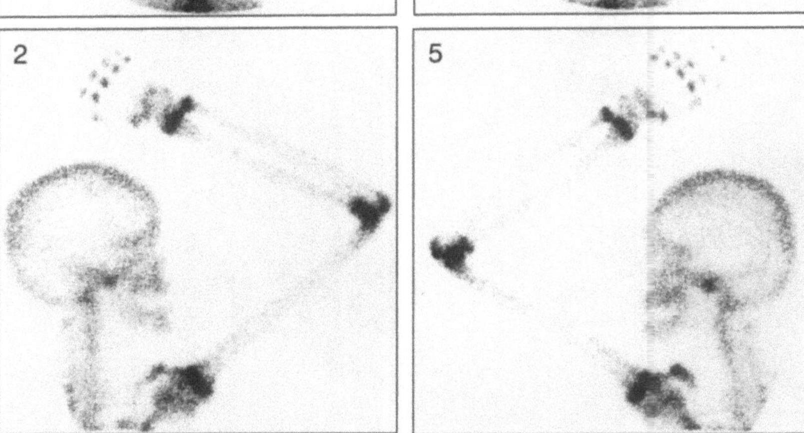

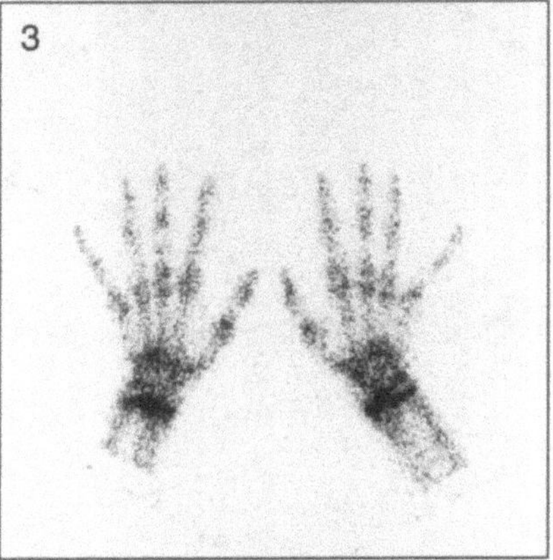

Technical Comments
- Note the significant difference between the activity in the epiphyseal
  plates of the wrist in Figs. 2, 3 and 5
- The lateral views of the skull (Figs. 2 and 5) were taken posteriorly
- The tips of the fingers have not been included in the field of view of the
  gamma camera in Figs. 2 and 5

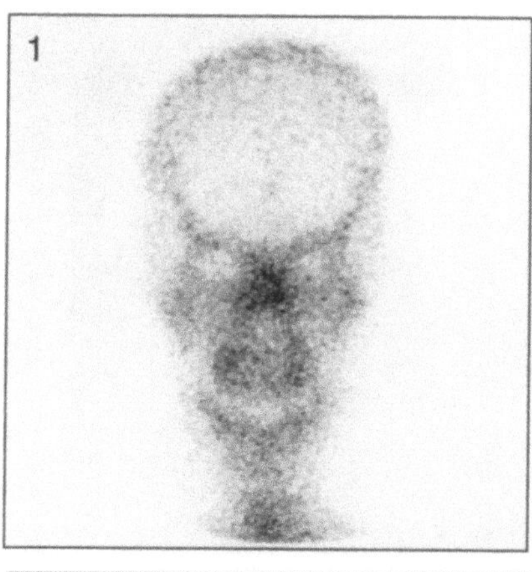

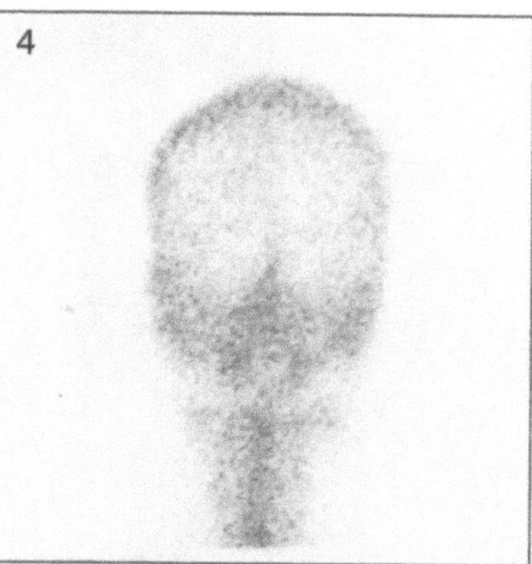

Fig. 1. Anterior view of skull

Fig. 4. Posterior view of skull

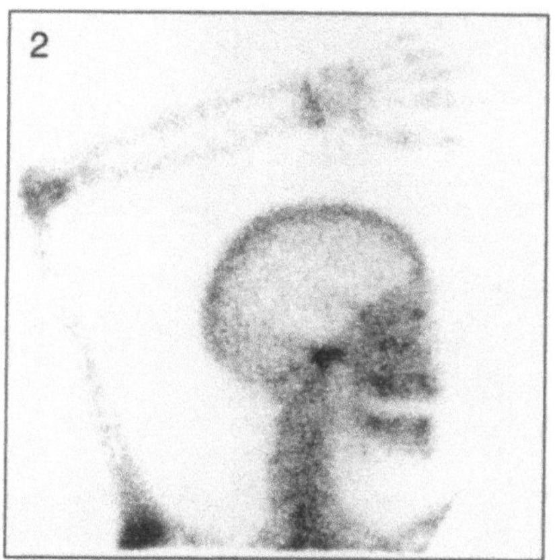

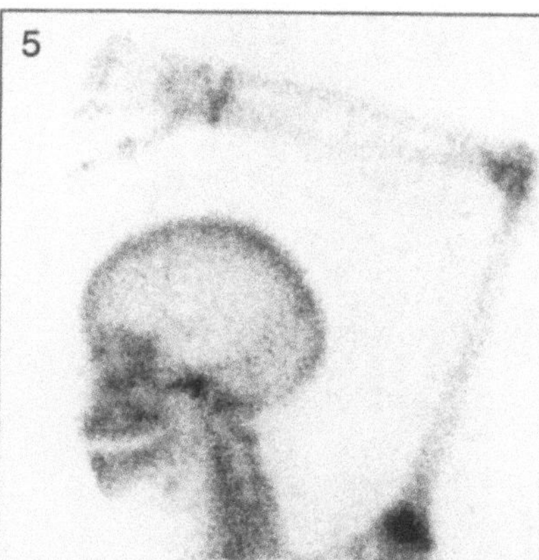

Fig. 2. Right lateral view of skull and right upper limb

Fig. 5. Left lateral view of skull and left upper limb

Technical Comments

– The lateral views of the skull (Figs. 2 and 5) were taken anteriorly
– The tips of the fingers have not been included in the field of view of the gamma camera in Figs. 2 and 5

Fig. 1. Anterior view of thorax and spine

Fig. 4. Anterior view of thorax and spine

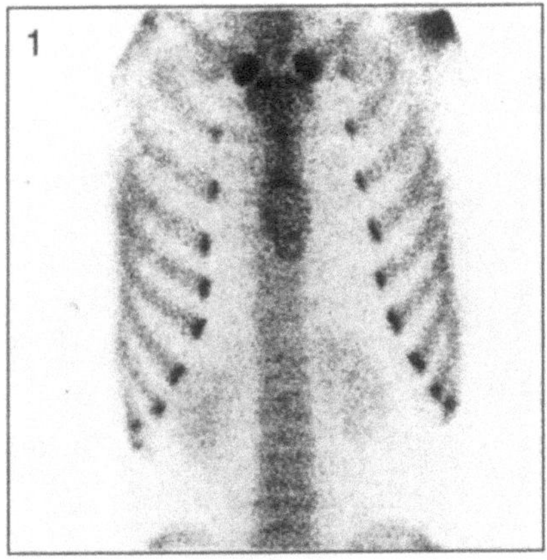

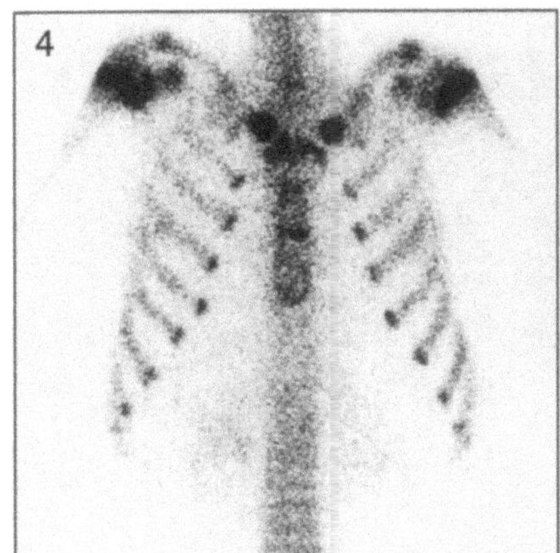

Fig. 2. Anterior view of thorax and spine

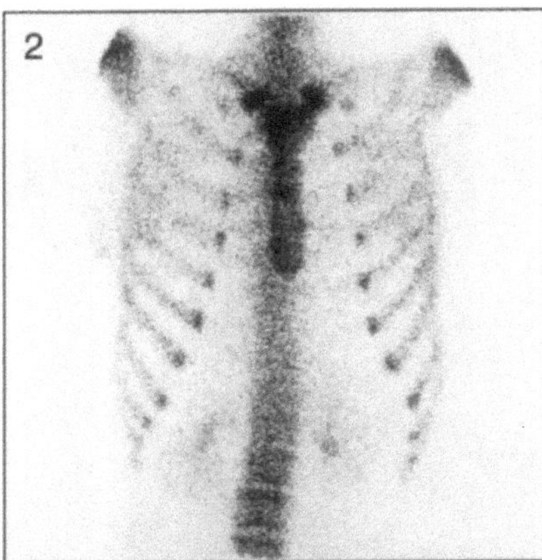

Fig. 3. Left anterior oblique view of thorax

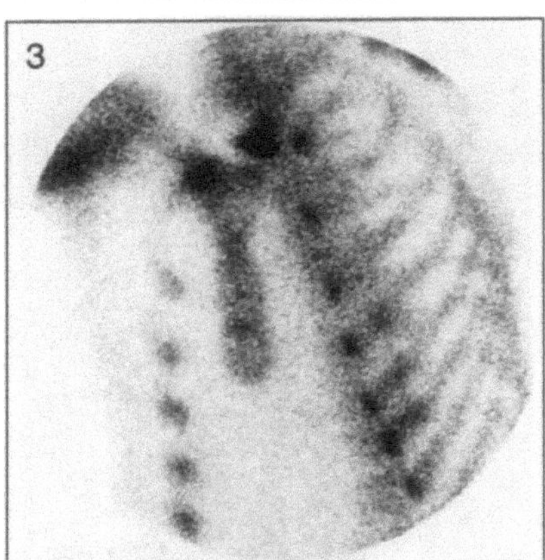

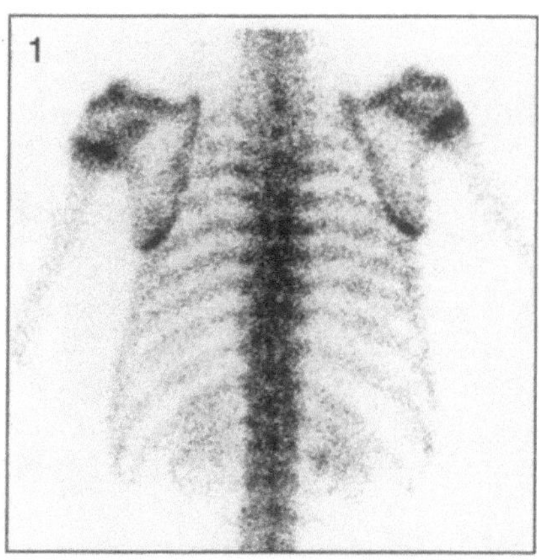

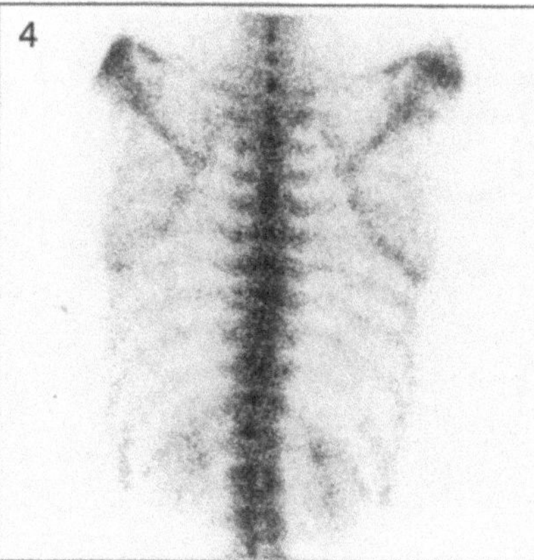

Fig. 1. Posterior view of thorax and spine

Fig. 4. Posterior view of thorax and spine

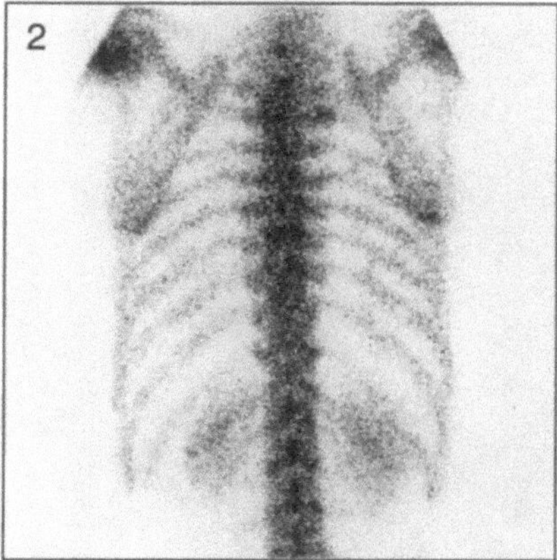

Fig. 2. Posterior view of thorax and spine

Fig. 1. Anterior view of spine, pelvis and femora

Fig. 4. Posterior view of spine, pelvis and femora

Fig. 2. Anterior view of spine, pelvis and femora

Fig. 5. Posterior view of spine and pelvis

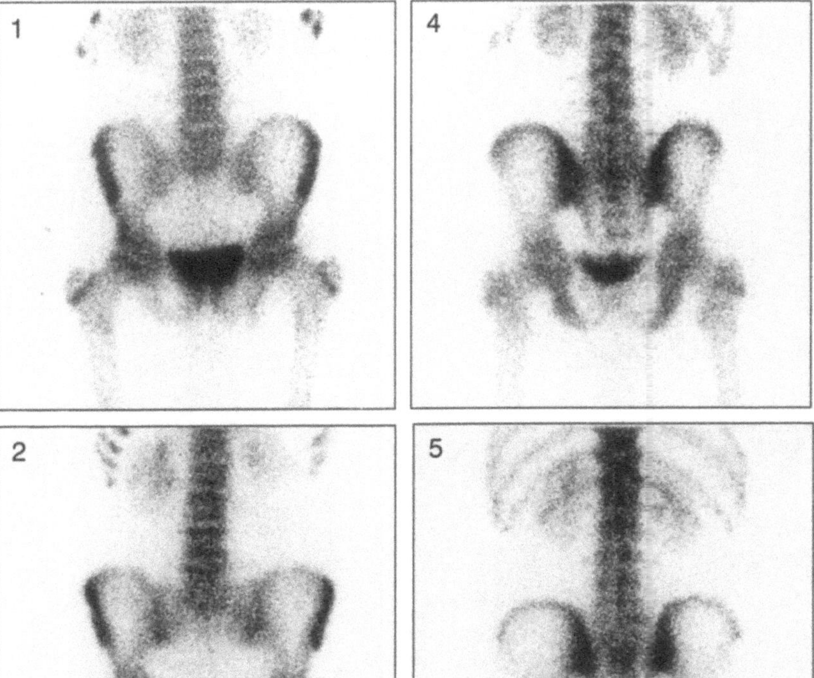

Technical Comment
– Urine contamination below the pelvis is seen in Fig. 2

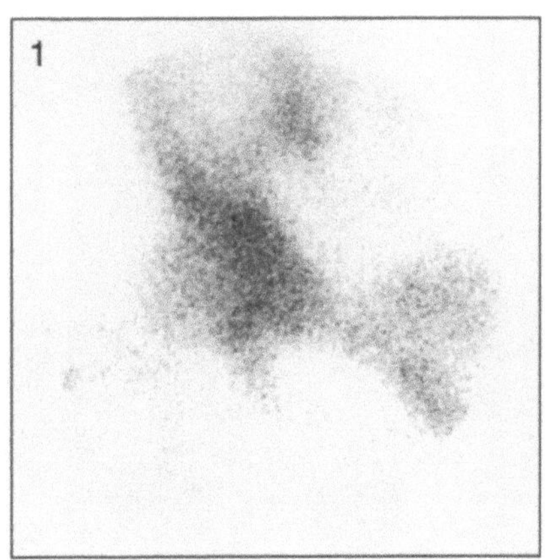
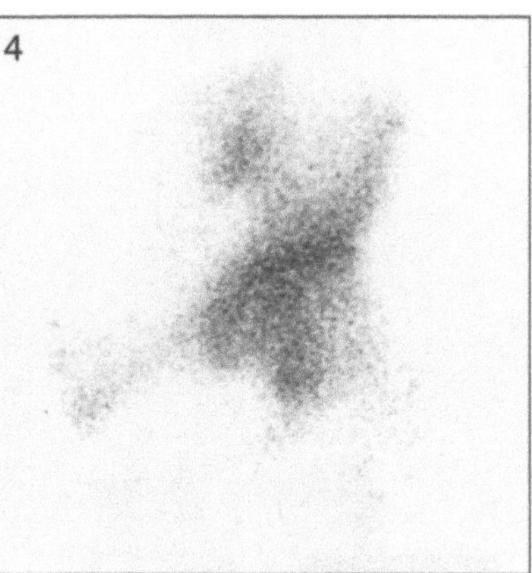

Fig. 1. Pinhole view of right hip

Fig. 4. Pinhole view of left hip

Fig. 1. Posterior view of femora

Fig. 4. Posterior view of femora

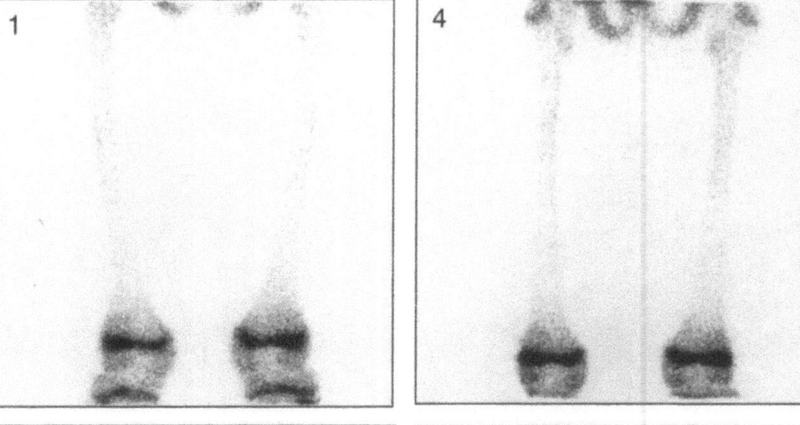

Fig. 2. Posterior view of tibia, fibula and ankles

Fig. 5. Posterior view of tibia, fibula and ankles

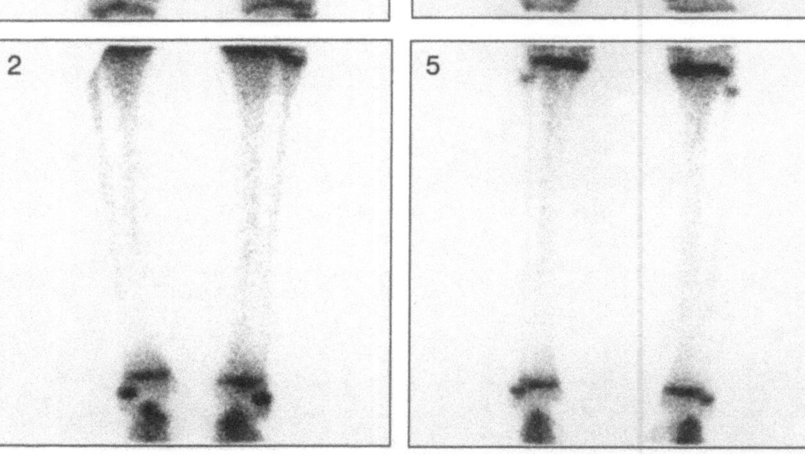

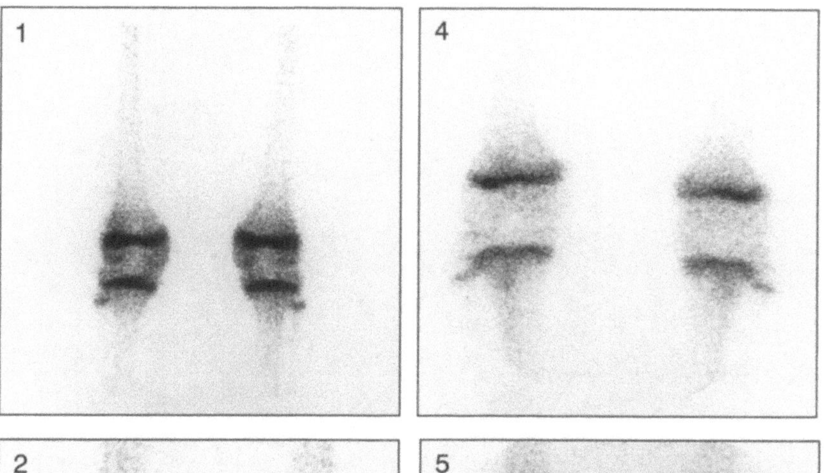

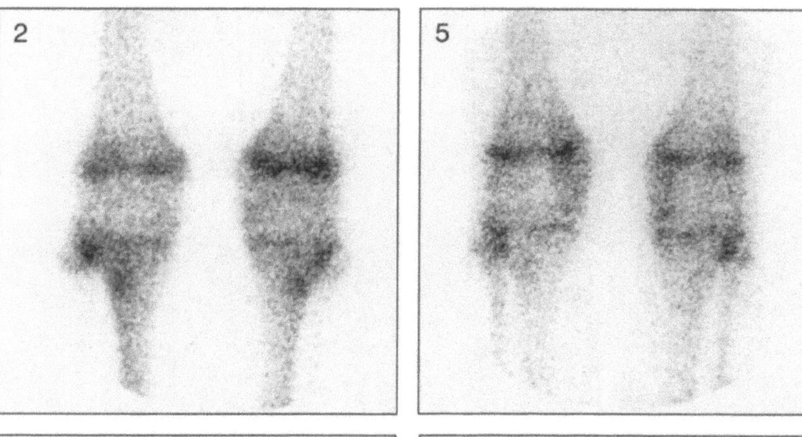

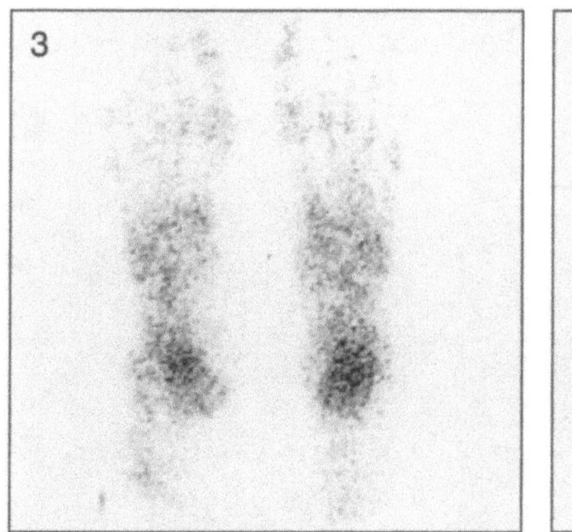

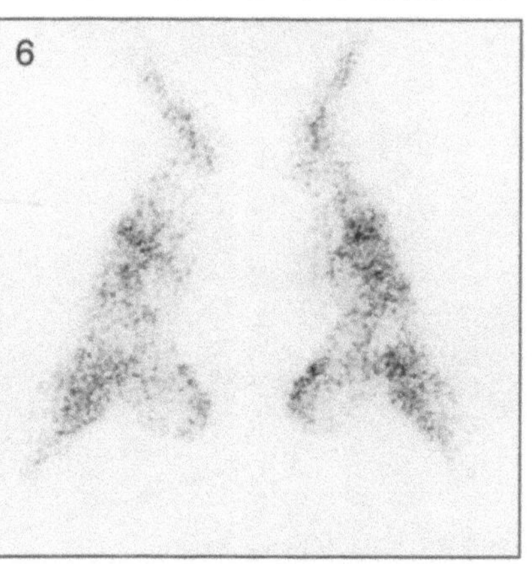

Fig. 1. Posterior view of knees

Fig. 4. Posterior view of knees

Fig. 2. Anterior view of knees

Fig. 5. Posterior view of knees

Fig. 3. Posterior view of feet

Fig. 6. Lateral view of feet

Technical Comment
– Note the differences in maturation of the epiphyseal plates in Figs. 1, 2, 4 and 5

18: Age 17–22 Years

- A double headed whole body
  gamma camera was used
- Left image is the anterior view
- Right image is the posterior
  view

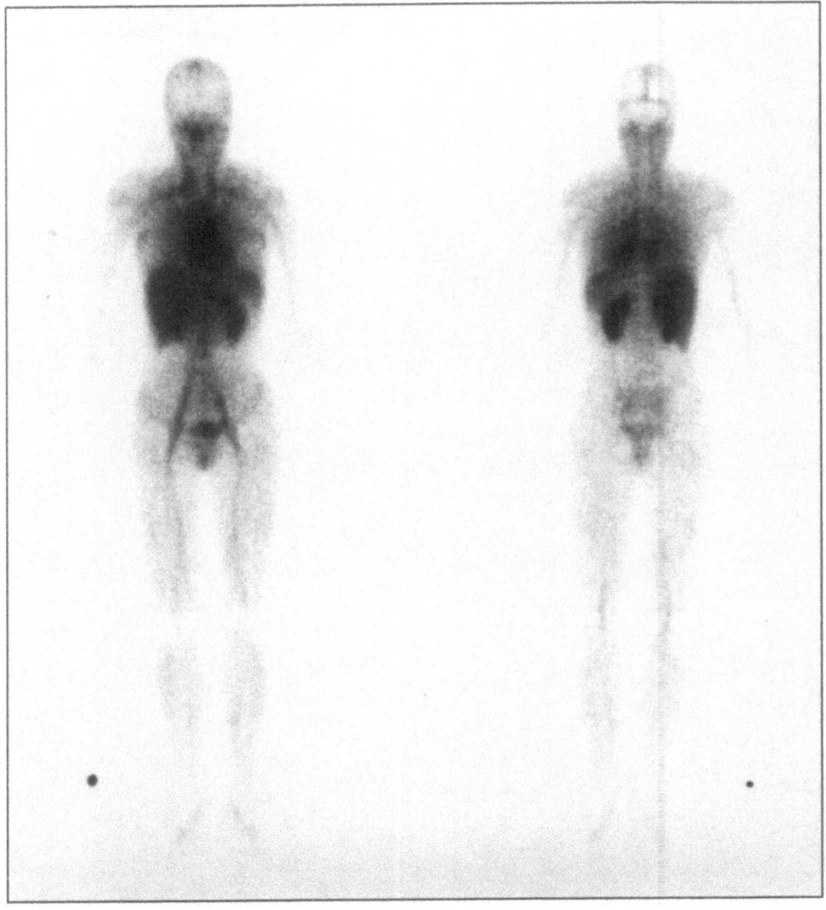

Technical Comments
- Note the mature skeleton with no visualization of the epiphyseal plates
- Marker on child's right side

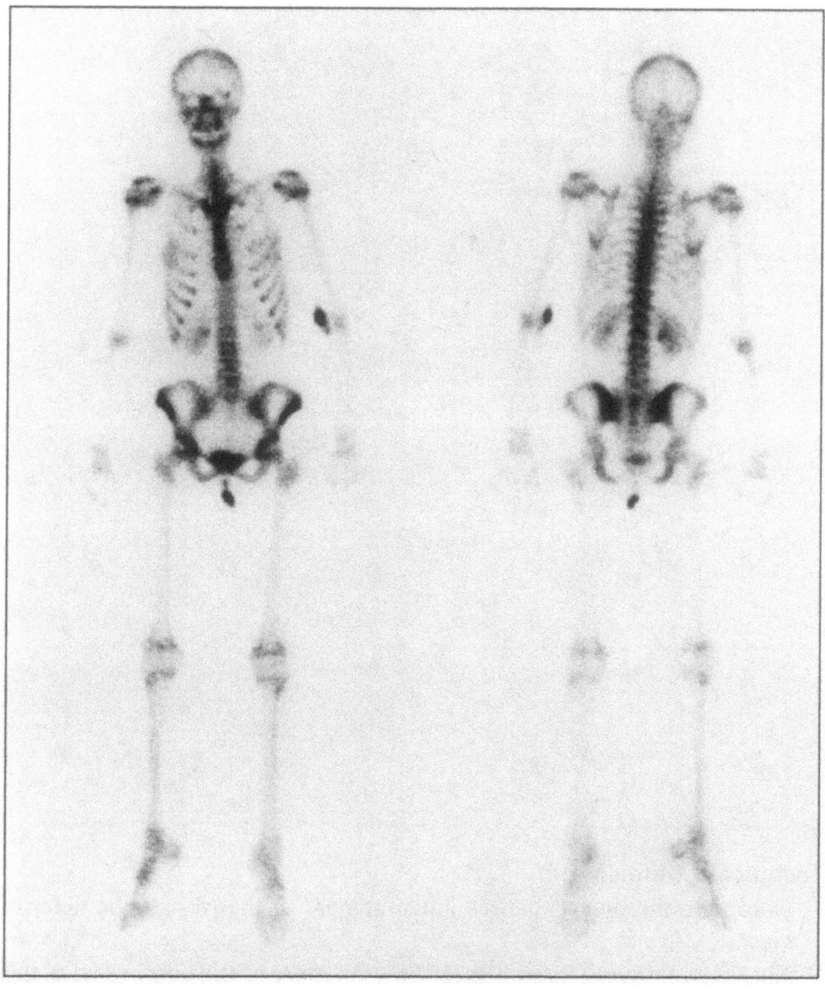

- A double headed whole body gamma camera was used
- Left image is the anterior view
- Right image is the posterior view

Technical Comments

- Isotope in the breasts causes indistinctness of the ribs on the anterior view
- Poor positioning of the feet is noted
- The focal area of apparent increased uptake in the left orbital region seen only on the anterior view is probably due to rotation of the head
- Note the mature skeleton with no visualization of the epiphyseal plates
- Note extravasation of isotope at the site of injection in the left elbow
- Urine contamination below the pelvis is seen

– A double headed whole body
  gamma camera was used
– Left image is the anterior view
– Right image is the posterior
  view

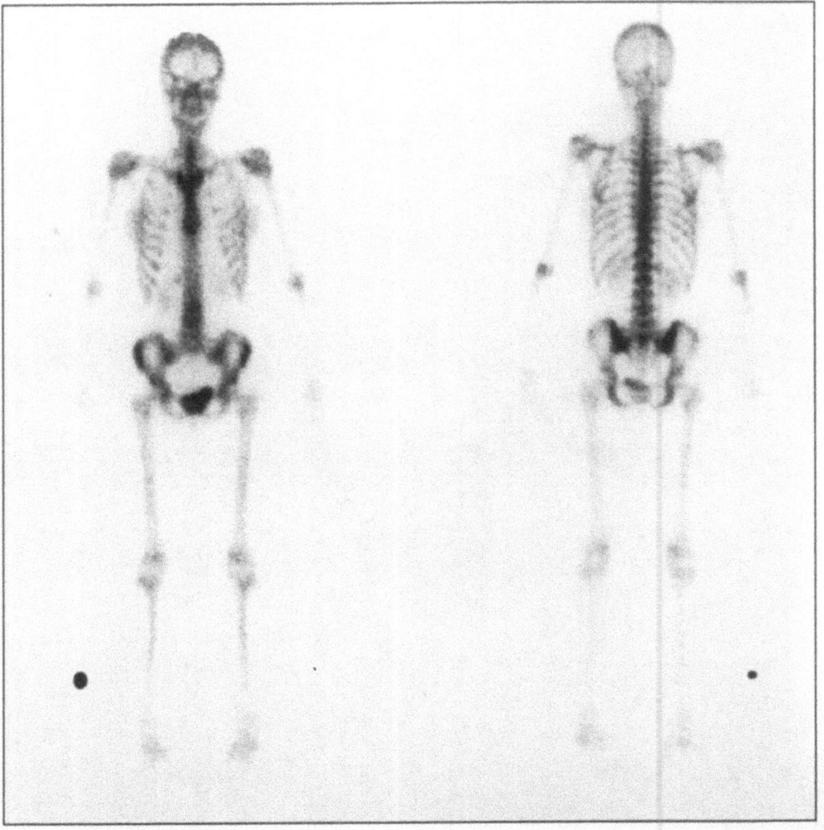

Technical Comments
– Isotope in the breasts causes indistinctness of the ribs on the anterior
  view
– Note the relatively mature skeleton with most of the epiphyseal plates
  virtually fused
– Marker on child's right side

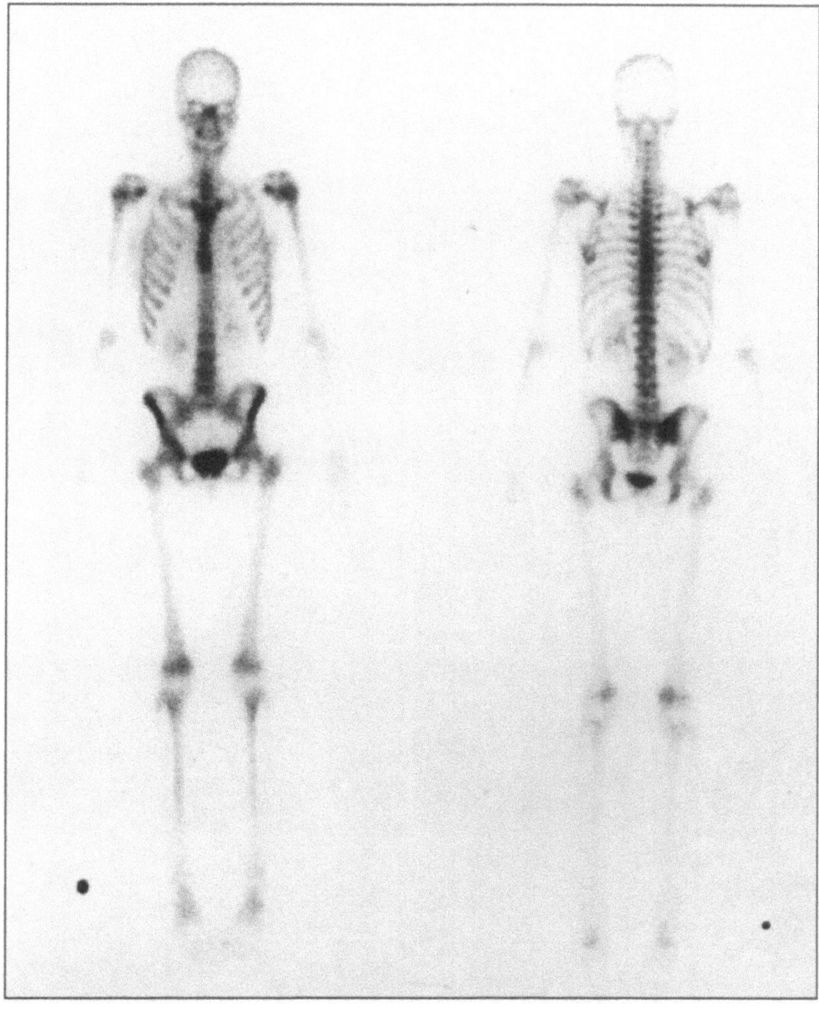

- A double headed whole body gamma camera was used
- Left image is the anterior view
- Right image is the posterior view

Technical Comments

- Isotope in the breasts causes indistinctness of the ribs on the anterior view
- Note the relatively mature skeleton with most of the epiphyseal plates virtually fused
- Marker on child's right side

Fig. 1. Anterior view of skull

Fig. 4. Posterior view of skull

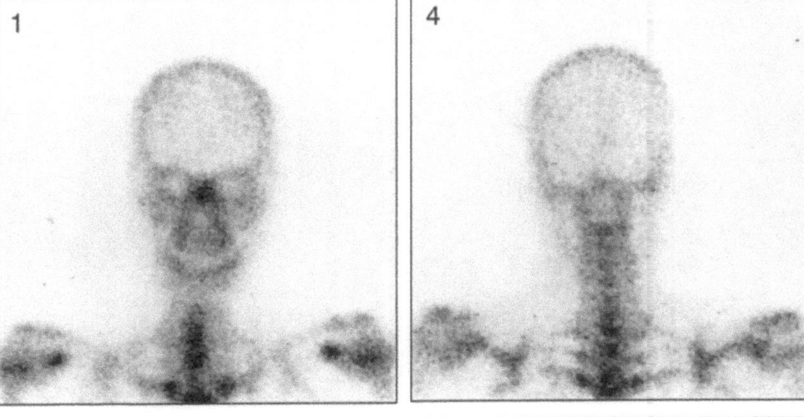

Fig. 2. Right lateral view of skull

Fig. 5. Left lateral view of skull

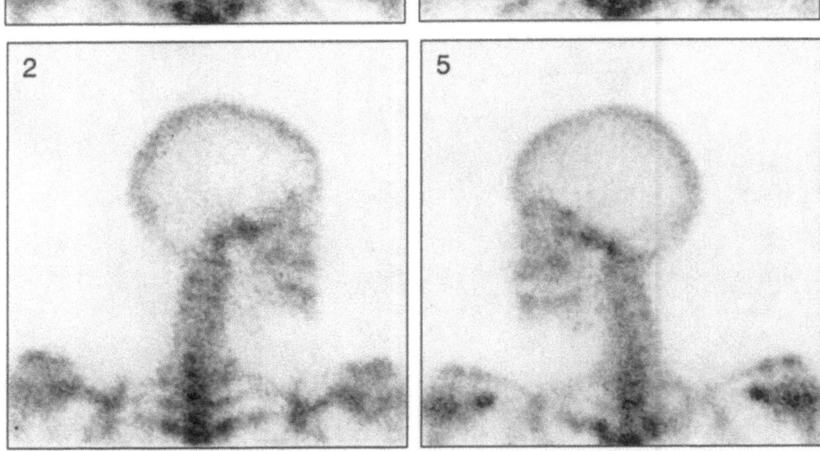

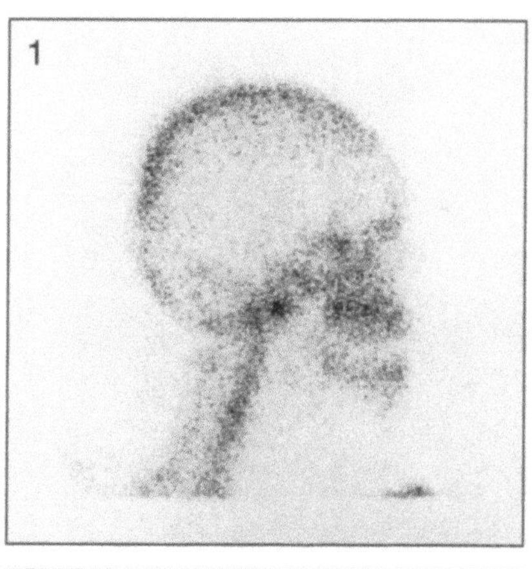

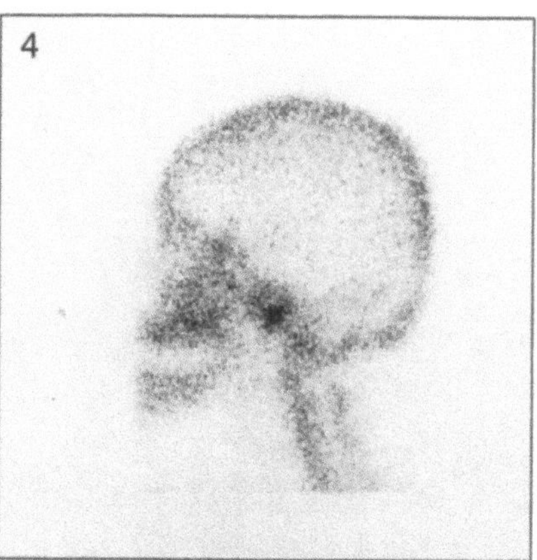

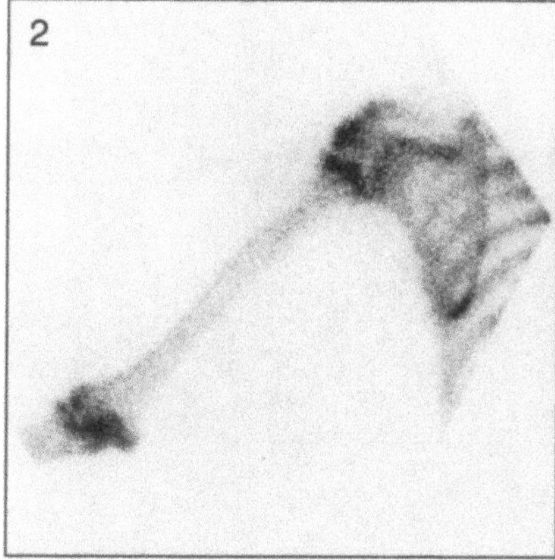

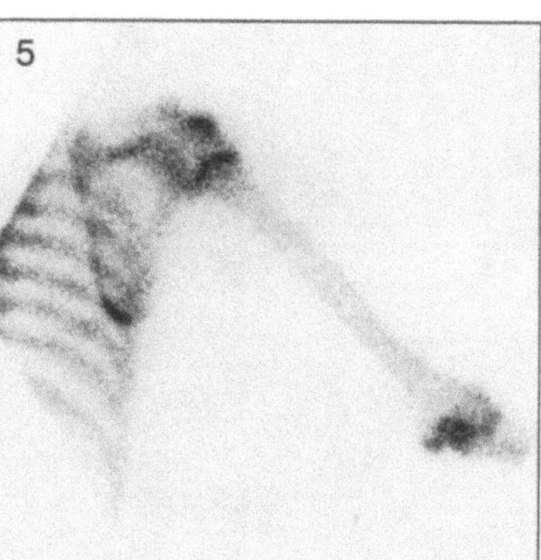

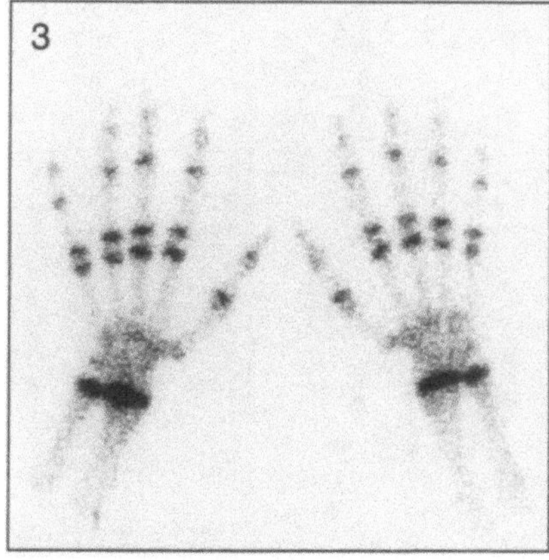

Fig. 1. Right lateral view of skull

Fig. 4. Left lateral view of skull

Fig. 2. Posterior view of left humerus

Fig. 5. Posterior view of right humerus

Fig. 3. Anterior view of hands

Technical Comment

– Note the variation in the maturity of the skeleton in this age group. The epiphysis around the hands and wrists are not yet fused in this particular child (Fig. 3)

Fig. 1. Anterior view of thorax

Fig. 4. Posterior view of thorax and spine

Fig. 2. Anterior view of spine and pelvis

Fig. 5. Posterior view of spine and pelvis

Fig. 3. Pinhole view of right hip

Fig. 6. Pinhole view of left hip

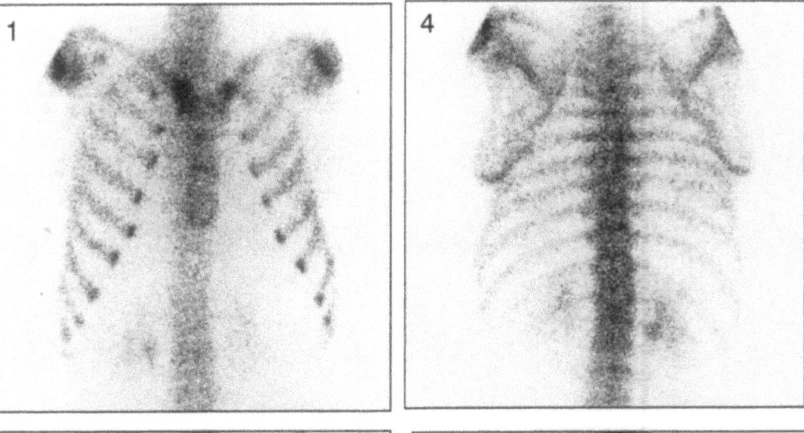

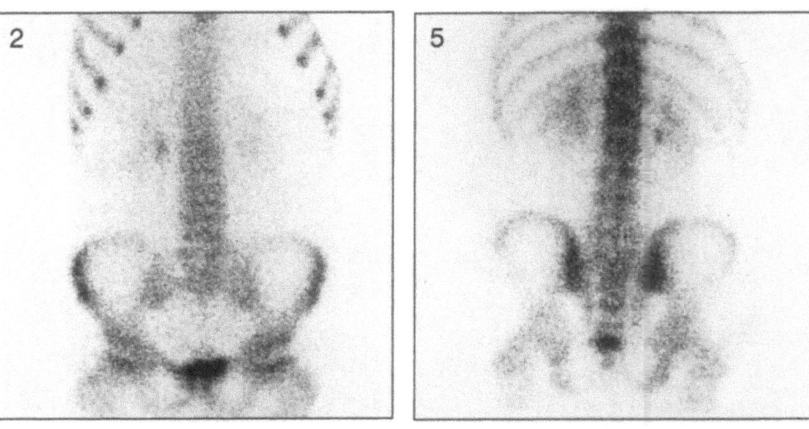

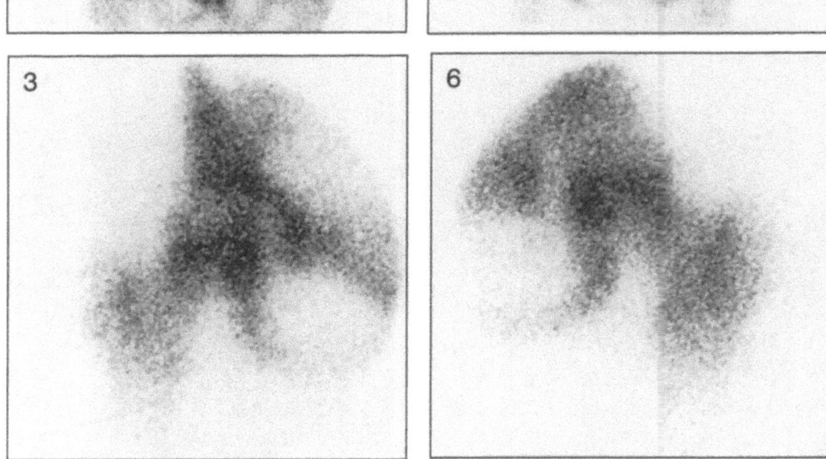

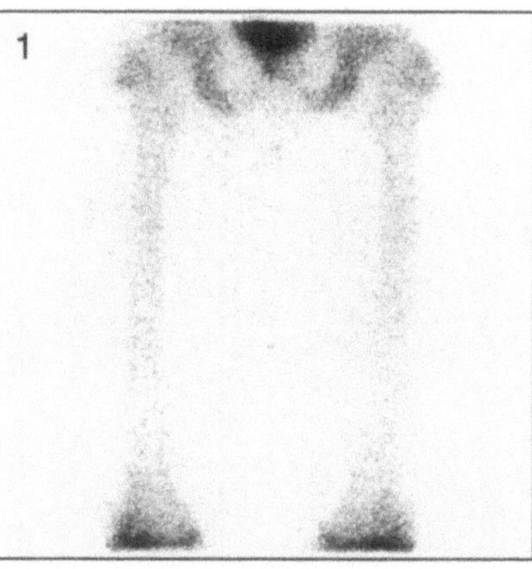

Fig. 1. Posterior view of femora

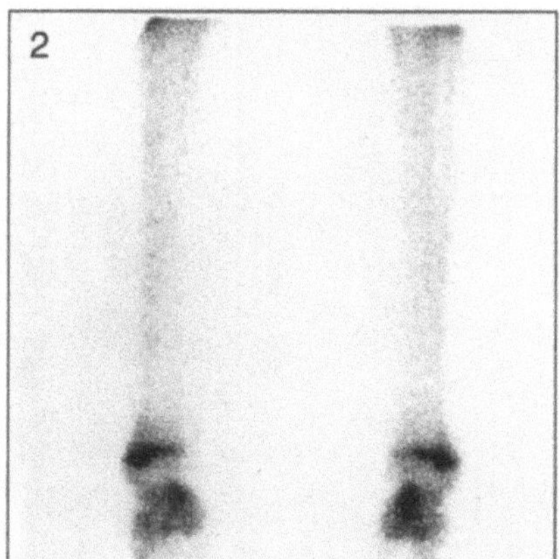

Fig. 2. Posterior view of tibia, fibula and ankles

Fig. 1. Anterior view of knees

Fig. 4. Posterior view of knees

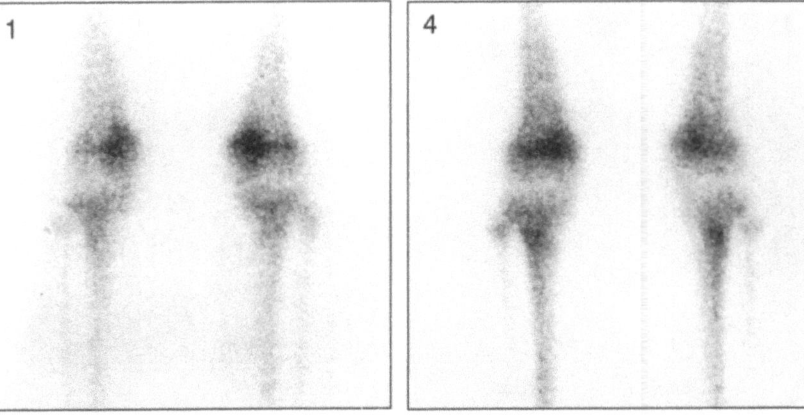

Fig. 2. Anterior view of feet

Fig. 5. Posterior view of feet

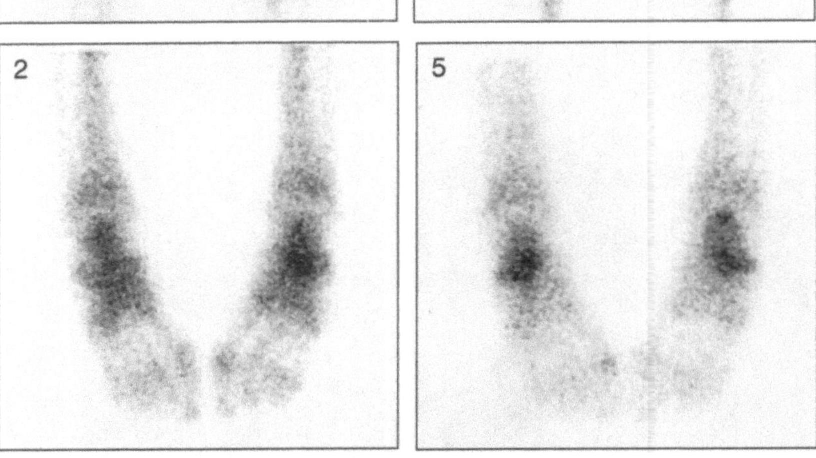

19: Knees

Fig. 1. Anterior view

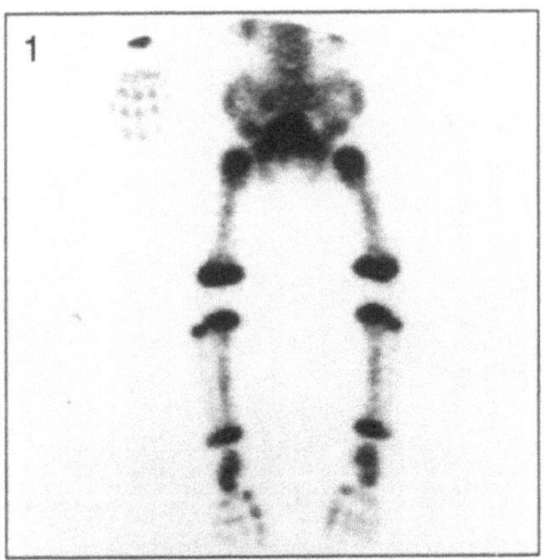

Technical Comment

– The toes are facing medially, "the radiographic neutral position". This
is the reason that the fibula is clearly seen

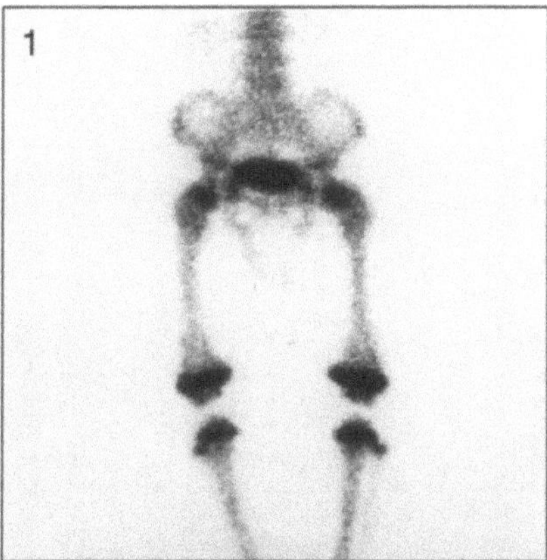

Fig. 1. Anterior view

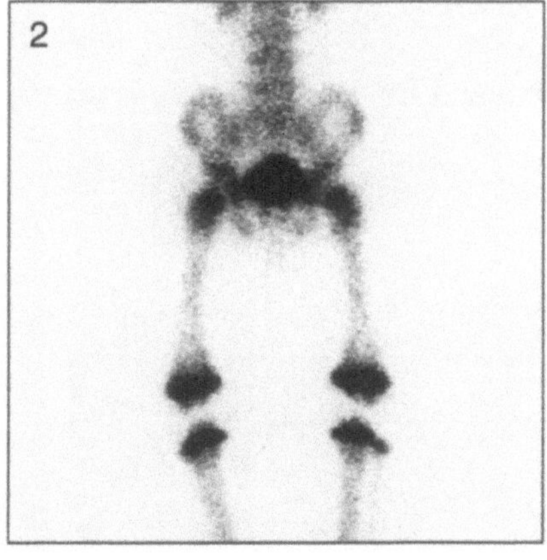

Fig. 2. Anterior view

Technical Comments

- Fig. 1 shows good positioning of the left knee and foot, this allows visualization of the left fibula. The fibula is not seen on the right due to poor positioning of both the knee and the foot
- Fig. 2 shows good positioning of the left knee and foot, this allows visualization of the left fibula
- Note the shape of the epiphyseal plates around the knees

Fig. 1. Anterior view

Fig. 4. Posterior view

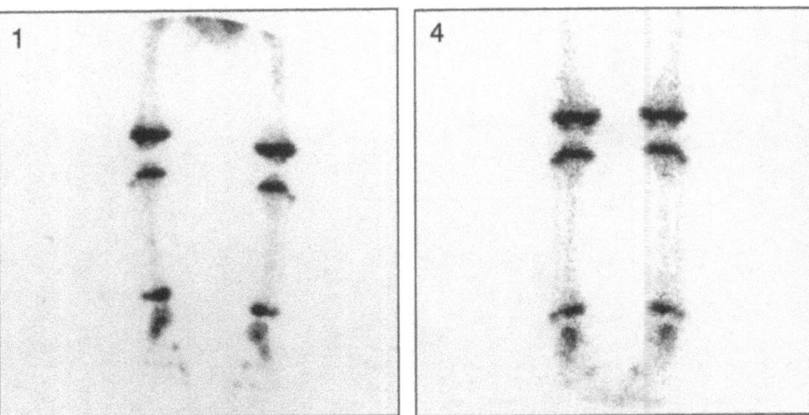

Technical Comment
– Good positioning for the knees, tibia and fibula. The fibula is clearly
  seen separate from the tibia

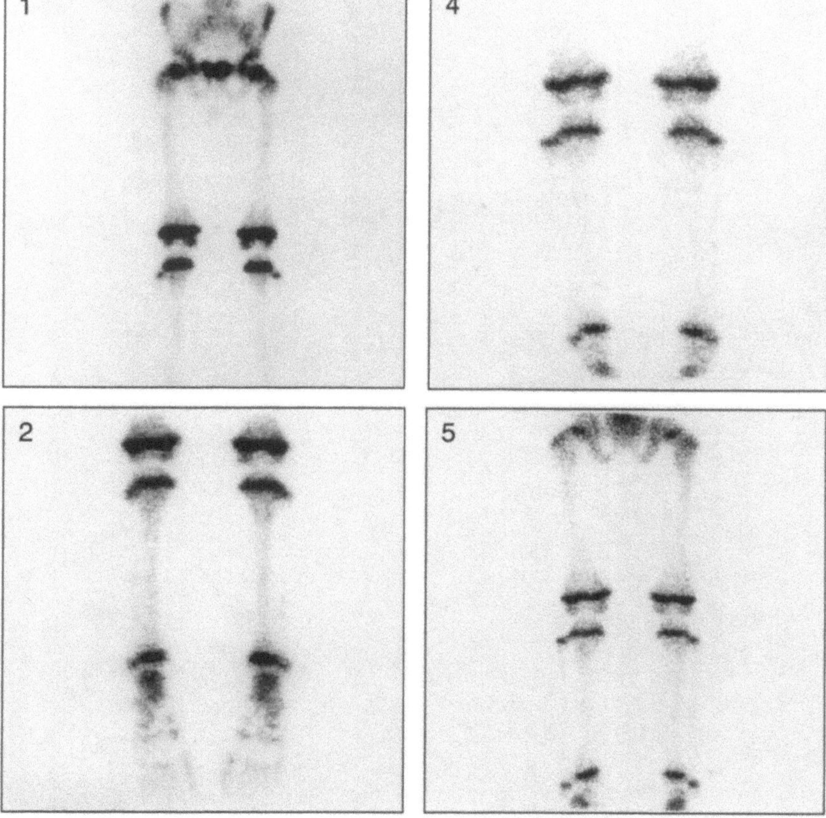

Fig. 1. Anterior view

Fig. 4. Posterior view

Fig. 2. Anterior view

Fig. 5. Posterior view

Technical Comments
- Note the clarity of the fibula separate from the tibia in Figs. 1, 2, 4 and 5. This is due to the "radiographic neutral position of the feet"
- Note the progressive changes in the epiphyseal plates with maturation

Fig. 1. Anterior view

Fig. 4. Posterior view

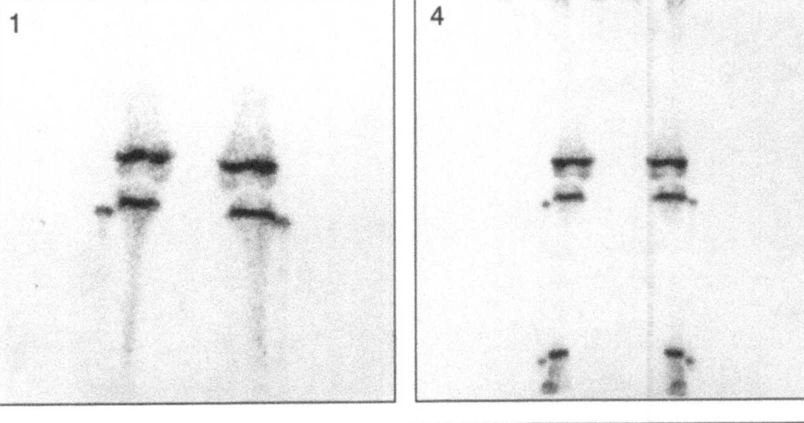

Fig. 5. Posterior view

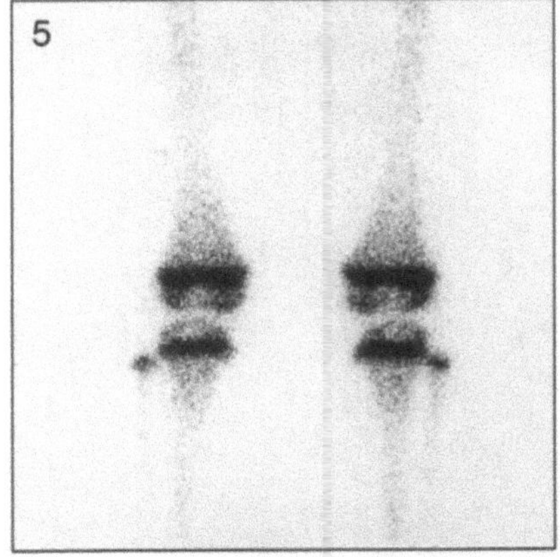

Technical Comment
– The clarity of the fibula on all the images suggests good positioning of
  the feet

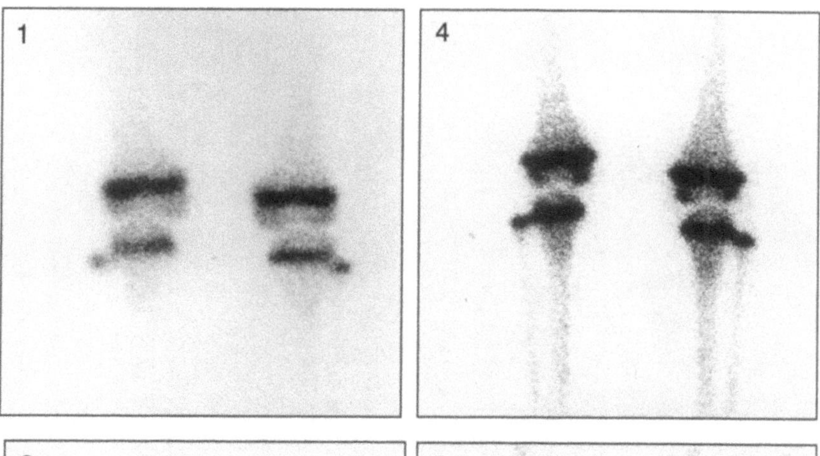

Fig. 1. Posterior view

Fig. 4. Posterior view

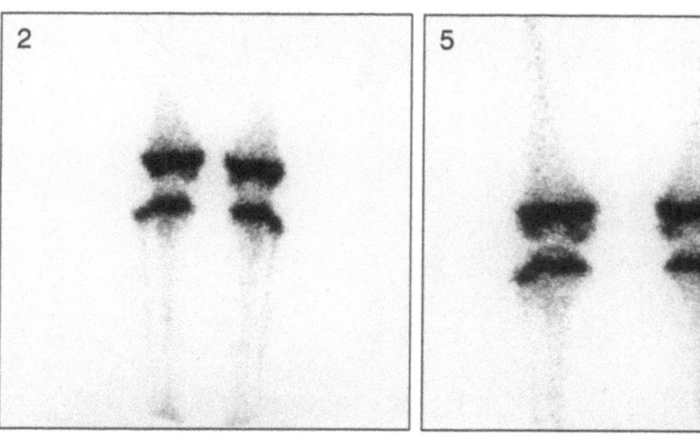

Fig. 2. Posterior view

Fig. 5. Posterior view

Technical Comment
– Note the clarity of the heads of the fibulae on all the images. Note the clarity of epiphyseal plates with their well defined margins

Fig. 1. Posterior view

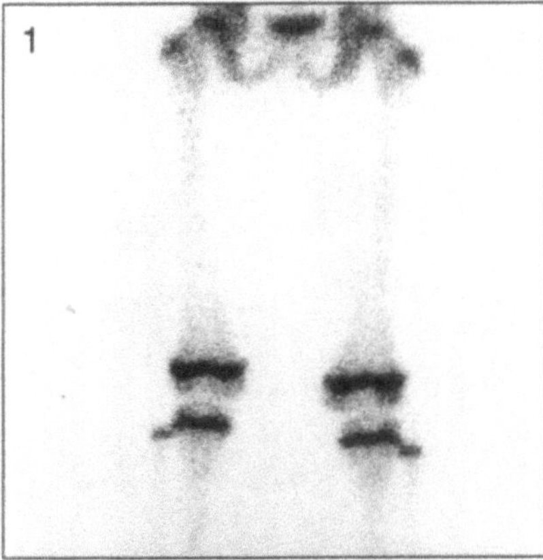

Fig. 2. Posterior magnified view

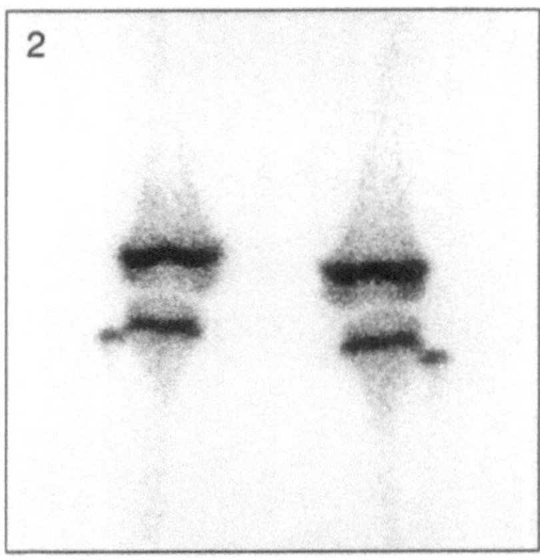

Technical Comments
- The magnified view (Fig. 2) shows the epiphyseal plates to best advantage
- The fibula is clearly seen in both figures because the toes were turned inward during image acquisition

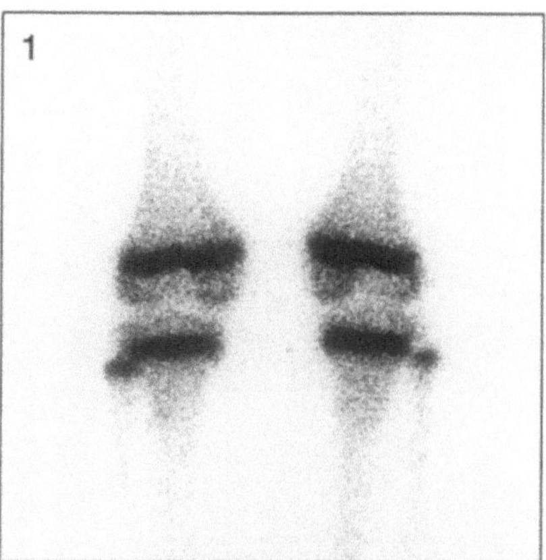

Fig. 1. Posterior magnified view

Fig. 2. Posterior view

Technical Comment
– Note the clear definition of the growth plate from the adjacent meta-
  physes

Fig. 1. Posterior magnified view

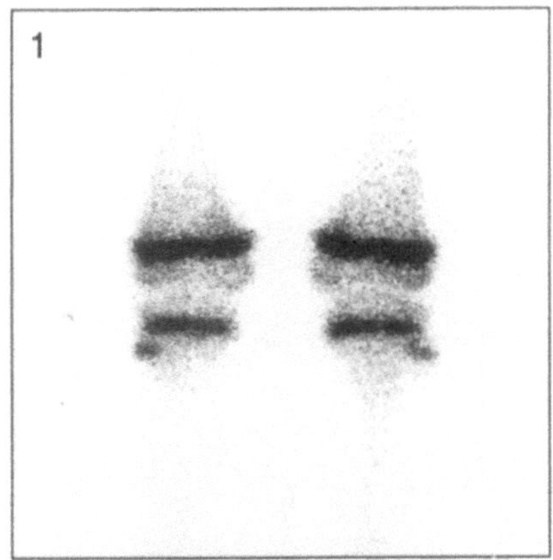

Fig. 2. Posterior magnified view

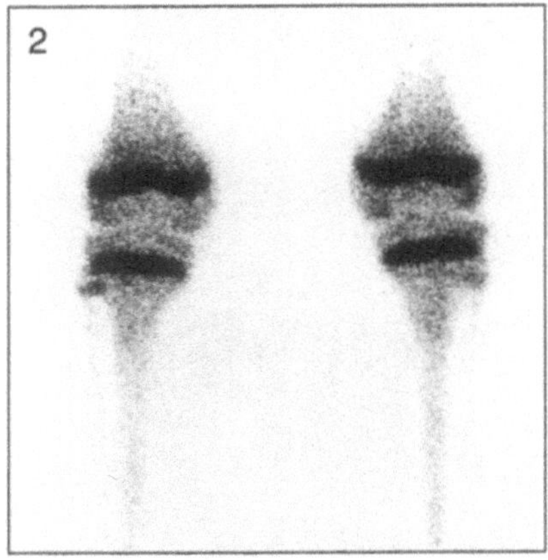

Fig. 3. Posterior view

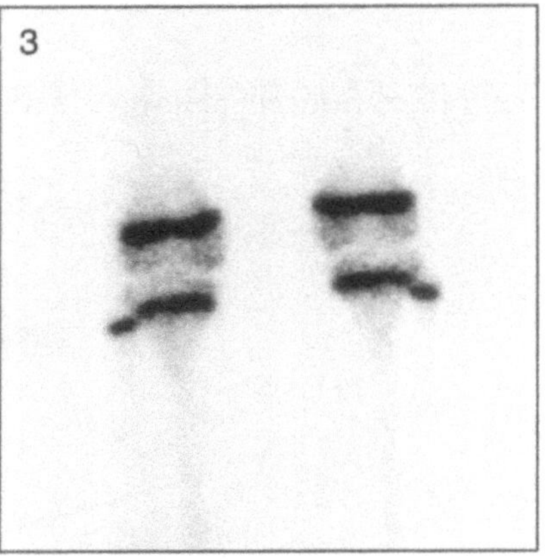

Technical Comments
- There is clear separation between the upper tibia and the fibula due to good positioning of the feet in all images
- Slight overexposure as seen in Fig. 2 shows the tibia better but causes overexposure of the epiphyseal plates. This is the reason for the different appearance of this image compared to Figs. 1 and 3

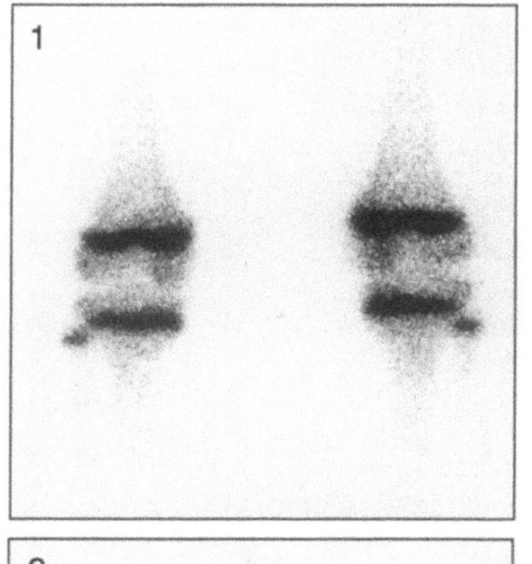

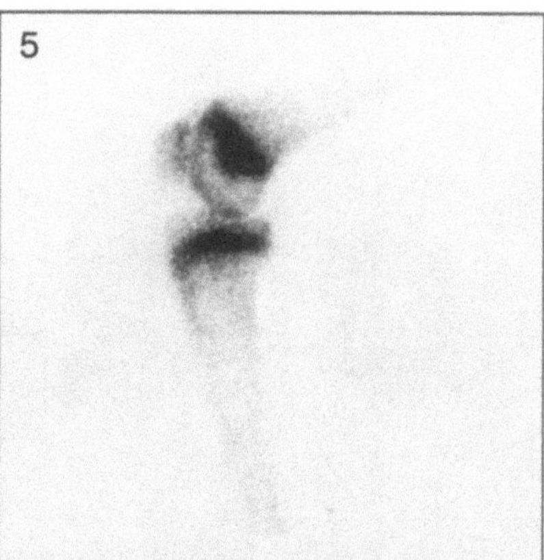

Fig. 1. Posterior view

Fig. 4. Anterior view

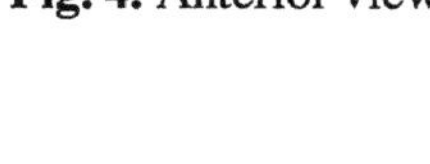

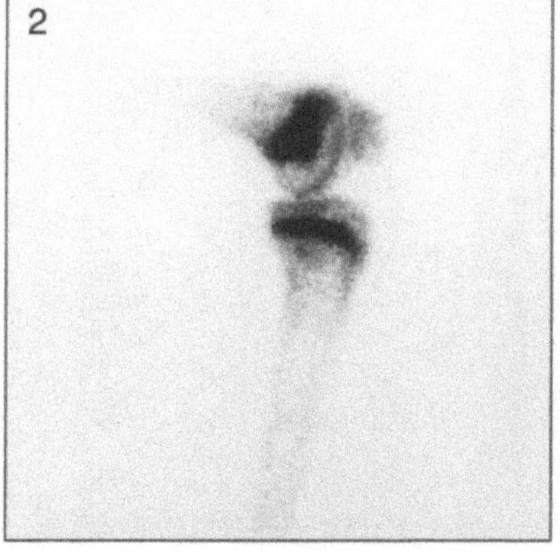

Fig. 2. Medial lateral view of left knee

Fig. 5. Medial lateral view of right knee

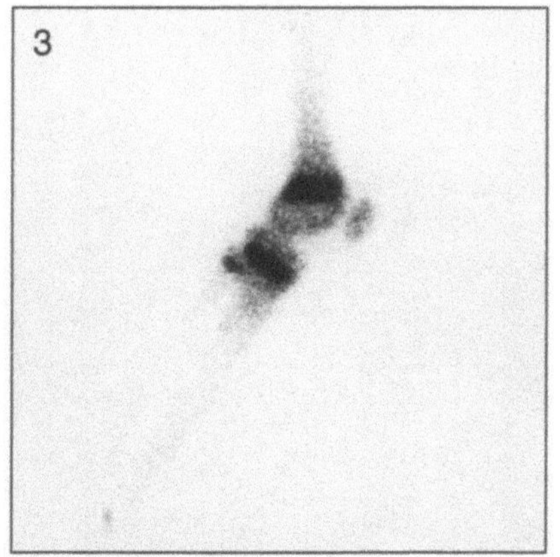

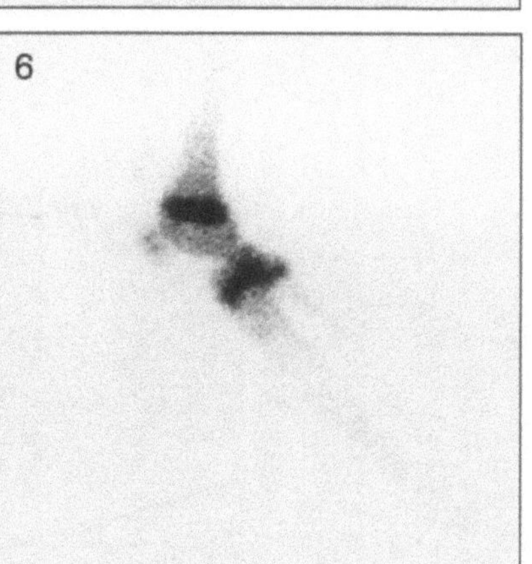

Fig. 3. Lateral lateral view of right knee

Fig. 6. Lateral lateral view of left knee

Technical Comment

– The pair of lateral knees in Figs. 2 and 5 represent medial lateral knees while Figs. 3 and 6 represent lateral knees in the lateral projection. This accounts for the clarity of the fibula in the lower series and the difficulty in seeing the fibula in the upper series

Fig. 1. Posterior view

Fig. 4. Posterior view

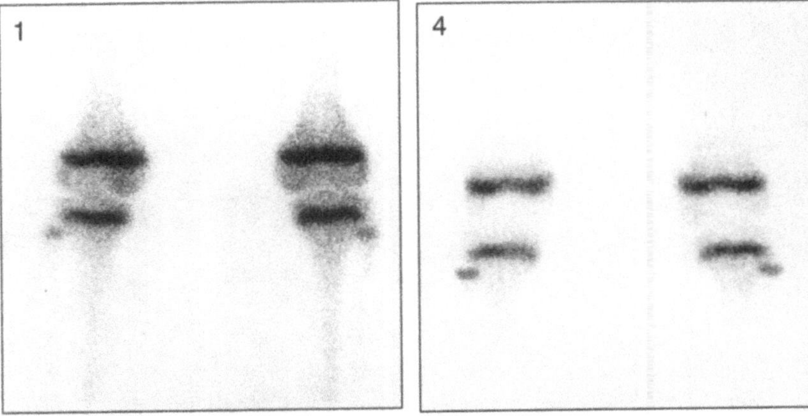

Fig. 2. Posterior view

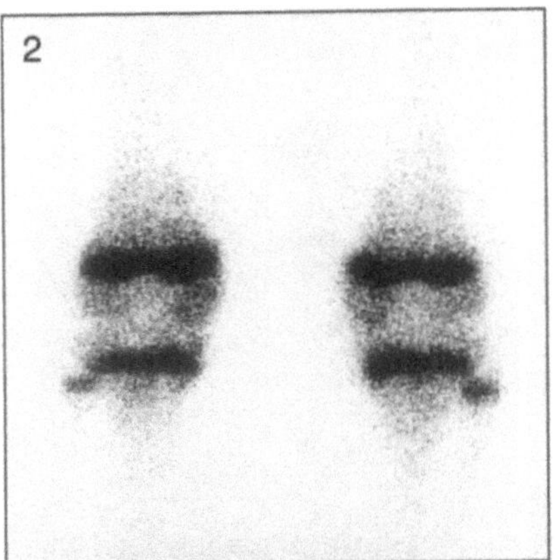

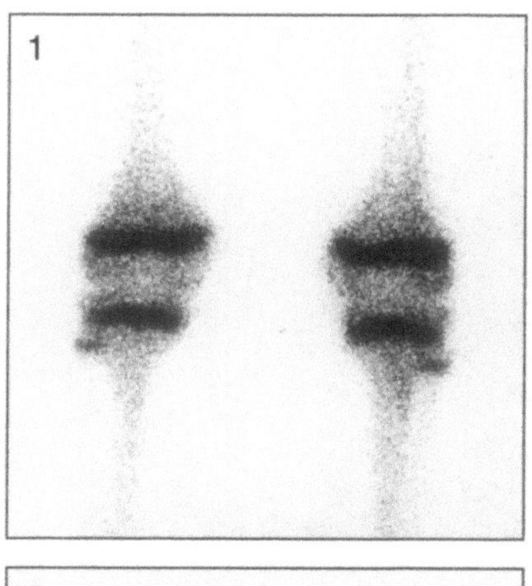

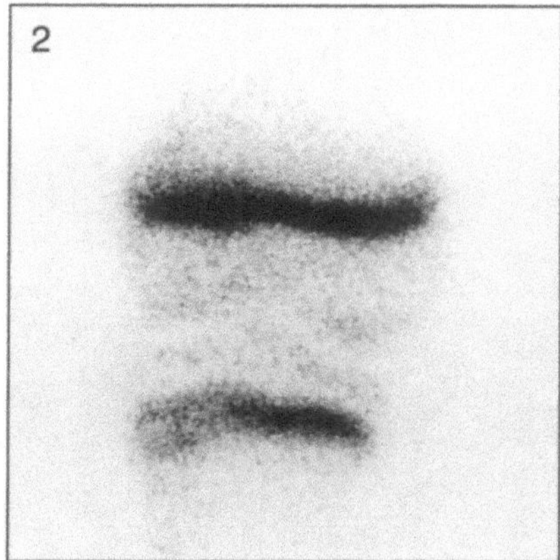

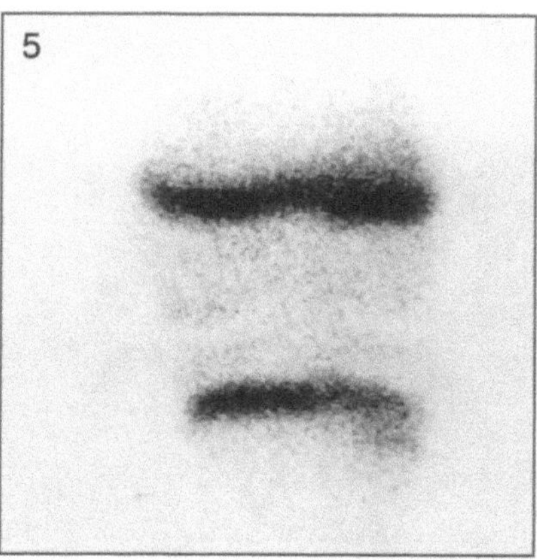

Fig. 1. Posterior magnified view

Fig. 4. Posterior view

Fig. 2. Anterior pinhole view of right knee

Fig. 5. Anterior pinhole view of left knee

Technical Comment

– Note that the fibula is better seen on the posterior view in Figs. 1 and 4 than on the anterior pinhole views in Figs. 2 and 5

Fig. 1. Posterior view

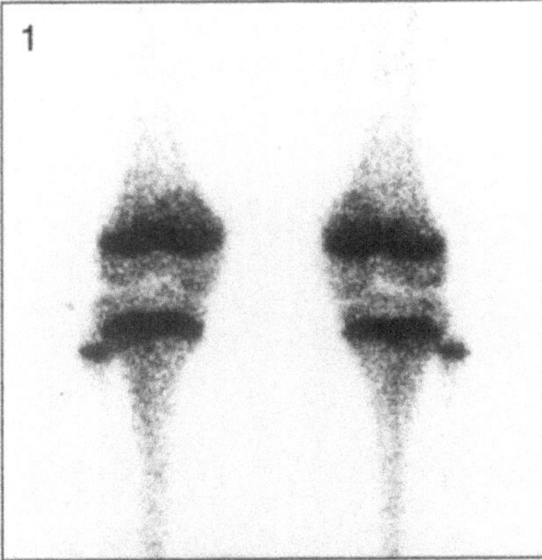

Fig. 2. Posterior view

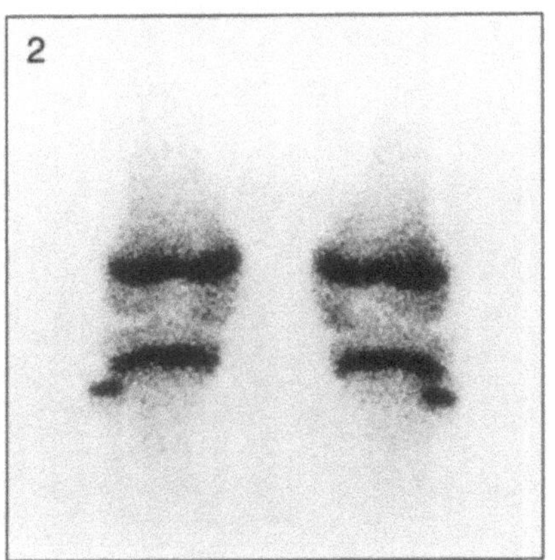

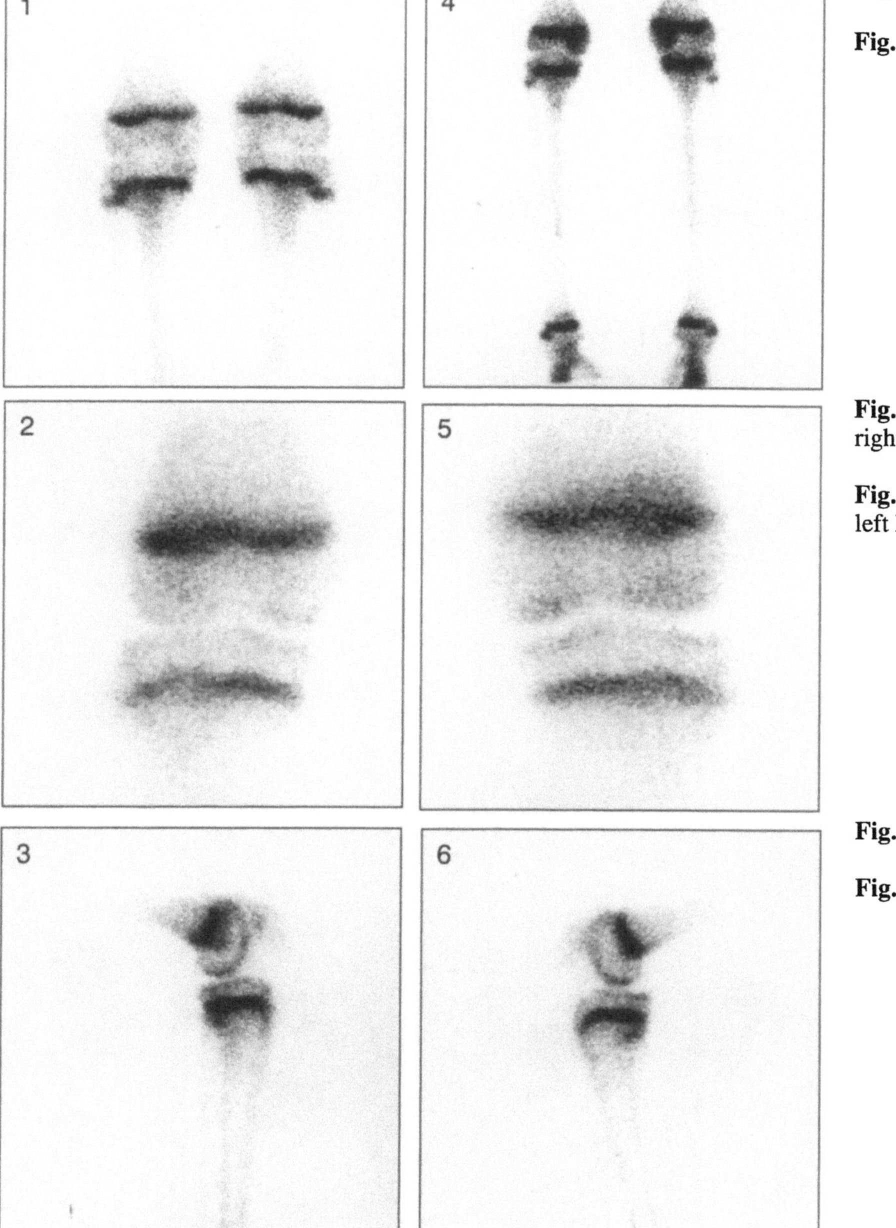

Fig. 1. Anterior view

Fig. 4. Posterior view

Fig. 2. Anterior pinhole view of right knee

Fig. 5. Anterior pinhole view of left knee

Fig. 3. Lateral view of left knee

Fig. 6. Lateral view of right knee

Technical Comment

– The fibula is not seen in Figs. 3 and 6 because the images are medial lateral views

Fig. 1. Anterior view

Fig. 4. Posterior view

Fig. 2. Anterior pinhole view of right knee

Fig. 5. Anterior pinhole view of left knee

Fig. 3. Lateral view

Fig. 6. Lateral view

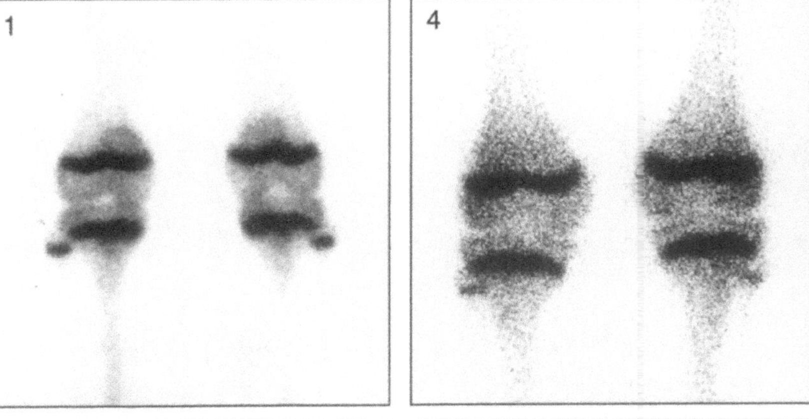

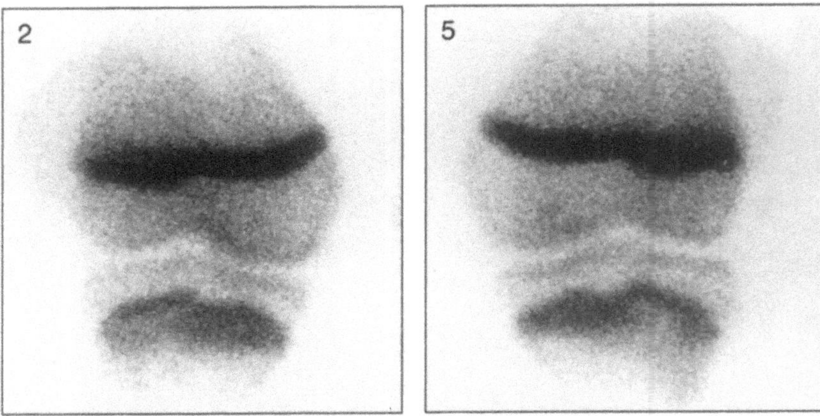

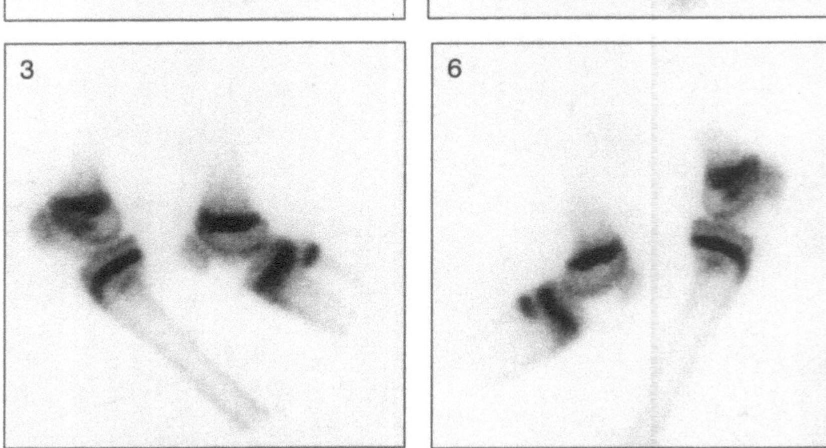

Technical Comment

– Figs. 3 and 6 represent lateral images of the knees, on the left there is a medial lateral and a lateral lateral while on the right it is the reverse order. The lateral lateral images show the head of the fibula clearly while the medial lateral show the head of the fibula within the tibia but below the epiphyseal plate

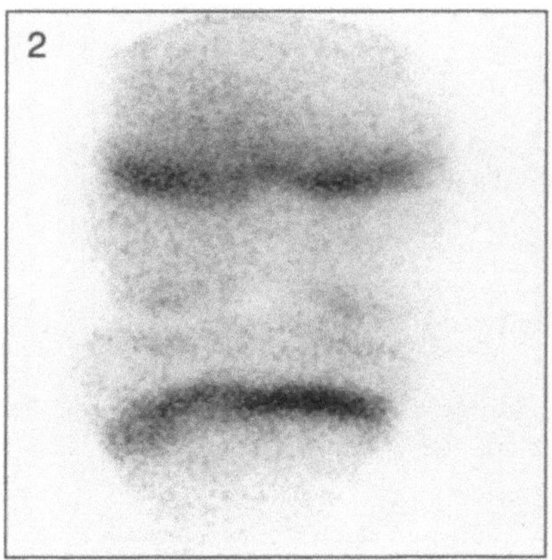

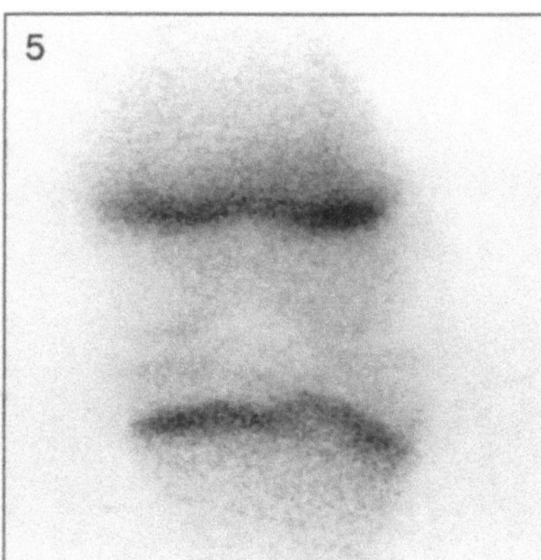

Fig. 1. Posterior view

Fig. 2. Anterior pinhole view of right knee

Fig. 5. Anterior pinhole view of left knee

Fig. 1. Posterior view

Fig. 4. Posterior view

Fig. 2. Anterior view

Fig. 5. Posterior view

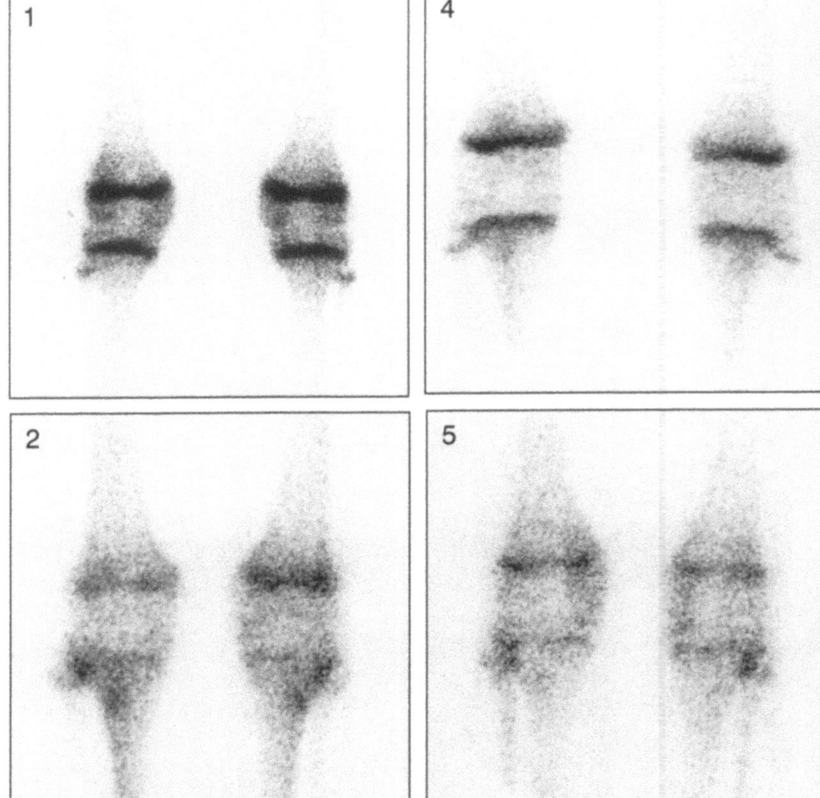

Technical Comment
– Note the differences in maturation of the epiphyseal plates in Figs. 1, 2, 4 and 5

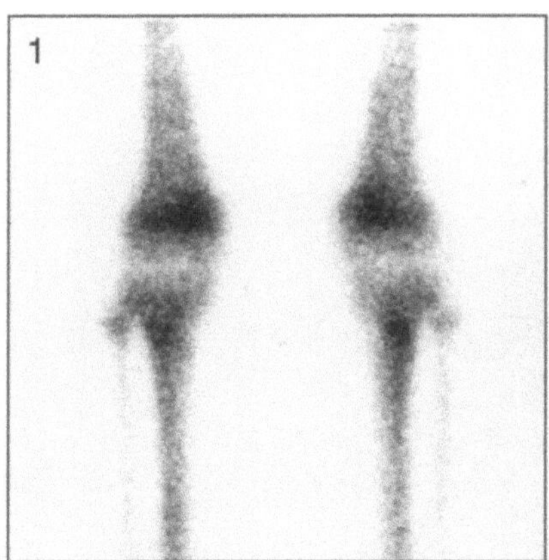

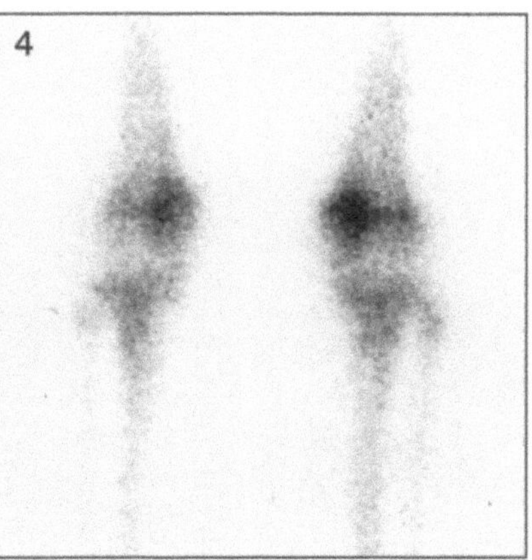

Fig. 1. Anterior view

Fig. 4. Posterior view

Technical Comment
– Note the epiphyseal plates are distinct as the skeleton approaches maturity

20: Hips

Fig. 1. Posterior view

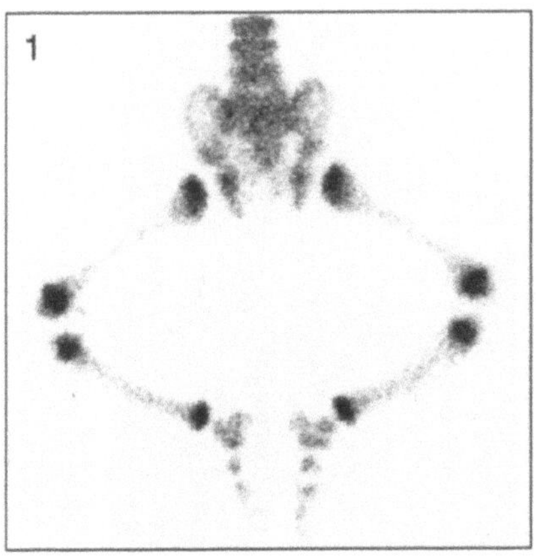

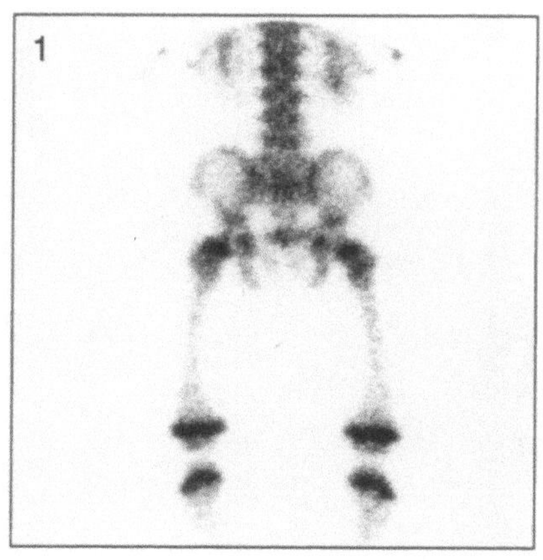

Fig. 1. Posterior view

Fig. 1. Posterior view

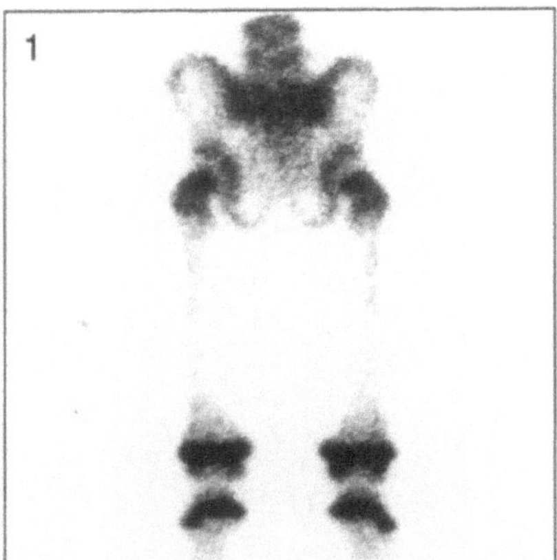

Fig. 2. Pinhole view of right hip

Fig. 5. Pinhole view of left hip

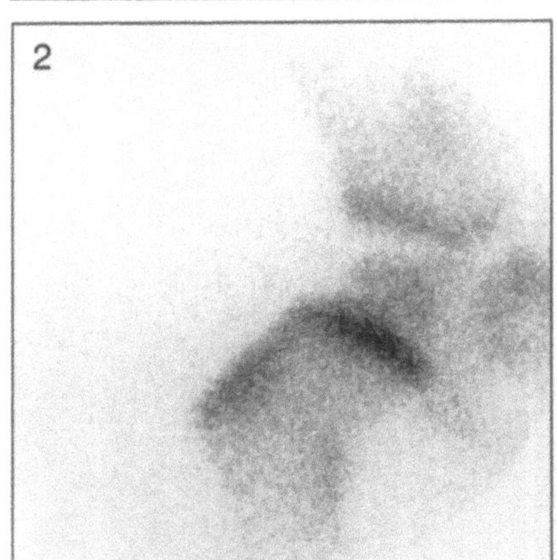
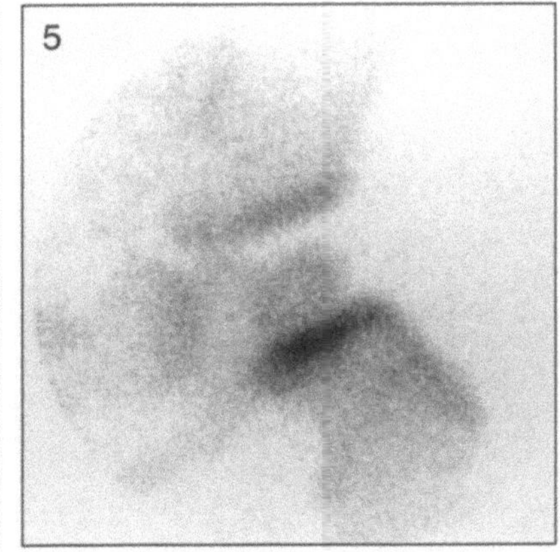

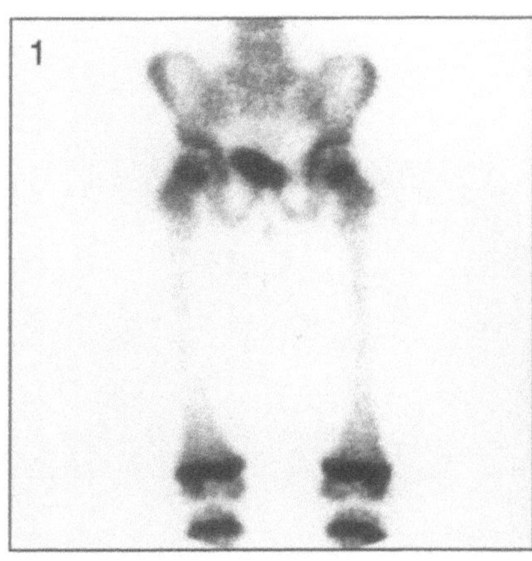

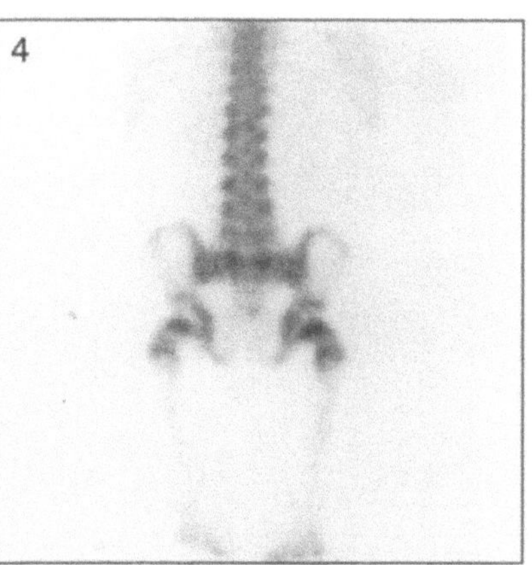

Fig. 1. Anterior view

Fig. 4. Posterior view

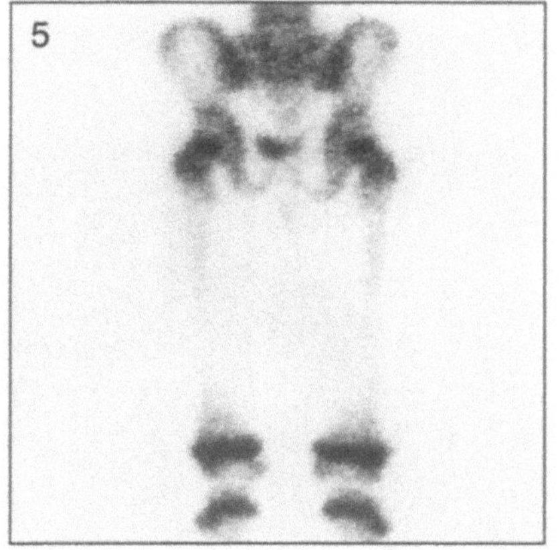

Fig. 5. Posterior view

Technical Comments

– Note that the bladder is virtually empty on all three images, an ideal situation
– Urine contamination below the pelvis is seen in Figs. 1 and 5

Fig. 1. Pinhole view of right hip

Fig. 4. Pinhole view of left hip

Fig. 2. Pinhole view of right hip

Fig. 5. Pinhole view of left hip

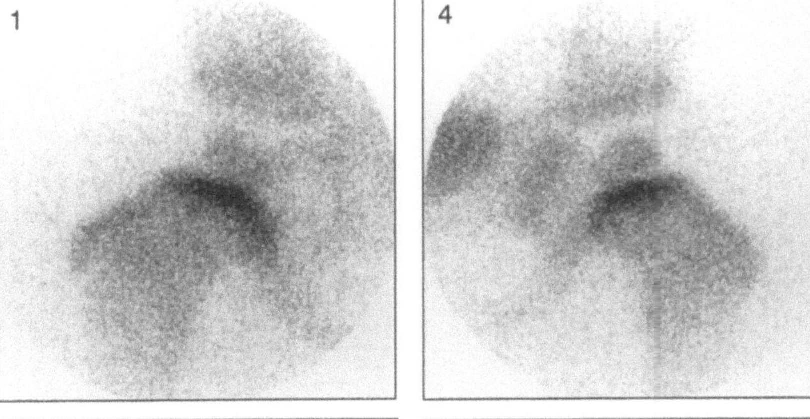

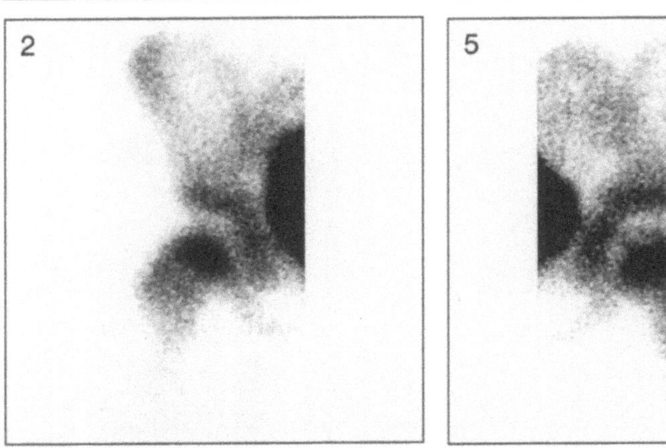

Technical Comment
- The femoral heads are well seen in Figs. 1 and 4 while poor visualization is noted in Figs. 2 and 5 because of poor positioning of the knees and feet

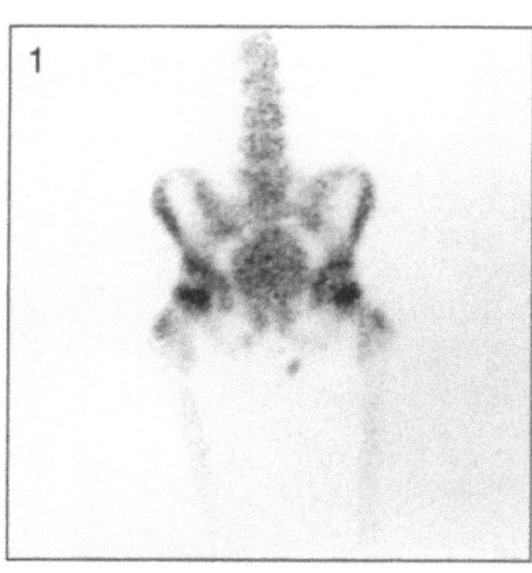

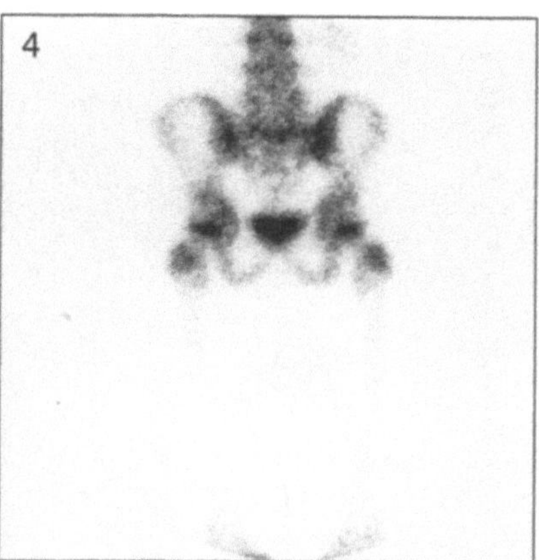

Fig. 1. Anterior view

Fig. 4. Posterior view

Technical Comments
- The bladder in Fig. 1 is of relatively large volume but has little activity suggesting that the child is well hydrated
- Urine contamination below the pelvis is seen in Fig. 1

Fig. 1. Pinhole view of right hip

Fig. 4. Pinhole view of left hip

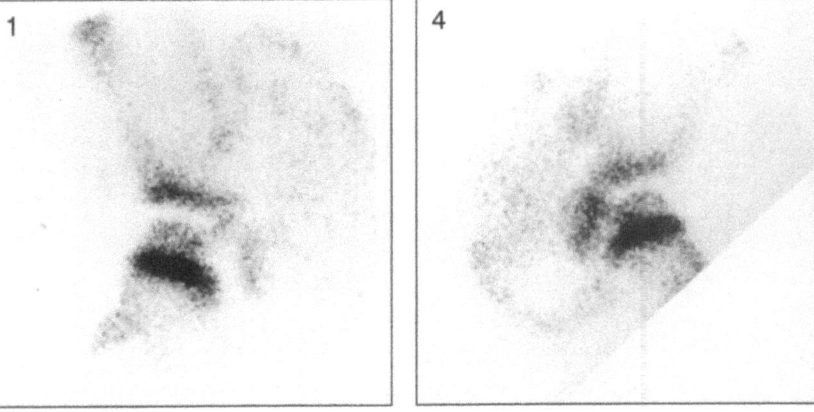

Fig. 2. Pinhole view of right hip

Fig. 5. Pinhole view of left hip

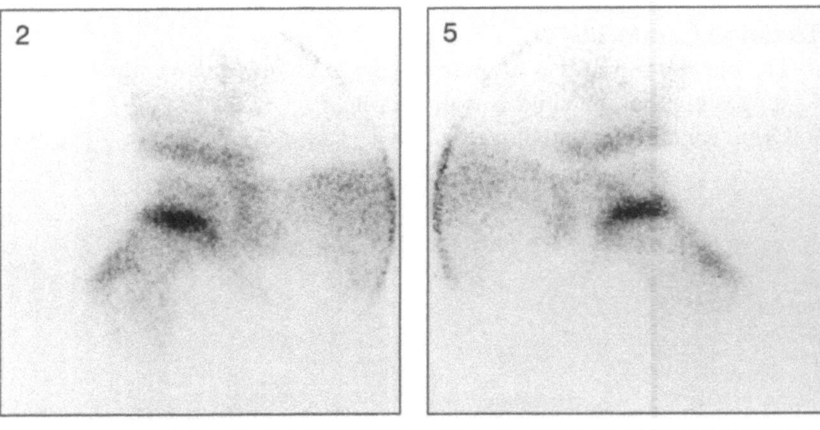

Fig. 3. Pinhole view of right hip

Fig. 6. Pinhole view of left hip

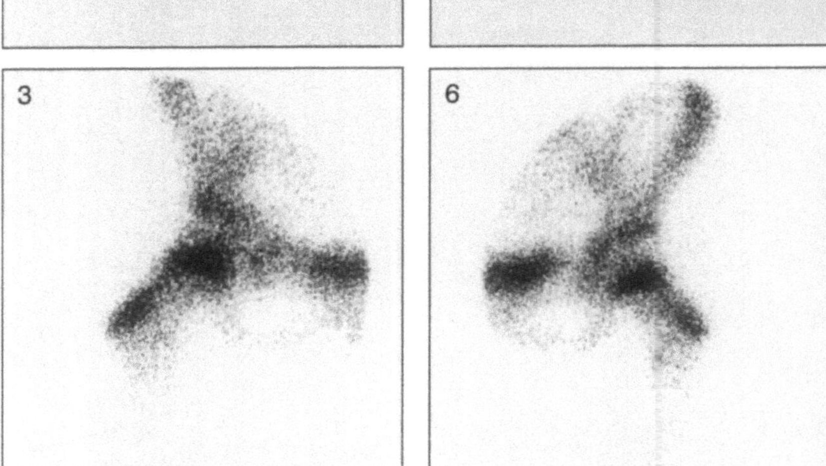

Technical Comments

- Different size of pinhole inserts give different degrees of magnification
- The position of the hips in the bottom series (Figs. 3 and 6) is not as good as in the top two. This is related to the positioning of the knees and the lack of in-turning of the feet. The "radiographic neutral position of the feet and knees" is essential when doing pinholes of the hips

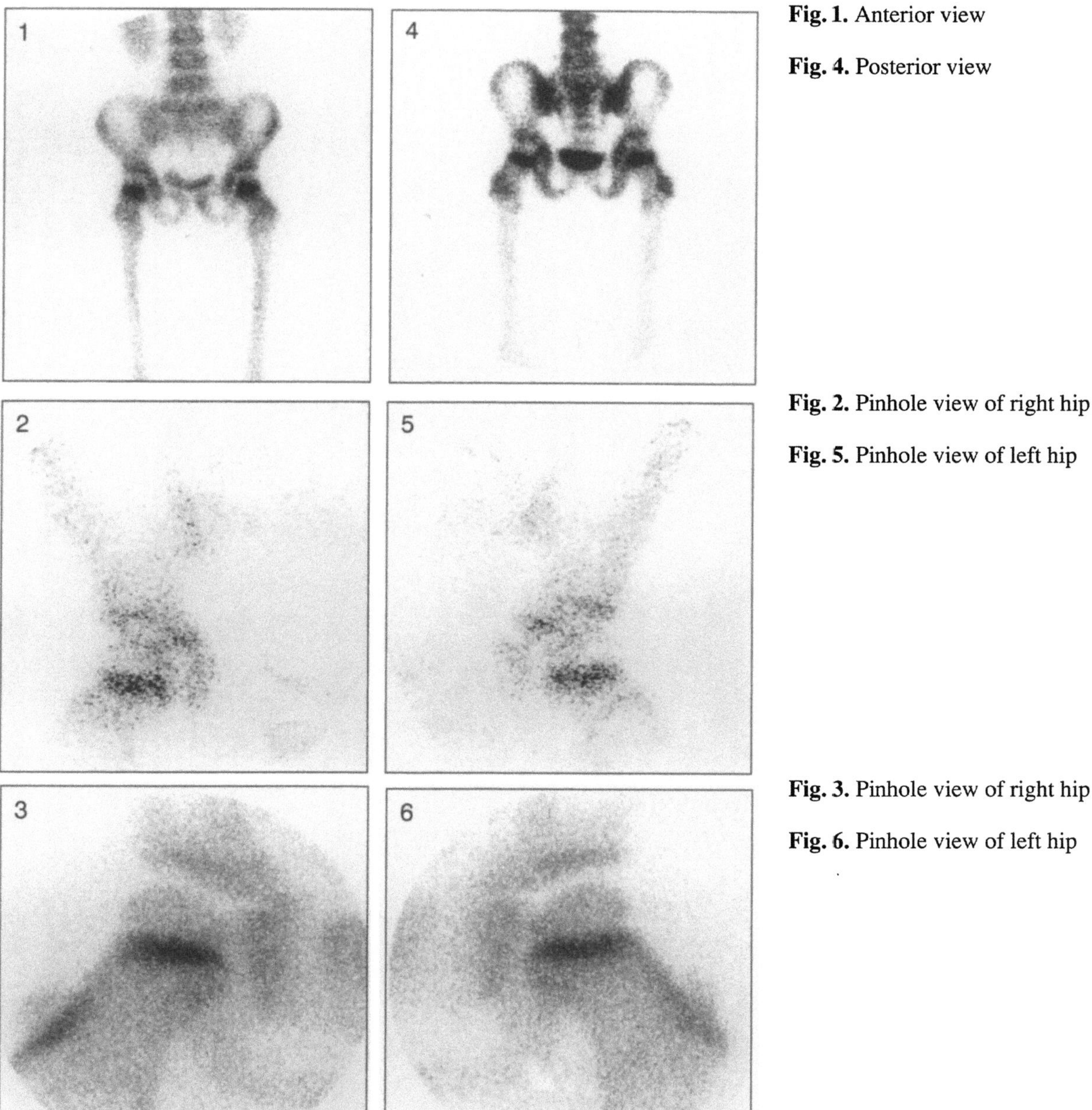

Fig. 1. Anterior view

Fig. 4. Posterior view

Fig. 2. Pinhole view of right hip

Fig. 5. Pinhole view of left hip

Fig. 3. Pinhole view of right hip

Fig. 6. Pinhole view of left hip

Fig. 1. Anterior view

Fig. 4. Posterior view

Fig. 2. Anterior view

Fig. 5. Posterior view

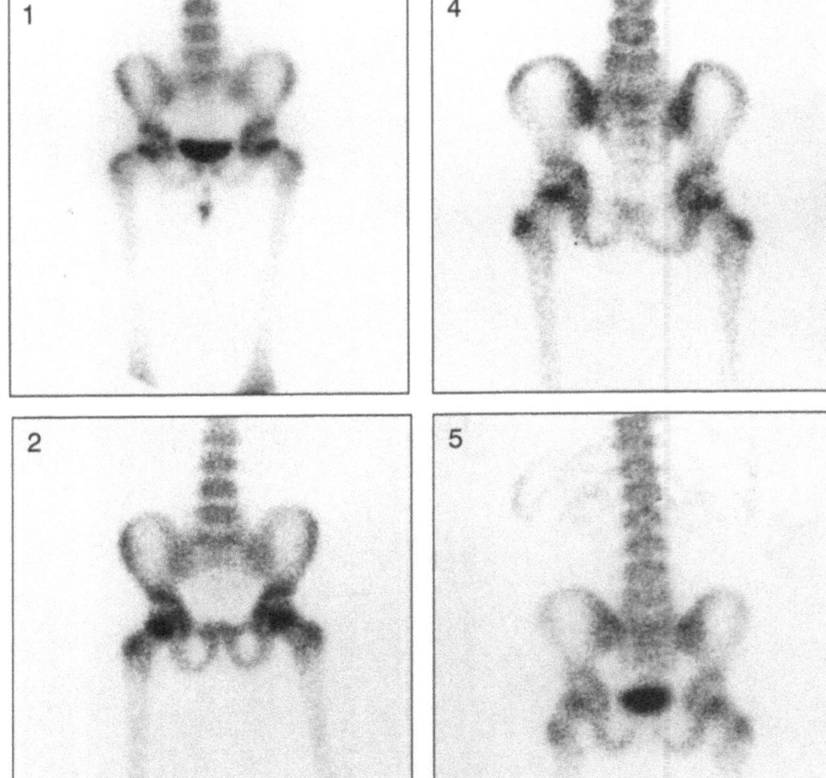

Technical Comment
– Urine contamination below the pelvis is seen in Fig. 1

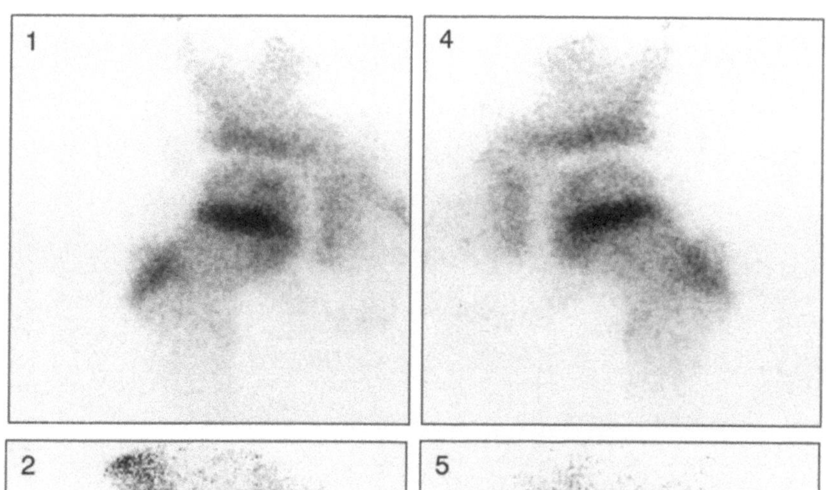

Fig. 1. Pinhole view of right hip

Fig. 4. Pinhole view of left hip

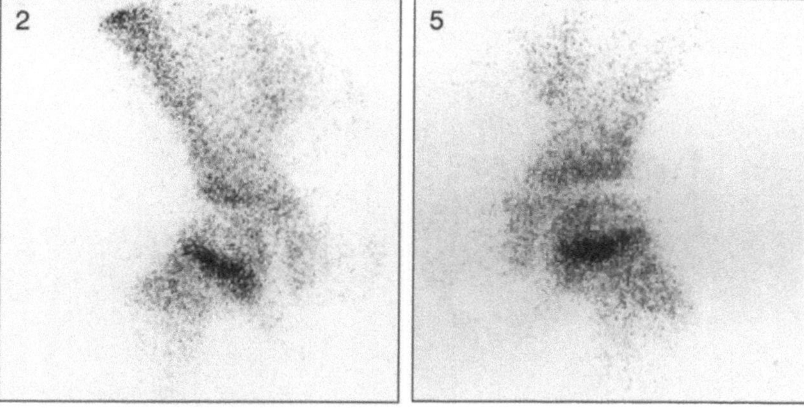

Fig. 2. Pinhole view of right hip

Fig. 5. Pinhole view of left hip

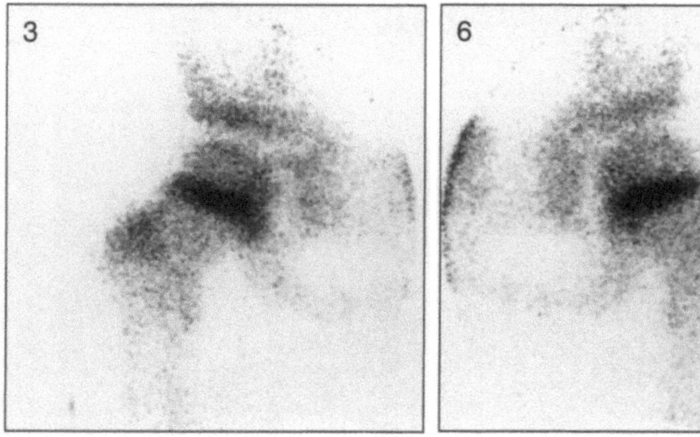

Fig. 3. Pinhole view of right hip

Fig. 6. Pinhole view of left hip

Fig. 1. Anterior view

Fig. 4. Posterior view

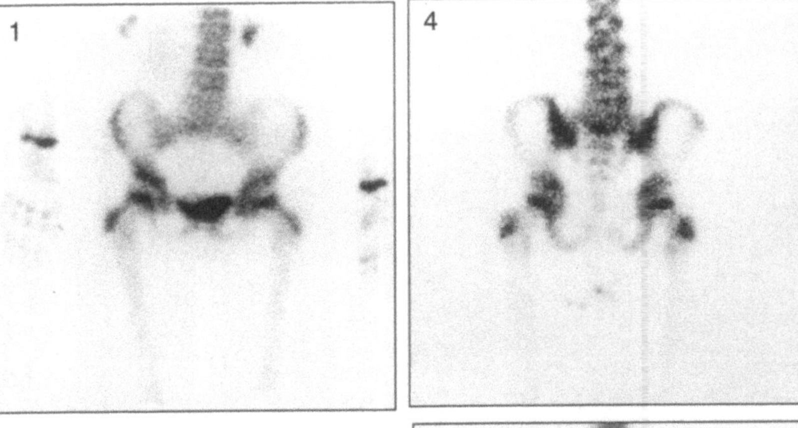

Fig. 5. Posterior view

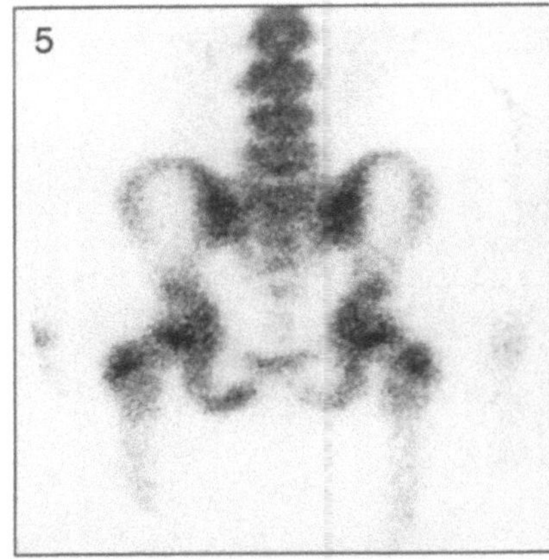

Fig. 3. Pinhole view of right hip

Fig. 6. Pinhole view of left hip

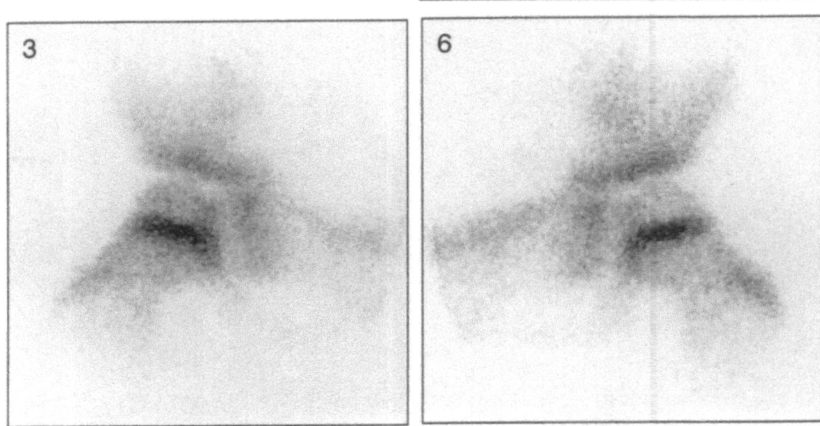

Technical Comment
- Urine contamination below the pelvis is seen in Fig. 4

▶ **Potential Pitfall**
- In Fig. 5 increased activity is noted at the junction of the left ischial tuberosity and the posterior pubic ramus due to the normal synchondrosis at this site

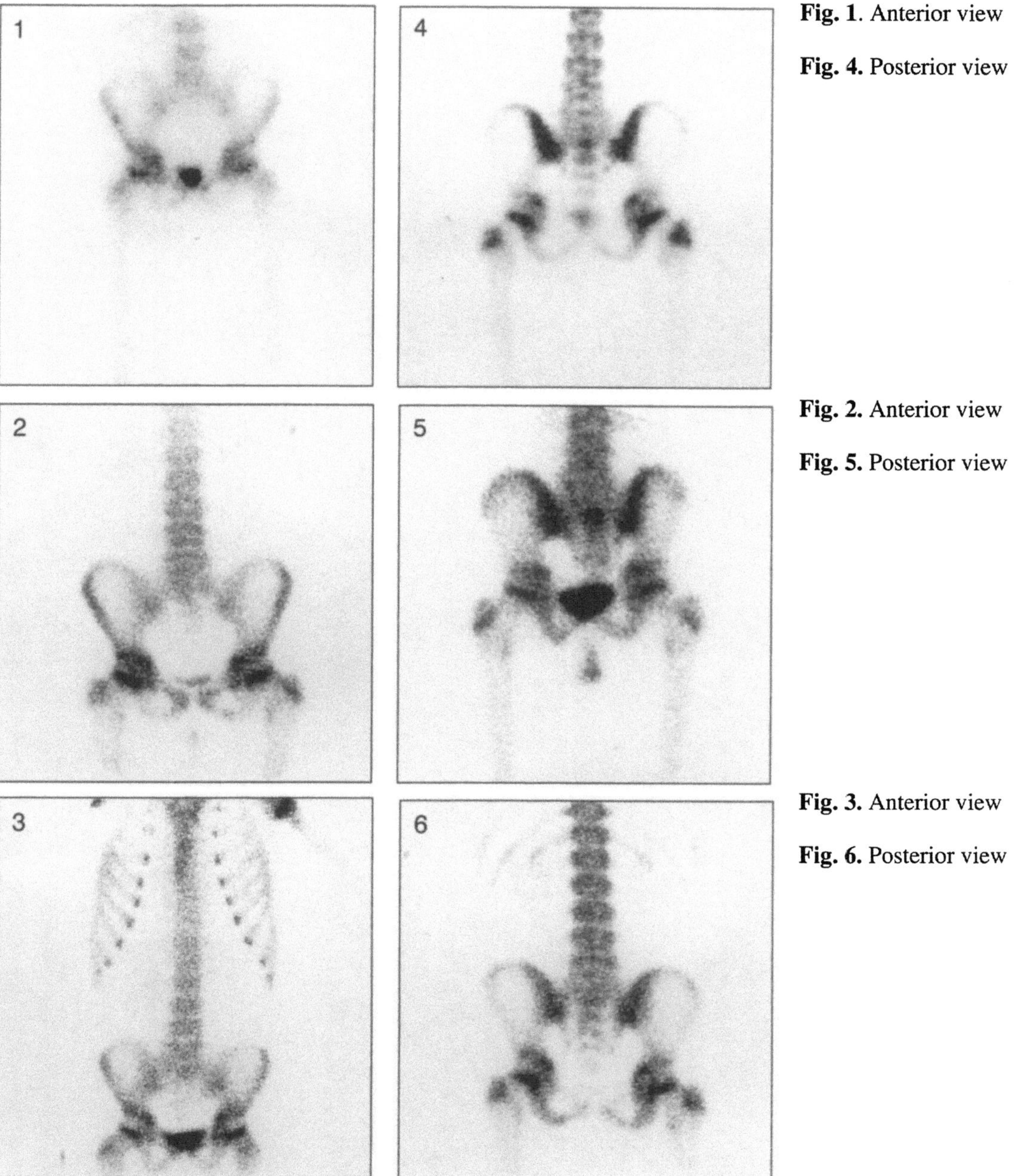

Fig. 1. Anterior view

Fig. 4. Posterior view

Fig. 2. Anterior view

Fig. 5. Posterior view

Fig. 3. Anterior view

Fig. 6. Posterior view

Technical Comment

– Urine contamination below the pelvis is seen in Figs. 2 and 5

Fig. 1. Pinhole view of right hip

Fig. 4. Pinhole view of left hip

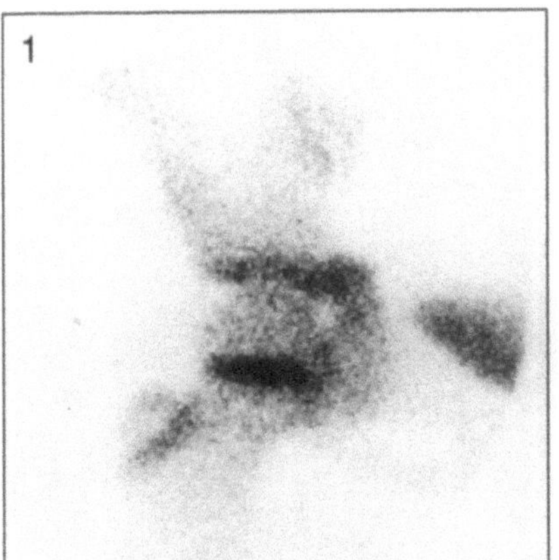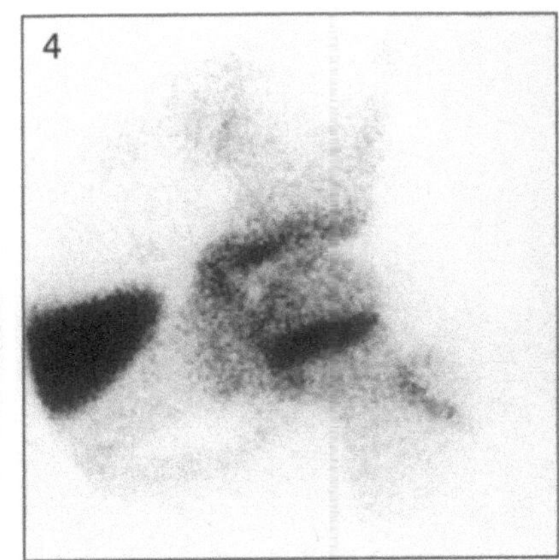

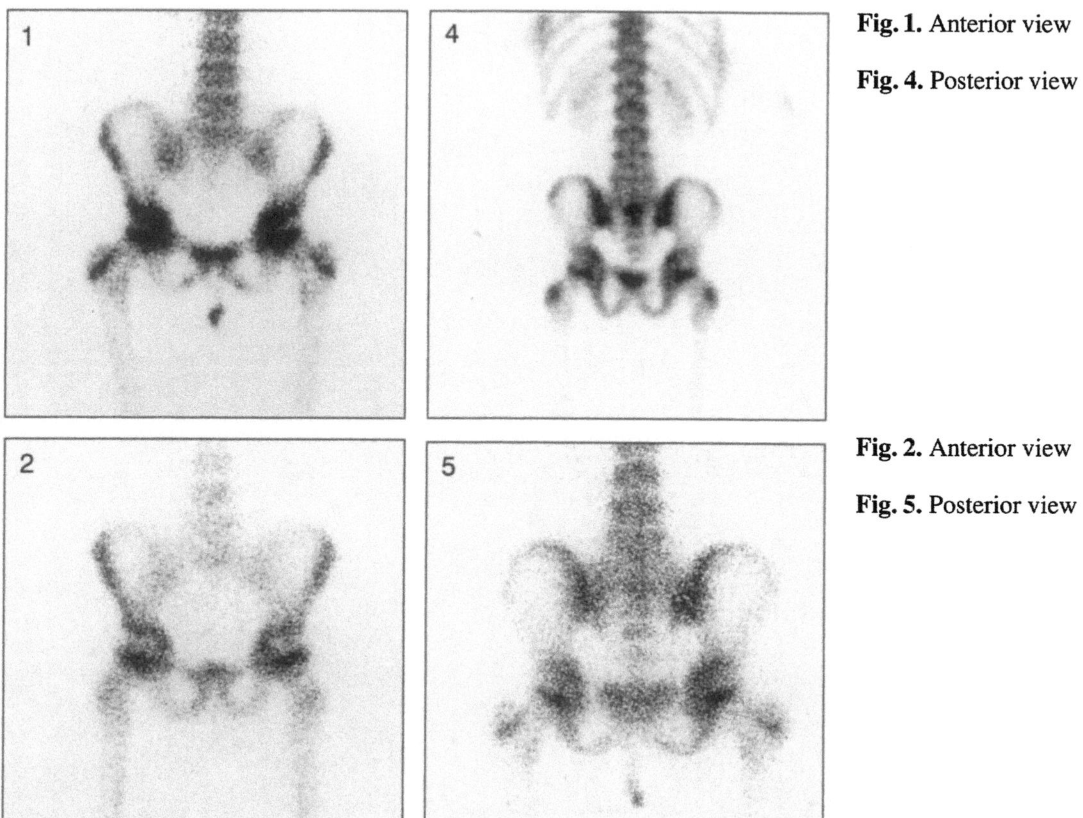

Fig. 1. Anterior view

Fig. 4. Posterior view

Fig. 2. Anterior view

Fig. 5. Posterior view

Technical Comment

– Urine contamination below the pelvis is seen in Figs. 1 and 5

Fig. 1. Pinhole view of right hip

Fig. 4. Pinhole view of left hip

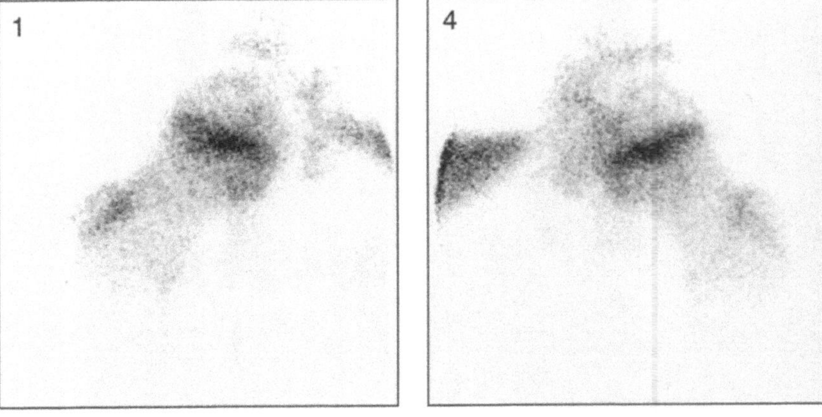

Fig. 2. Pinhole view of right hip

Fig. 5. Pinhole view of left hip

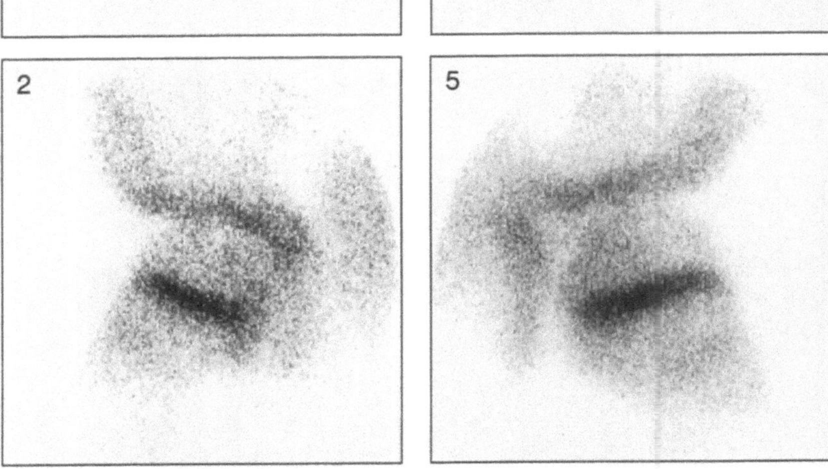

Technical Comment
- Figs. 2 and 5 show a change in magnification creating an apparent difference between the sizes of the two hips.

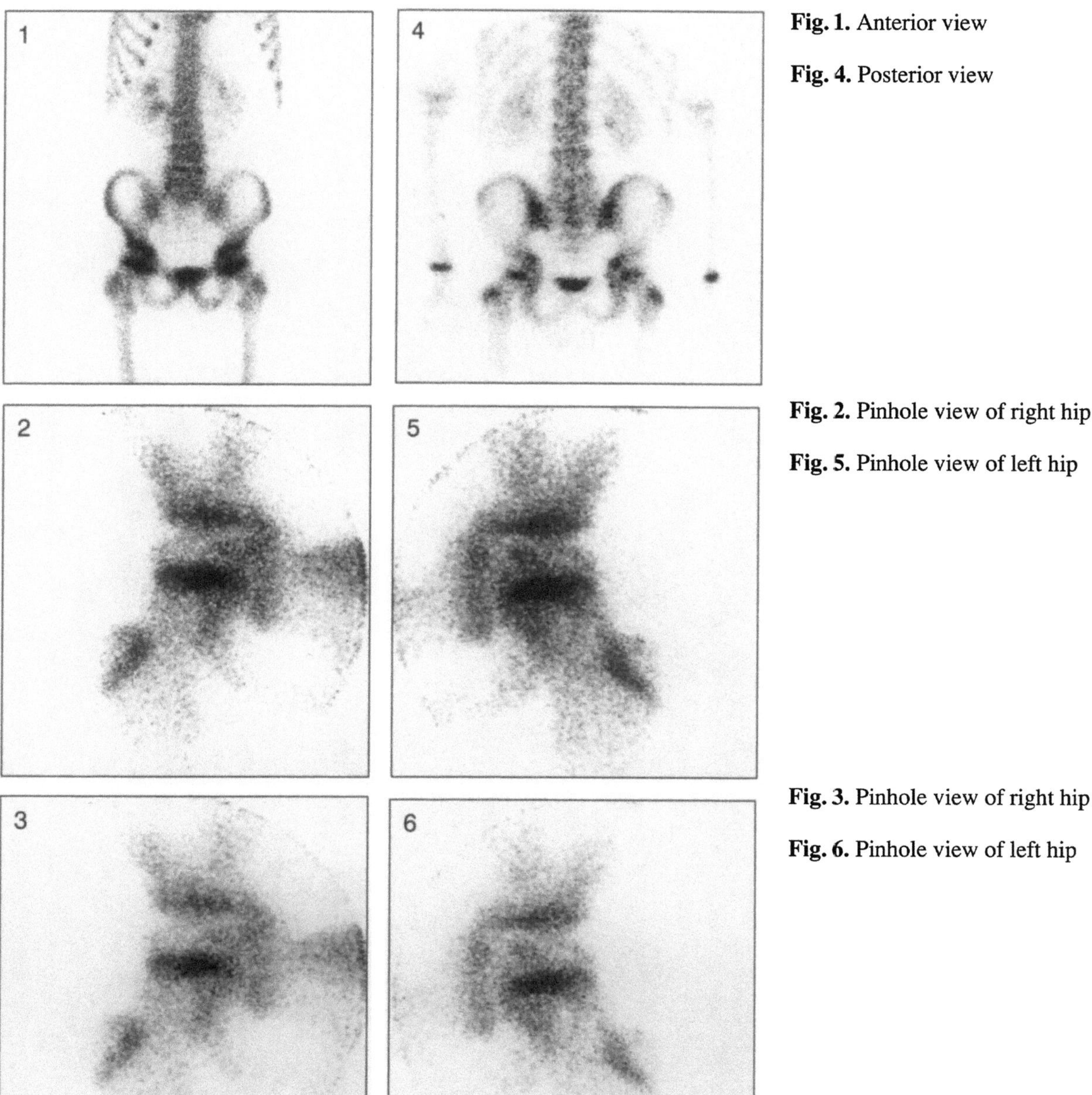

Fig. 1. Anterior view

Fig. 4. Posterior view

Fig. 2. Pinhole view of right hip

Fig. 5. Pinhole view of left hip

Fig. 3. Pinhole view of right hip

Fig. 6. Pinhole view of left hip

Fig. 1. Anterior view

Fig. 4. Posterior view

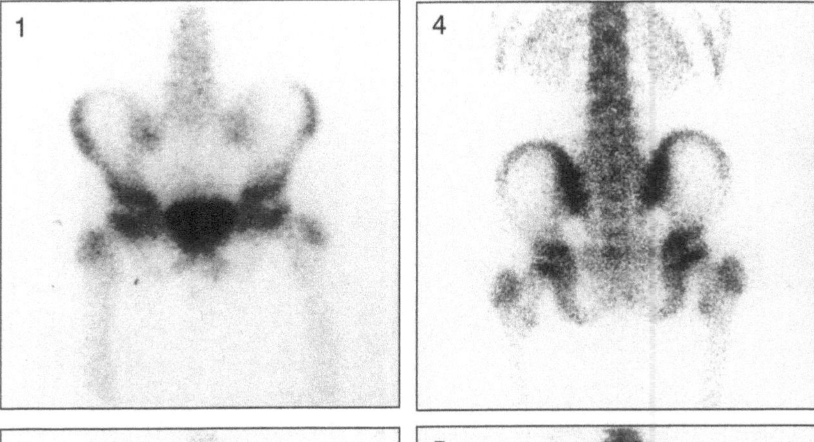

Fig. 2. Anterior view

Fig. 5. Posterior view

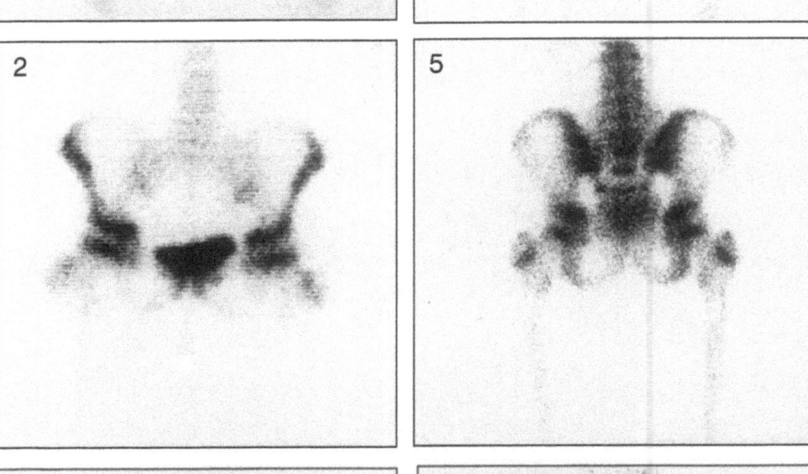

Fig. 3. Pinhole view of right hip

Fig. 6. Pinhole view of left hip

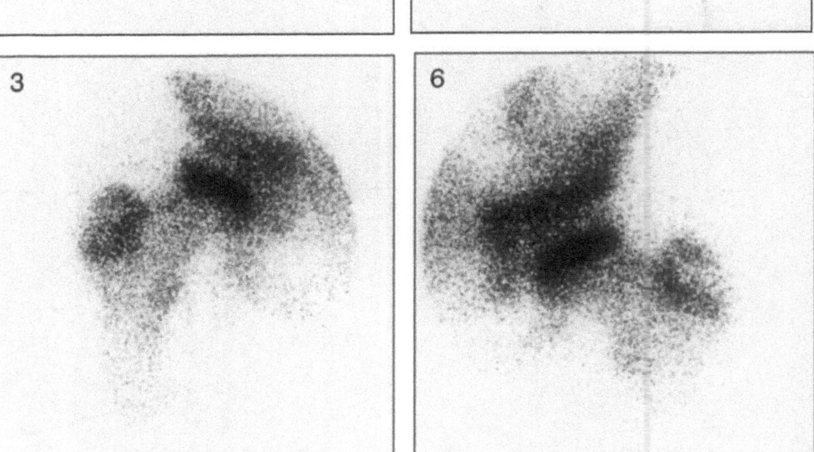

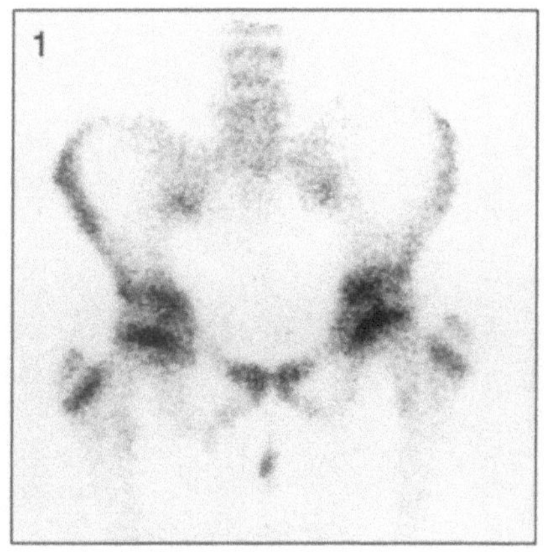

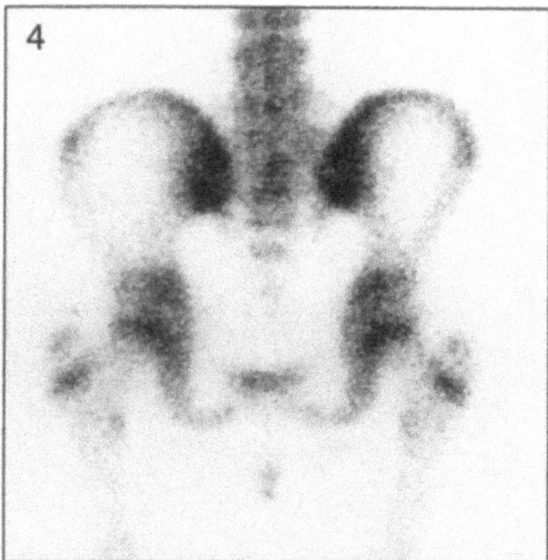

Fig. 1. Anterior view

Fig. 4. Posterior view

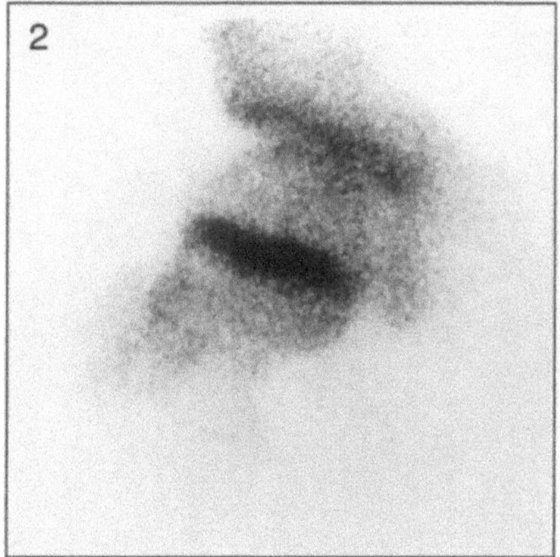

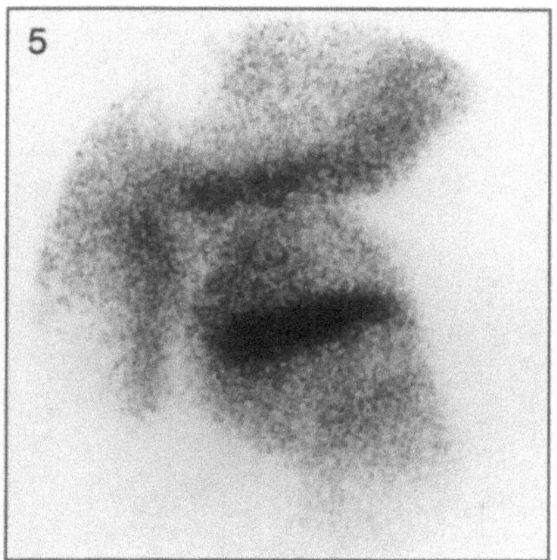

Fig. 2. Pinhole view of right hip

Fig. 5. Pinhole view of left hip

Technical Comment

- Urine contamination below the pelvis is seen in Figs. 1 and 4
- Figs. 2 and 5 show a change in magnification creating an apparent difference between the sizes of the two hips

Fig. 1. Anterior view

Fig. 4. Posterior view

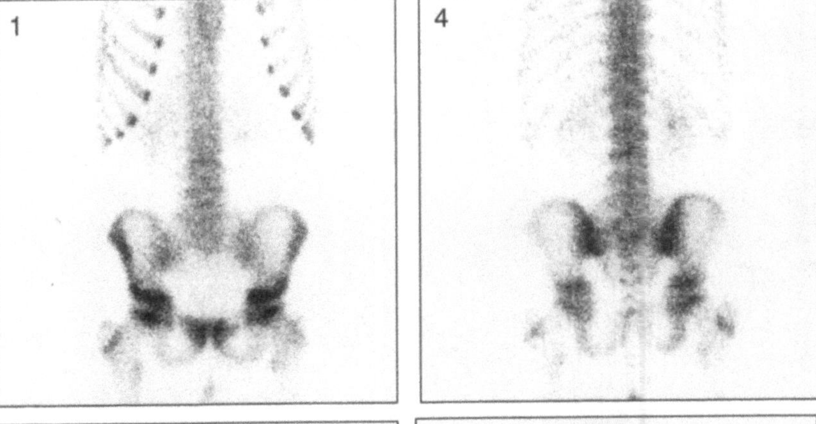

Fig. 2. Anterior view

Fig. 5. Posterior view

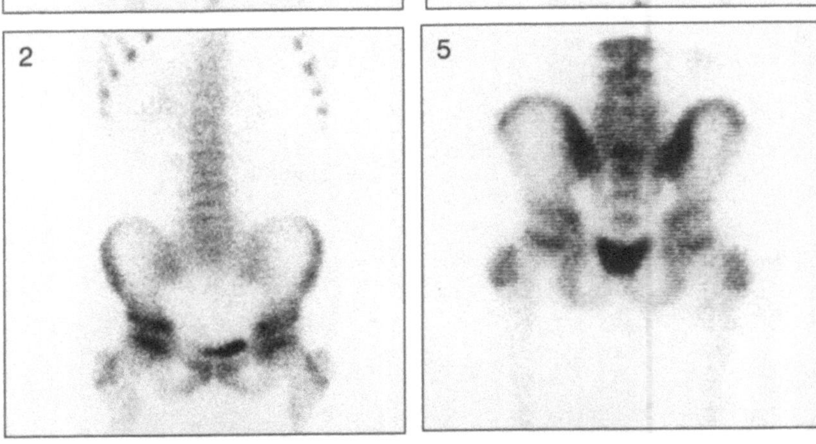

Fig. 3. Pinhole view of right hip

Fig. 6. Pinhole view of left hip

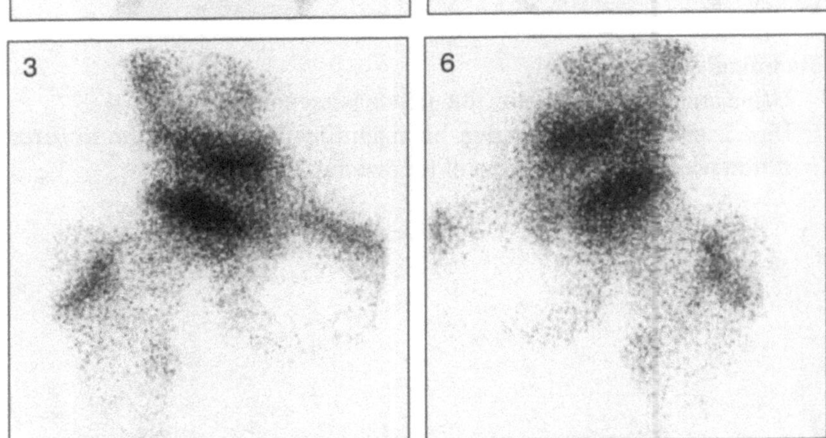

Technical Comment
- Urine contamination below the pelvis is seen in Figs. 1 and 4

▶ **Potential Pitfall**
- The pubic rami show normal increased uptake of tracer in Figs. 1 and 2

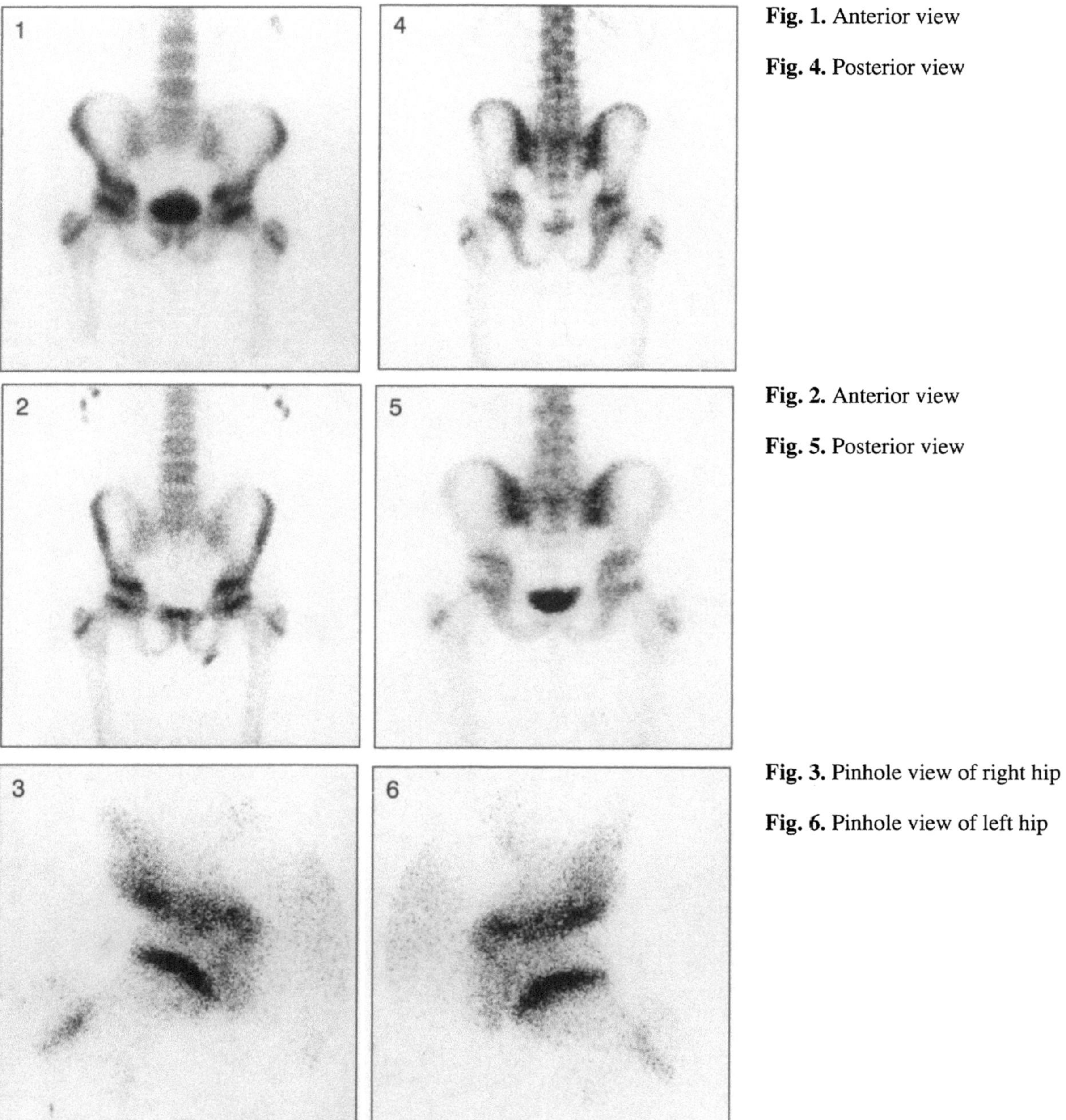

Fig. 1. Anterior view

Fig. 4. Posterior view

Fig. 2. Anterior view

Fig. 5. Posterior view

Fig. 3. Pinhole view of right hip

Fig. 6. Pinhole view of left hip

Technical Comment
– Urine contamination adjacent to the left ischium is seen in Fig. 2

Fig. 1. Anterior view

Fig. 4. Posterior view

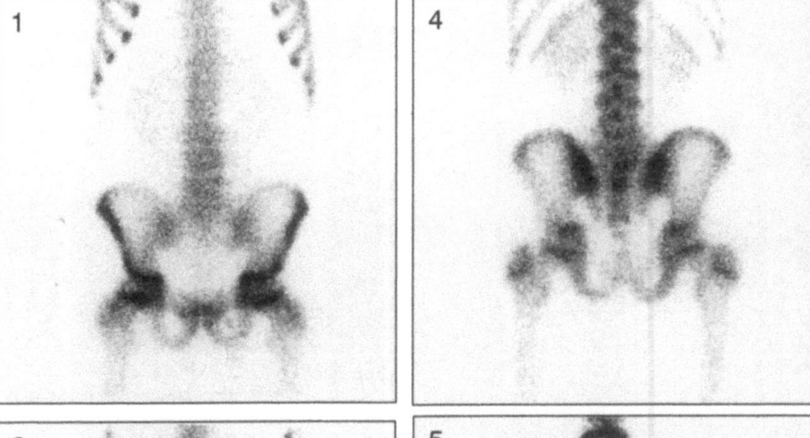

Fig. 2. Anterior view

Fig. 5. Posterior view

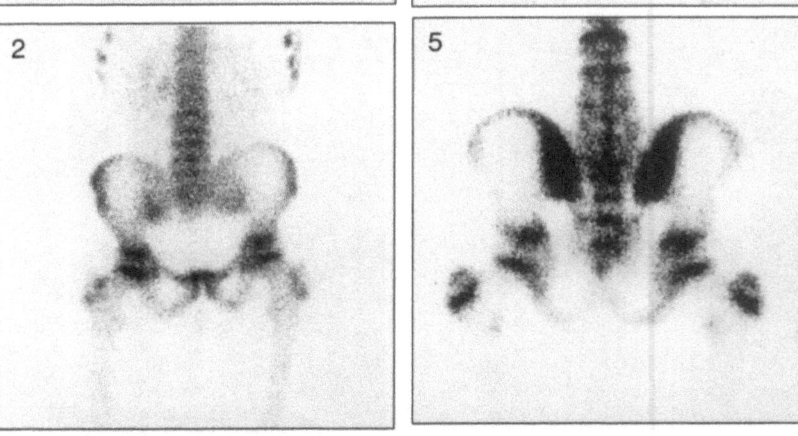

Fig. 3. Pinhole view of right hip

Fig. 6. Pinhole view of left hip

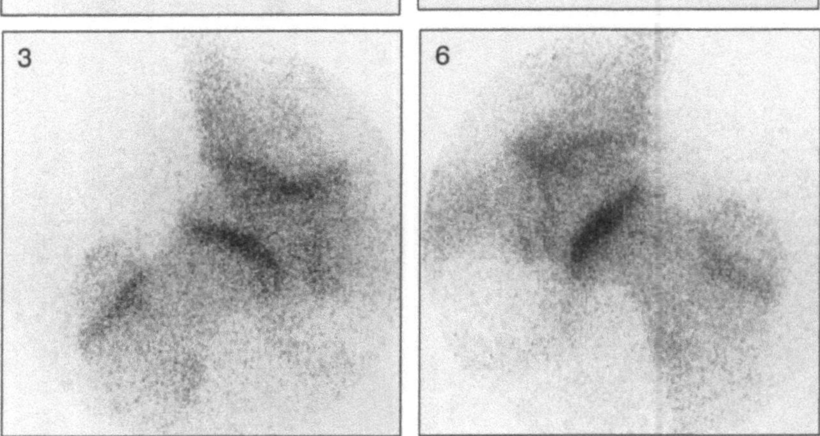

Technical Comment
- Urine contamination to the left of the pubic bone is seen in Fig. 1

▶ **Potential Pitfall**
- The apparent increased activity in the right sacro-iliac joint in Fig. 2 is due to rotation of the pelvis and is not pathological

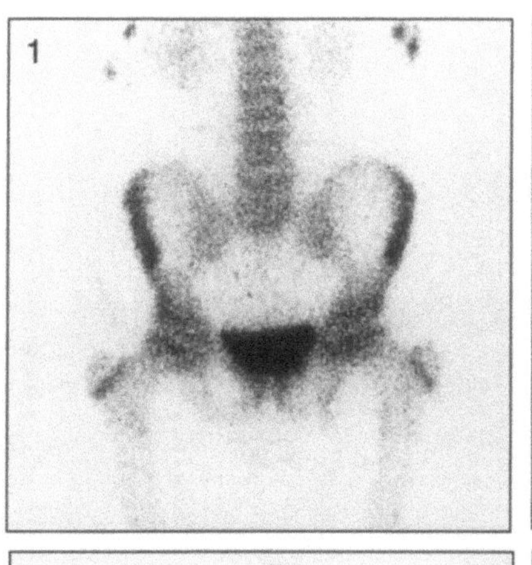

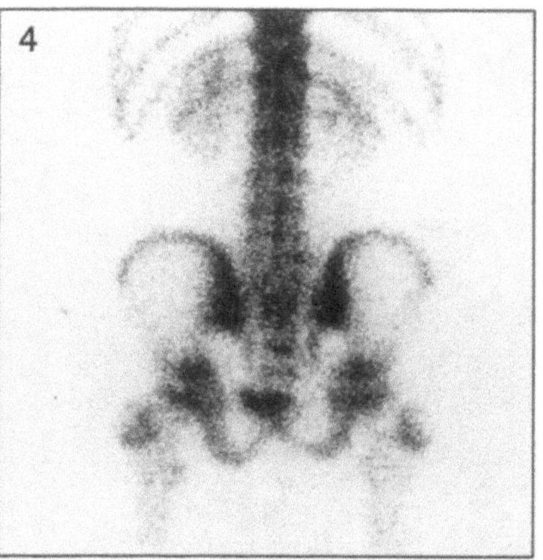

Fig. 1. Anterior view

Fig. 4. Posterior view

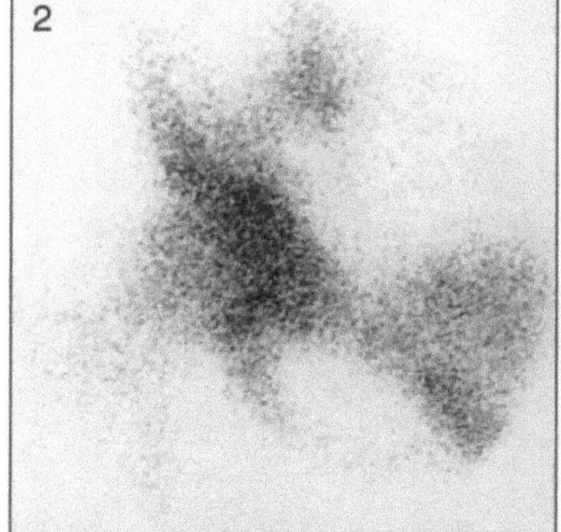

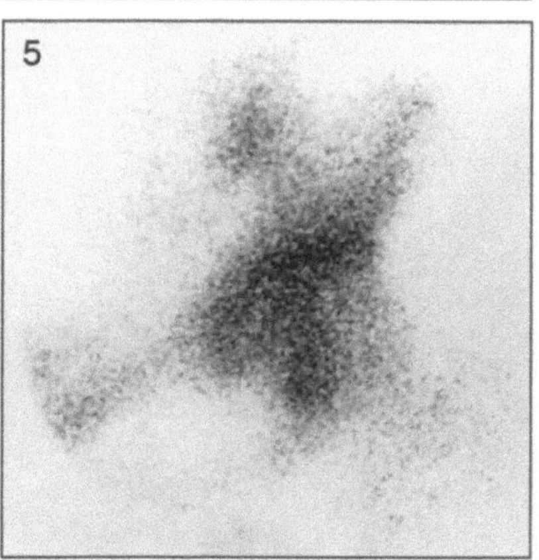

Fig. 2. Pinhole view of right hip

Fig. 5. Pinhole view of left hip

Technical Comment

– The maturity of the skeleton has resulted in decreased uptake of the epiphyseal plates

Fig. 1. Anterior view

Fig. 4. Posterior view

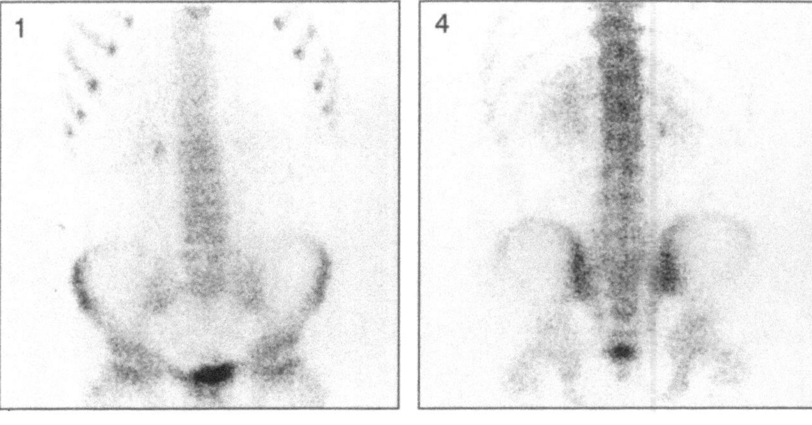

Fig. 2. Pinhole view of right hip

Fig. 5. Pinhole view of left hip

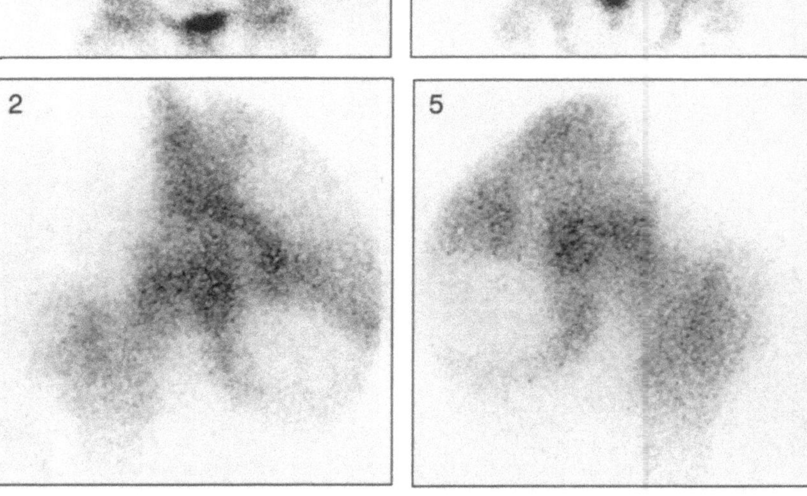

Springer-Verlag and the Environment

We at Springer-Verlag firmly believe that an international science publisher has a special obligation to the environment, and our corporate policies consistently reflect this conviction.

We also expect our business partners – paper mills, printers, packaging manufacturers, etc. – to commit themselves to using environmentally friendly materials and production processes.

The paper in this book is made from low- or no-chlorine pulp and is acid free, in conformance with international standards for paper permanency.